Grundlehren der mathematischen Wissenschaften 259

A Series of Comprehensive Studies in Mathematics

Grundlehren der mathematischen Wissenschaften

A Series of Comprehensive Studies in Mathematics

A Selection

162. Nevanlinna: Analytic Functions
163. Stoer/Witzgall: Convexity and Optimization in Finite Dimensions I
164. Sario/Nakai: Classification Theory of Riemann Surfaces
165. Mitrinovic/Vasic: Analytic Inequalities
166. Grothendieck/Dieudonné: Eléments de Géometrie Algébrique I
167. Chandrasekharan: Arithmetical Functions
168. Palamodov: Linear Differential Operators with Constant Coefficients
169. Rademacher: Topics in Analytic Number Theory
170. Lions: Optimal Control of Systems Governed by Partial Differential Equations
171. Singer: Best Approximation in Normed Linear Spaces by Elements of Linear Subspaces
172. Bühlmann: Mathematical Methods in Risk Theory
173. Maeda/Maeda: Theory of Symmetric Lattices
174. Stiefel/Scheifele: Linear and Regular Celestial Mechanic, Perturbed Two-body Motion—Numerical Methods—Canonical Theory
175. Larsen: An Introduction to the Theory of Multipliers
176. Grauert/Remmert: Analytische Stellenalgebren
177. Flügge: Practical Quantum Mechanics I
178. Flügge: Practical Quantum Mechanics II
179. Giraud: Cohomologie non abélienne
180. Landkof: Foundations of Modern Potential Theory
181. Lions/Magenes: Non-Homogeneous Boundary Value Problems and Applications I
182. Lions/Magenes: Non-Homogeneous Boundary Value Problems and Applications II
183. Lions/Magenes: Non-Homogeneous Boundary Value Problems and Applications III
184. Rosenblatt: Markov Processes, Structure and Asymptotic Behavior
185. Rubinowicz: Sommerfeldsche Polynommethode
186. Handbook for Automatic Computation. Vol. 2. Wilkinson/Reinsch: Linear Algebra
187. Siegel/Moser: Lectures on Celestial Mechanics
188. Warner: Harmonic Analysis on Semi-Simple Lie Groups I
189. Warner: Harmonic Analysis on Semi-Simple Lie Groups II
190. Faith: Algebra: Rings, Modules, and Categories I
191. Faith: Algebra II, Ring Theory
192. Mallcev: Algebraic Systems
193. Pólya/Szegö: Problems and Theorems in Analysis I
194. Igusa: Theta Functions
195. Berberian: Baer*-Rings
196. Athreya/Ney: Branching Processes
197. Benz: Vorlesungen über Geometric der Algebren
198. Gaal: Linear Analysis and Representation Theory
199. Nitsche: Vorlesungen über Minimalflächen
200. Dold: Lectures on Algebraic Topology
201. Beck: Continuous Flows in the Plane
202. Schmetterer: Introduction to Mathematical Statistics
203. Schoeneberg: Elliptic Modular Functions
204. Popov: Hyperstability of Control Systems
205. Nikollskii: Approximation of Functions of Several Variables and Imbedding Theorems

Continued after Index

Peter L. Duren

Univalent Functions

With 12 Illustrations

Springer-Verlag
New York Berlin Heidelberg Tokyo

Peter L. Duren
Department of Mathematics
University of Michigan
Ann Arbor, MI 48109
U.S.A.

AMS Subject Classification (1980): 30CXX

Library of Congress Cataloging in Publication Data
Duren, Peter L., 1935–
Univalent functions.
(Grundlehren der mathematischen Wissenschaften; 259)
Includes bibliographies and index.
1. Univalent functions. I. Title. II. Series.
QA331.D96 1983 515.9 83-9091

ISBN 978-1-4419-2816-0

Printed in the United States of America.

9 8 7 6 5 4 3 2 1

To my wife and children . . .

To Gay

and to Betsy and Bill

Preface

The theory of univalent functions is an old subject, born around the turn of the century, yet it remains an active field of current research. Progress has been especially rapid in recent years. The purpose of this book is to present a modern overview, to give a full account of the more classical aspects of the subject while emphasizing recent developments and open problems. The book is designed to serve both as an introductory text and as a reference for research workers. Assuming only a basic acquaintance with real and complex analysis, it offers an exposition of the principal methods of the field and their applications.

Most of the book is concerned with the class S of functions analytic and univalent in the unit disk and normalized in a standard way. One of the major problems of the field is the Bieberbach conjecture, dating from the year 1916, which asserts that the Taylor coefficients of each function of class S satisfy the inequality $|a_n| \leq n$. For many years this famous problem has stood as a challenge and has inspired the development of ingenious methods which now form the backbone of the entire subject. A substantial part of this book is directed toward various partial solutions of the Bieberbach conjecture.

The organization of the subject seems to proceed more naturally by methods than by results. The methods come from diverse areas of mathematics. Each has its peculiar advantages for solving certain kinds of problems. The plan of the book is to develop each method in turn, together with its main applications. Thus there are individual chapters devoted to elementary methods, Loewner's parametric method, area methods and the Grunsky inequalities, Milin's and FitzGerald's methods of exponentiating the Grunsky inequalities, Baernstein's method of maximal functions, and variational methods. This structure inevitably produces some overlapping of results. For example, four different methods are used to prove

$$|a_n| < en, \quad |a_n| < 1.243n, \quad |a_n| < \sqrt{7/6}\,n, \quad \text{and} \quad |a_n| < (e/2)n.$$

The choice of topics was guided to some extent by my own interests, but it was also influenced by Christian Pommerenke's book of the same title. In order to avoid duplication, I steered away from certain topics covered there. His treatment of the Garabedian–Schiffer inequalities, for instance, is so elegant that I felt no regret in omitting this important topic myself. A more serious omission is Jenkins' general coefficient theorem and its applications, a method which complements the variational method in a remarkable way. One special case, Teichmüller's theorem, is used without proof in the final chapter. This material is covered in Jenkins' book and (by G. Jensen) in Pommerenke's book. In general, I have emphasized the full class S and have given relatively little attention to special subclasses. Fortunately, the forthcoming book by A. W. Goodman will do justice to this aspect of the subject.

In an effort to make the book accessible to a wide audience, I have tried to minimize the prerequisites and to keep the exposition at an elementary level wherever possible. A preliminary chapter is included to review the basic principles of real and complex analysis which are invoked repeatedly throughout the book. The discussion of univalent functions then begins with a detailed introduction (Chapter 2) to the more elementary and classical part of the theory. This material is particularly elegant and is easily accessible to the beginning student. It already illustrates the interplay of geometry and analysis which pervades the whole subject and gives it a special fascination.

Each chapter concludes with a list of exercises. Some are very easy, but a good many are too difficult to be considered exercises in the usual sense. They should be taken rather as invitations to consult the original papers (as cited) for certain topics not fully treated in the text. Thus no author should feel insulted to find his theorem offered as an exercise, nor should anyone feel defeated if unable to prove that theorem independently. On the other hand, some references accompany relatively simple exercises and are cited only for historical reasons.

The book was written over a period of about twelve years. During that time I was able to lecture on selected topics in several graduate courses at the University of Michigan. The students' criticism led to numerous improvements in the exposition. For several years I also had the benefit of an active seminar composed of enthusiastic groups of students and colleagues at Michigan. Together we learned many of the topics presented in this book. Among the participants were Renate McLaughlin, George Leeman, Jack Quine, Y. J. Leung, Johnny Brown, Anna Tsao, and Brad Osgood, all of whom wrote Ph.D. theses on univalent functions. They and many other workers in the field made observations and suggestions which ultimately enhanced the book. In addition, several people read parts of the manuscript and pointed out errors, omissions, and confusions. I am particularly indebted to Russell Lyons, who read most of the final typescript and made a large number of helpful criticisms.

The direction of my own research in univalent functions was decisively influenced by several colleagues and collaborators, including Max Schiffer, Glenn Schober, Dov Aharonov, Y. J. Leung, and Johnny Brown. I owe a special debt to Professor Schiffer, who introduced me to the field through his lectures at Stanford University in 1961 and through our subsequent collaboration. Over the years he has been a continual source of inspiration.

My work was supported in part by the National Science Foundation through summer research grants, and by two sabbatical leaves from the University of Michigan. While in Ann Arbor I had easy access to the literature through the excellent mathematical library, which contains virtually every item cited in the bibliography. Special thanks go to Mike Ciapa for helping to track down many an obscure reference. Toward the end of the project I had the benefit of mathematical facilities at the University of Maryland, at the Université de Paris-Sud (Orsay), and at the Institut Mittag-Leffler in Djursholm, Sweden. I am grateful for the friendly hospitality extended to me at those institutions. In addition, my friends and colleagues at the University of Michigan, especially Fred Gehring and Allen Shields, lent approval and support to the writing effort over a long period of time. My father, William L. Duren, Jr., generously read early versions of the manuscript and gave valuable criticism and encouragement. But above all my effort was sustained by my wife and children, who not only endured with patience a seemingly interminable project, but supported it with enthusiasm. The book is dedicated to them.

Large portions of the manuscript were skillfully typed by Arlett Gibbons, Dorothy Lentz, and Marsha Spickenagel. Their help is much appreciated. Finally, my thanks go to the staff at Springer-Verlag for patiently awaiting a long-overdue manuscript and for efficiently converting it to type. Their expertise in mathematical publishing will be readily apparent.

Djursholm, Sweden
February, 1983

PETER L. DUREN

Contents

Chapter 1

Geometric Function Theory

The purpose of this preliminary chapter is to review and assemble for later reference some of the general principles of complex analysis which underlie the theory of univalent functions. In many instances the statements of theorems are supported by bare indications of a proof, or by no proof at all. Detailed proofs and further discussion may be found in standard introductory texts such as Ahlfors [5] or Titchmarsh [1]. The reader with a good knowledge of complex analysis is advised to proceed directly to Chapter 2 and to refer back to Chapter 1 as the need arises.

§1.1. Basic Principles

A *domain* is an open connected set in the complex plane $\mathbb{C}$. A domain is said to be *simply connected* if its complement is connected. A *neighborhood* of a set $E \subset \mathbb{C}$ is an open set which contains E.

A *compact set* is a closed bounded subset of $\mathbb{C}$. The *closure* $\bar{E}$ of a set $E \subset \mathbb{C}$ is the smallest closed set which contains E. It is equivalent to describe $\bar{E}$ as the union of E with its set of limit points. The *interior* $\mathring{E}$ of the set E is the largest open set contained in E. It may well happen that $\mathring{E}$ is empty. The *boundary* of E is the set $\partial E = \bar{E} - \mathring{E}$ consisting of all points in the closure of E which are not interior points.

The *unit disk* $\mathbb{D}$ consists of all points $z \in \mathbb{C}$ of modulus $|z| < 1$. Its boundary, the unit circle, is denoted by $\mathbb{T}$.

An *arc* in the complex plane is the continuous image of a line segment. Strictly speaking, an arc must be viewed as a continuous mapping $z = \varphi(t)$ of an interval $a \leq t \leq b$ into $\mathbb{C}$. A *rectifiable arc* is an arc which has a length; that is, for which the mapping function φ is of bounded variation. A *Jordan arc* is an arc without self-intersections. A *closed curve* is the continuous image of a circle, or an arc whose endpoints coincide. A *simple closed curve*, or a *Jordan curve*, is a closed curve without self-intersections. The Jordan curve theorem asserts that every Jordan curve divides the plane into two regions, the interior and the exterior of the curve. The interior of a Jordan curve is called a *Jordan domain*.

A complex-valued function f of a complex variable is *differentiable* at a point $z_0 \in \mathbb{C}$ if it has a derivative

$$f'(z_0) = \lim_{z \to z_0} \frac{f(z) - f(z_0)}{z - z_0}$$

at z_0. Such a function f is *analytic* at z_0 if it is differentiable at every point in some neighborhood of z_0. It is one of the "miracles" of complex analysis that f must then have derivatives of all orders at z_0, and that f has a *Taylor series* expansion

$$f(z) = \sum_{n=0}^{\infty} a_n(z - z_0)^n, \qquad a_n = f^{(n)}(z_0)/n!,$$

convergent in some open disk centered at z_0.

The Taylor series development is easily obtained from the *Cauchy integral formula*

$$f^{(n)}(z) = \frac{n!}{2\pi i} \int_C \frac{f(\zeta)}{(\zeta - z)^{n+1}} \, d\zeta,$$

where C is a rectifiable Jordan curve, f is analytic inside and on C, and z is inside C. The Cauchy integral formula is derived for $n = 0$ from the *Cauchy integral theorem*:

$$\int_C f(\zeta) \, d\zeta = 0.$$

The more general formula (for $n \geq 1$) is then obtained by a standard process of "differentiation under the integral sign." All of these results remain valid for functions analytic inside C and continuous in the closure.

The *uniqueness principle* for analytic functions is a direct consequence of the Taylor series representation. It asserts that whenever two analytic functions agree on a sequence of points which cluster inside their common domain of analyticity, they must agree everywhere. Equivalently, if f is analytic at a point z_0 and $f(z_n) = 0$ on a sequence of distinct points z_n converging to z_0, then $f(z) \equiv 0$.

Cauchy's theorem has a converse known as *Morera's theorem*: if f is a complex-valued function which is continuous in a simply connected domain D and has the property that

$$\int_C f(\zeta) \, d\zeta = 0$$

for every rectifiable Jordan curve C lying in D, then f is analytic in D.

The modulus of an analytic function describes a peakless landscape. This is the content of the *maximum modulus theorem*, which asserts that the modulus of a function f analytic in a domain D can have no weak local maximum in D unless f is constant. A function analytic in a bounded domain and continuous in the closure must therefore attain its maximum modulus on the boundary.

A simple but important consequence of the maximum modulus theorem is the Schwarz lemma, which may be stated as follows.

Schwarz Lemma. *Let f be analytic in the unit disk $\mathbb{D}$, with $f(0) = 0$ and $|f(z)| < 1$ in $\mathbb{D}$. Then $|f'(0)| \leq 1$ and $|f(z)| \leq |z|$ in $\mathbb{D}$. Strict inequality holds in both estimates unless f is a rotation of the disk: $f(z) = e^{i\theta}z$.*

Proof. Apply the maximum modulus theorem to the analytic function $g(z) = f(z)/z$.

A function f analytic and single-valued in an annular region $0 < |z - z_0| < \rho$ has a *Laurent series* expansion

$$f(z) = \sum_{n=-\infty}^{\infty} a_n(z - z_0)^n$$

there. The *principal part* of the expansion is the sum $\sum_{n=-\infty}^{-1} a_n(z - z_0)^n$. The number a_{-1} is called the *residue* of f at z_0. The famous *residue theorem* asserts that if f is analytic and single-valued inside and on a rectifiable Jordan curve C, aside from a finite number of singular points $z_1, z_2, \ldots, z_n$ inside C, where f has residues $\alpha_1, \alpha_2, \ldots, \alpha_n$, then

$$\int_C f(z)\,dz = 2\pi i(\alpha_1 + \alpha_2 + \cdots + \alpha_n).$$

The direction of integration is understood to be the positive or "counter-clockwise" direction.

The argument principle is often a useful device for determination of the range and valence of an analytic function. It is proved by a simple application of the residue theorem.

Argument Principle. *Let f be analytic in the closure of a domain D bounded by a rectifiable Jordan curve C, and suppose that $f(z) \neq 0$ on C. Then the number of zeros of f in D, counted according to multiplicity, is equal to $1/2\pi$ times the net change in the argument of $f(z)$ as z traverses C in the positive direction.*

Proof. Let $\Delta_C \arg f$ denote the net change in the argument of f over C. Then by the residue theorem,

$$i\Delta_C \arg f = \Delta_C \log f = \int_C \frac{f'(z)}{f(z)}\,dz$$
$$= 2\pi i\{m_1 + m_2 + \cdots + m_n\},$$

where m_k is the order of the zero of f at the point $z_k \in D$.

In more picturesque language, the argument principle asserts that the number of zeros of f inside C is equal to the winding number of the image curve $f(C)$ about the origin.

An especially useful form of the argument principle is known as Rouché's theorem.

Rouché's Theorem. *Let f and g be analytic inside and on a rectifiable Jordan curve C, with $|g(z)| < |f(z)|$ on C. Then f and $(f + g)$ have the same number of zeros, counted according to multiplicity, inside C.*

Proof.

$$\Delta_C \arg(f + g) = \Delta_C \arg f + \Delta_C \arg(1 + g/f) = \Delta_C \arg f.$$

If a sequence $\{f_n\}$ of functions analytic in a domain D converges uniformly on each compact subset of D to a function f, then f is also analytic in D. This is easily proved with the aid of the Cauchy integral formula. Hurwitz's theorem establishes a close connection between the zeros of f and the zeros of the functions f_n.

Hurwitz's Theorem. *Let f_n be analytic in a domain D, and suppose $f_n(z) \to f(z)$ as $n \to \infty$, uniformly on each compact subset of D. Then either $f(z) \equiv 0$ in D, or every zero of f is a limit-point of a sequence of zeros of the functions f_n.*

Proof. Suppose $f(z_0) = 0$ but $f(z) \not\equiv 0$. It is enough to show that every neighborhood of z_0 contains a zero of some function f_n. Choose $\delta > 0$ so small that the disk $|z - z_0| \leq \delta$ lies in D and $f(z) \neq 0$ on the circle C defined by $|z - z_0| = \delta$. Let m be the minimum of $|f(z)|$ on C. Then for all $n \geq N$,

$$|f_n(z) - f(z)| < m \leq |f(z)|$$

on C. Thus by Rouché's theorem, f_n has the same number of zeros as f does inside C. In other words, $f_n(z)$ must vanish at least once inside C whenever $n \geq N$.

A function f analytic in a domain D is said to be *univalent* there if it does not take the same value twice: $f(z_1) \neq f(z_2)$ for all pairs of distinct points z_1 and z_2 in D. In other words, f is a one-to-one (or injective) mapping of D onto another domain. Alternate terms in common use are *schlicht* and *simple*. It is a remarkable fact, fundamental to the theory of univalent functions, that univalence is essentially preserved under uniform convergence.

Theorem. *Let f_n be analytic and univalent in a domain D, and suppose $f_n(z) \to f(z)$ as $n \to \infty$, uniformly on each compact subset of D. Then f is either univalent or constant in D.*

Proof. Suppose, on the contrary, that $f(z_1) = f(z_2) = \alpha$ for some pair of distinct points z_1 and z_2 in D. Then if $f(z) \not\equiv \alpha$, it follows from Hurwitz's theorem, or rather from its proof, that for $n \geq N$ the function $f_n(z) - \alpha$ vanishes in prescribed (disjoint) neighborhoods of both z_1 and z_2. This violates the univalence of f_n, so $f(z) \equiv \alpha$.

Alternatively, the theorem can be proved by direct appeal to Rouché's theorem. It should be remarked that the limit function can actually be constant. For example, let $f_n(z) = z/n$.

§1.2. Local Mapping Properties

A complex function $w = f(z)$ may be viewed geometrically as a mapping from a region in the z-plane to a region in the w-plane, defined by $u = u(x, y)$ and $v = v(x, y)$, where $z = x + iy$ and $w = u + iv$. Wherever f is analytic, its real and imaginary parts satisfy the *Cauchy–Riemann equations*

$$\frac{\partial u}{\partial x} = \frac{\partial v}{\partial y}, \qquad \frac{\partial u}{\partial y} = -\frac{\partial v}{\partial x}.$$

It follows that $|f'(z)|^2$ is the Jacobian of the mapping. Thus by the inverse mapping theorem, f is *locally univalent* wherever $f'(z) \neq 0$. More precisely, if f is analytic at z_0 and $f'(z_0) \neq 0$, then f is univalent in some neighborhood of z_0. Conversely, if f is locally univalent at z_0, then $f'(z_0) \neq 0$. Both statements can be proved by appeal to Rouché's theorem (see Titchmarsh [1], p. 198). For analytic mappings, therefore, the nonvanishing of the Jacobian is necessary and sufficient for local univalence. For more general smooth mappings this condition is sufficient but not necessary.

It must be emphasized that an analytic function may be locally univalent throughout a domain and yet fail to be univalent. For example, the function $f(z) = z^2$ is locally univalent in the domain

$$D = \{z: 1 < |z| < 2, 0 < \arg z < 3\pi/2\}$$

but is not univalent there.

It is well known that the Jacobian of a smooth mapping may be viewed as the local magnification factor of area. Thus if a function f is analytic and univalent in a domain D, the area of the image domain $\Delta = f(D)$ is given by the double integral

$$\iint_D |f'(z)|^2 \, dx \, dy.$$

If D is a Jordan domain bounded by a rectifiable Jordan curve C, with f analytic and univalent in its closure, an application of Green's theorem (see §1.7) shows that the area of Δ can be expressed by the contour integral

$$\frac{1}{2i}\int_\Gamma \bar{w}\, dw = \frac{1}{2i}\int_C \overline{f(z)} f'(z)\, dz,$$

where $\Gamma = f(C)$ is the image of C.

The derivative of an analytic function has further geometric significance. Its modulus is the local magnification factor of arclength, or the local measure of distortion. Thus if f is analytic on a rectifiable arc C and $\Gamma = f(C)$ is the image of C under f, the arclength of Γ is given by

$$\int_C |f'(z)||dz|.$$

Wherever $f'(z) \neq 0$, its argument represents the local rotation under the mapping f. Thus if C is a smooth arc having parametrization $z = \varphi(t)$, with $z_0 = \varphi(t_0)$ and $f'(z_0) \neq 0$, then the tangent vector of the image arc $\Gamma = f(C)$ at $w_0 = f(z_0)$ has inclination

$$\arg\{f'(z_0)\varphi'(t_0)\} = \arg f'(z_0) + \arg \varphi'(t_0).$$

In other words, the mapping f rotates the arc C by an angle $\arg f'(z_0)$ at the point z_0. In particular, the angle between two arcs intersecting at z_0 will be preserved under any analytic mapping f with $f'(z_0) \neq 0$. Because of this angle-preserving property, an analytic univalent function is known as a *conformal mapping.* This term is sometimes extended to locally univalent functions, but in this book it will be reserved for globally univalent functions.

Every *linear fractional transformation*

$$w = f(z) = \frac{az + b}{cz + d}, \qquad ad - bc \neq 0,$$

where a, b, c, d are complex constants, provides a conformal mapping of the extended complex plane $\hat{\mathbb{C}}$ onto itself. The extended complex plane is obtained from the complex plane $\mathbb{C}$ by adjoining the point at infinity. It can be identified by stereographic projection with the surface of a sphere, called the *Riemann sphere*.

Two points are said to be *inverse points* (or symmetric points) of a circle if they lie on the same radial half-line from the center and the product of their distances from the center is equal to the square of the radius. Thus α and β are inverse points of a circle $|z - z_0| = \rho$ if $\arg\{\alpha - z_0\} = \arg\{\beta - z_0\}$ and $|\alpha - z_0||\beta - z_0| = \rho^2$. The center of the circle and the point at infinity are taken to be inverse points. Each point on the circle is self-inverse. Two points are said to be inverse points of a line if the line is the perpendicular bisector of the segment joining the two points. Each point on the line is self-inverse.

A "*circle*" is either a circle or a line. Every linear fractional transformation maps "circles" to "circles" and inverse points to inverse points. Because of this property, it is easily seen that the most general linear fractional transformation of the unit disk onto itself has the form

$$f(z) = e^{i\theta} \frac{z - \alpha}{1 - \bar{\alpha}z}, \qquad |\alpha| < 1.$$

With the aid of the Schwarz lemma, it can be shown that in fact every conformal mapping of the unit disk onto itself has this form.

§1.3. Normal Families

A family $\mathscr{F}$ of functions analytic in a domain D is called a *normal family* if every sequence of functions $f_n \in \mathscr{F}$ has a subsequence which converges uniformly on each compact subset of D. A family $\mathscr{F}$ is *compact* if whenever $f_n \in \mathscr{F}$ and $f_n(z) \to f(z)$ uniformly on compact subsets of D, it follows that $f \in \mathscr{F}$. The defining property of a normal family is analogous to the Bolzano–Weierstrass property of a bounded set of points in Euclidean space. Compact families are analogous to closed sets.

A family $\mathscr{F}$ of functions analytic in D is said to be *locally bounded* if the functions are uniformly bounded on each closed disk $B \subset D$; that is, if $|f(z)| \leq M$ for all $z \in B$ and for every $f \in \mathscr{F}$, where the bound M depends only on B. In view of the Heine–Borel theorem, it then follows that the functions are uniformly bounded on each compact subset of D. If $\mathscr{F}$ is a locally bounded family of analytic functions, then by the Cauchy integral formula the family of derivatives $\{f' : f \in \mathscr{F}\}$ is also locally bounded. This remark is the key to Montel's theorem, which provides a very useful criterion for normality.

Montel's Theorem. *Every locally bounded family of analytic functions is normal.*

Proof. The theorem can be established by appeal to the closely related Arzela–Ascoli theorem on equicontinuous families of functions, but it is preferable to give an independent proof. Let $\mathscr{F}$ be a locally bounded family of functions analytic in a domain D, and let $\{f_n\}$ be an arbitrary sequence of functions in $\mathscr{F}$. Let $\{\zeta_j\}$ be a dense sequence of points in D. (Choose, for instance, the points with rational coordinates.) Because $\mathscr{F}$ is locally bounded, the numerical sequence $\{f_n(\zeta_j)\}$ is bounded for each fixed j. Extract a subsequence of $\{f_n\}$ which converges at ζ_1, then extract a further subsequence which converges at ζ_2, etc. The "diagonal" subsequence $\{g_n\}$ then converges at every point ζ_j, $j = 1, 2, \ldots$.

We claim that $\{g_n(z)\}$ converges at every point $z \in D$ and that the convergence is uniform on each compact set $K \subset D$. To show this, first use the Heine–Borel theorem to produce a bounded open set E containing K, with closure $\bar{E} \subset D$. Because $\mathscr{F}$ is locally bounded, it follows from the Cauchy formula that $|g_n'(z)| \leq M$ on E for every n. Given $\varepsilon > 0$, apply the Heine–Borel theorem again to construct a finite number of open disks $\Delta_k \subset E$, with diameters $d(\Delta_k) < \varepsilon/3M$, whose union contains K. Observe that for each pair of points α and β in the same disk Δ_k, the identity

$$g_n(\beta) - g_n(\alpha) = \int_\alpha^\beta g'(z)\, dz$$

shows that

$$|g_n(\alpha) - g_n(\beta)| \leq M|\alpha - \beta| \leq Md(\Delta_k) < \frac{\varepsilon}{3}.$$

For each index k, select a point $\omega_k \in \Delta_k$ from the dense sequence $\{\zeta_j\}$ where $\{g_n\}$ converges. Now each point $z \in K$ belongs to some disk Δ_k and so

$$\begin{aligned} |g_n(z) - g_m(z)| &\leq |g_n(z) - g_n(\omega_k)| \\ &\quad + |g_n(\omega_k) - g_m(\omega_k)| + |g_m(\omega_k) - g_m(z)| \\ &\leq \frac{2\varepsilon}{3} + |g_n(\omega_k) - g_m(\omega_k)| < \varepsilon \end{aligned}$$

for all $n, m \geq N$, where N is independent of z because there are only finitely many of the points ω_k. This proves that $\{g_n\}$ is a uniform Cauchy sequence on each compact set $K \subset D$, and so is uniformly convergent there.

The converse of Montel's theorem is also true.

Theorem. *Every normal family is locally bounded.*

Proof. Let $\mathscr{F}$ be a family of functions analytic in a domain D. If $\mathscr{F}$ is not locally bounded, it fails to be uniformly bounded on some compact set $K \subset D$. Thus

$f_n(z_n) \to \infty$ for some functions $f_n \in \mathscr{F}$ and points $z_n \in K$. But then no subsequence of $\{f_n\}$ can converge uniformly on K, so $\mathscr{F}$ is not a normal family.

For functions in a normal family, pointwise convergence actually implies uniform convergence on each compact subset. This remarkable fact is known as Vitali's theorem.

Vitali's Theorem. *Let the functions f_n be analytic and locally bounded in a domain D, and suppose that $\{f_n(z)\}$ converges at each point of a set which has a cluster point in D. Then $\{f_n(z)\}$ converges uniformly on each compact subset of D.*

Proof. Because the functions f_n are locally bounded, they form a normal family. Extract a subsequence $\{g_n\}$ which converges uniformly on each compact subset of D to an analytic function g. If $\{f_n\}$ does not converge uniformly on compact sets to g, then there exist $\varepsilon > 0$, a compact set $K \subset D$, a subsequence $\{f_{n_k}\}$, and a sequence of points $z_k \in K$ such that

$$|f_{n_k}(z_k) - g(z_k)| \geq \varepsilon, \qquad k = 1, 2, \ldots .$$

Extract a further subsequence of $\{f_{n_k}\}$ which converges uniformly on compact sets to a function h. Then $h = g$ because the two analytic functions agree on the set of points where $\{f_n\}$ converges, which has a cluster point in D. This contradiction completes the proof.

The theory of univalent functions is largely concerned with the family S of functions f analytic and univalent in the unit disk $\mathbb{D}$, and satisfying the conditions $f(0) = 0$ and $f'(0) = 1$. One of the basic results of the theory is the growth theorem (see §2.3), which asserts in part that

$$|f(z)| \leq |z|(1 - |z|)^{-2}, \qquad z \in \mathbb{D},$$

for each $f \in S$. In particular, the functions $f \in S$ are uniformly bounded on each compact subset of $\mathbb{D}$. Thus the family S is locally bounded, and so by Montel's theorem it is a normal family. Furthermore, we know (§1.1) that if $f_n \in S$ and $f_n(z) \to f(z)$ uniformly on each compact subset of $\mathbb{D}$, then f is analytic in $\mathbb{D}$ and is either univalent or constant. But it cannot be constant, because the uniform convergence implies (by the Cauchy integral formula) that $f_n'(0) \to f'(0)$, so $f'(0) = 1 \neq 0$. This shows that f is univalent in $\mathbb{D}$. In fact, it is clear that $f \in S$, because the normalizations $f(0) = 0$ and $f'(0) = 1$ are preserved under uniform convergence. Thus by appeal to the growth theorem, which will be proved independently in Chapter 2, we have obtained the following basic result.

Theorem. *The class S of univalent functions is a compact normal family.*

§1.4. Extremal Problems

Let $\mathcal{F}$ be a family of functions analytic in a domain D, and let ϕ be a complex-valued functional defined on $\mathcal{F}$. Thus $\phi(f)$ is a certain complex number for each $f \in \mathcal{F}$. One simple example is the point-evaluation functional $\phi(f) = f(\zeta)$ for fixed $\zeta \in D$. A functional ϕ is said to be *continuous* if $\phi(f_n) \to \phi(f)$ whenever a sequence of functions $f_n \in \mathcal{F}$ converges uniformly on each compact subset of D to a function $f \in \mathcal{F}$. For instance, the point-evaluation functionals are continuous, and so are the derivative functionals $\phi(f) = f'(\zeta)$, by the Cauchy integral formula.

For a given functional ϕ defined on a family $\mathcal{F}$, it is natural to pose the *extremal problem* of finding the supremum of $\text{Re}\{\phi\}$ over $\mathcal{F}$. Extremal problems play a central role in the geometric theory of functions, for two main reasons. On the one hand, they formulate problems of finding sharp estimates for certain geometric quantities. Obviously, the explicit knowledge of

$$\sup_{f \in \mathcal{F}} \text{Re}\{\phi(f)\} = M < \infty$$

will give the sharp inequality $\text{Re}\{\phi(f)\} \leq M$ for all $f \in \mathcal{F}$. On the other hand, it often happens that the supremum is attained for some function $f \in \mathcal{F}$, which is then called an *extremal function*. As a general rule, extremal functions are distinguished by elegant properties not enjoyed by other members of $\mathcal{F}$. Extremal problems therefore provide an effective method for establishing the existence of functions with certain natural properties. This is a standard method, for example, in the construction of canonical conformal mappings of multiply connected domains. As an important illustration, the method will be applied in the next section to prove the Riemann mapping theorem.

The basic principle is that whenever an extremal problem is posed for a continuous functional ϕ over a compact normal family $\mathcal{F}$, there must be an extremal function. In particular, $\text{Re}\{\phi\}$ is bounded on $\mathcal{F}$. Indeed, let $M \leq \infty$ denote the supremum of $\text{Re}\{\phi\}$ over $\mathcal{F}$. Choose a sequence of functions $f_n \in \mathcal{F}$ for which $\text{Re}\{\phi(f_n)\} \to M$. Since $\mathcal{F}$ is a compact normal family, some subsequence $\{f_{n_k}\}$ converges uniformly on compact subsets of D to a function $f \in \mathcal{F}$. But ϕ is continuous, so $\phi(f_{n_k}) \to \phi(f)$. This shows that $M < \infty$ and $\text{Re}\{\phi(f)\} = M$ for some $f \in \mathcal{F}$.

A family $\mathcal{G} \subset \mathcal{F}$ is called a *dense subfamily* of $\mathcal{F}$ if for each $f \in \mathcal{F}$ there is a sequence of functions $g_n \in \mathcal{G}$ which converges to f uniformly on each compact subset of D. If ϕ is a continuous functional on $\mathcal{F}$, the supremum of $\text{Re}\{\phi\}$ over any dense subfamily is equal to its supremum over $\mathcal{F}$. This simple remark can be enormously useful for solving an extremal problem, because it may restrict attention to functions with nice properties.

§1.5. The Riemann Mapping Theorem

As early as 1851, Riemann enunciated the basic theorem that every simply connected domain can be mapped conformally onto the unit disk. Riemann's proof tacitly assumed the existence of a solution to a certain extremal problem for the Dirichlet integral and was therefore incomplete. Half a century later, Koebe found a proof which avoids the difficulty by posing an extremal problem over a compact normal family, where the existence of an extremal function is assured. Koebe's proof has become a standard model for existence proofs in geometric function theory. Our presentation follows Ahlfors [5].

Riemann Mapping Theorem. *Let D be a simply connected domain which is a proper subset of the complex plane. Let ζ be a given point in D. Then there is a unique function f which maps D conformally onto the unit disk and has the properties $f(\zeta) = 0$ and $f'(\zeta) > 0$.*

Proof. The hypothesis that D not be the whole plane is essential because of Liouville's theorem that every bounded entire function is constant. The uniqueness assertion is easily established. Indeed, if g is another mapping with the given properties, the function $h = g \circ f^{-1}$ is a conformal mapping of the unit disk onto itself and is therefore a linear fractional mapping of the form displayed at the end of §1.2. But $h(0) = 0$ and $h'(0) > 0$, so h is the identity. Thus $f = g$ and the mapping is unique.

We now turn to the proof of existence. Consider the family $\mathscr{F}$ of all functions f analytic and univalent in D, with $f(\zeta) = 0$, $f'(\zeta) > 0$, and $|f(z)| < 1$ for all $z \in D$. This is the family of all normalized conformal mappings of D *into* the unit disk. According to Montel's theorem, $\mathscr{F}$ is a normal family. To see that $\mathscr{F}$ is nonempty, choose a finite point $\alpha \notin D$ and consider the function $g(z) = (z - \alpha)^{1/2}$. Since D is simply connected, g has a single-valued branch. This function g is analytic and univalent in D, and $g(z_1) \neq -g(z_2)$ for all points z_1 and z_2 in D. Thus because g assumes all values in some disk $|w - g(\zeta)| \leq \varepsilon$, it must omit the entire disk $|w + g(\zeta)| \leq \varepsilon$. Let ψ be the linear fractional mapping of the region $|w + g(\zeta)| > \varepsilon$ onto the unit disk with $\psi(g(\zeta)) = 0$ and $\psi'(g(\zeta)) > 0$. Then $\psi \circ g \in \mathscr{F}$.

Now let $\sup_{f \in \mathscr{F}} f'(\zeta) = M \leq \infty$, and choose a sequence of functions $f_n \in \mathscr{F}$ for which $f_n'(\zeta) \to M$. Since $\mathscr{F}$ is a normal family, some subsequence converges uniformly on compact sets to an analytic function f which is either univalent or constant. The limit function has the properties $f(\zeta) = 0$ and $f'(\zeta) = M > 0$. In particular, $M < \infty$ and f is not constant, so $f \in \mathscr{F}$.

The extremal function f is actually the required conformal mapping of D onto the unit disk. If not, then f omits some point $\omega \in \mathbb{D}$, and some branch of

$$F(z) = \left\{\frac{f(z) - \omega}{1 - \bar{\omega}f(z)}\right\}^{1/2}$$

is analytic and single-valued in D. Furthermore, F is univalent in D and $|F(z)| < 1$ there. The function

$$G(z) = e^{-i\theta} \frac{F(z) - F(\zeta)}{1 - \overline{F(\zeta)}F(z)},$$

where $e^{i\theta} = F'(\zeta)/|F'(\zeta)|$, therefore belongs to $\mathscr{F}$. However, a straightforward calculation gives

$$G'(\zeta) = \frac{|F'(\zeta)|}{1 - |F(\zeta)|^2} = \frac{1+|\omega|}{2\sqrt{|\omega|}} f'(\zeta),$$

and so $G'(\zeta) > f'(\zeta)$. This contradiction to the extremal property of f shows that f cannot omit any point in the unit disk. The proof is complete.

If D is a Jordan domain, the Riemann mapping can be extended continuously to the boundary, and the extended function maps the boundary curve in one-to-one fashion onto the unit circle. This important result is due to Carathéodory [3].

Carathéodory Extension Theorem. *Let D be a domain bounded by a Jordan curve C, and let f map D conformally onto the unit disk $\mathbb{D}$. Then f can be extended to a homeomorphism of $\bar{D} = D \cup C$ onto the closed disk $\bar{\mathbb{D}}$.*

For a proof, the reader is referred to the book of Goluzin [28]. One obvious corollary is that a conformal mapping of a Jordan domain D onto a Jordan domain Δ can be extended to a homeomorphism of $\bar{D}$ onto $\bar{\Delta}$.

§1.6. Analytic Continuation

According to the uniqueness principle, if two functions f and g are analytic in a domain D and $f(z) = g(z)$ on an arc Γ in D, then $f(z) = g(z)$ everywhere in D. Another version of this phenomenon is the *principle of analytic continuation*: if f and g are analytic in respective disjoint domains D and Δ which have a common boundary arc Γ, if f and g are both defined and analytic on Γ, and if $f(z) = g(z)$ on Γ, then the function

$$F(z) = \begin{cases} f(z), & z \in D \cup \Gamma, \\ g(z), & z \in \Delta \cup \Gamma, \end{cases}$$

is analytic in $D \cup \Gamma \cup \Delta$. The function F is called an *analytic continuation* of f (or g) across Γ to the larger domain.

It is an important fact that the principle of analytic continuation remains valid under the weaker hypothesis that f and g are continuous in $D \cup \Gamma$ and in $\Delta \cup \Gamma$, respectively, but are not assumed *a priori* to be analytic on Γ. This is the content of the following theorem.

Theorem. *Let D and Δ be disjoint domains with a common rectifiable boundary arc Γ. Let f be analytic in D and continuous in $D \cup \Gamma$, and let g be analytic in Δ and continuous in $\Delta \cup \Gamma$. Suppose $f(z) = g(z)$ for all $z \in \Gamma$. Then f has an analytic continuation F to $D \cup \Gamma \cup \Delta$ with $F(z) = g(z)$ in Δ.*

Proof. It must be shown that the function F defined as above is actually analytic on Γ. Given a point $z_0 \in \Gamma$, choose Jordan arcs $C_1 \subset D$ and $C_2 \subset \Delta$ with common endpoints on Γ whose union $C = C_1 \cup C_2$ is a Jordan curve enclosing z_0. Let γ be the subarc of Γ inside C. Let D^* and Δ^* be the subdomains of D and Δ enclosed by the Jordan curves $J_1 = C_1 \cup \gamma$ and $J_2 = C_2 \cup \gamma$, respectively. (See Figure 1.1.) By the extended version of Cauchy's theorem,

$$\frac{1}{2\pi i} \int_{J_1} \frac{f(\zeta)}{\zeta - z}\, d\zeta = \begin{cases} f(z), & z \in D^*, \\ 0, & z \in \Delta^*. \end{cases}$$

Similarly,

$$\frac{1}{2\pi i} \int_{J_2} \frac{g(\zeta)}{\zeta - z}\, d\zeta = \begin{cases} 0, & z \in D^*, \\ g(z), & z \in \Delta^*. \end{cases}$$

Because $f(\zeta) = g(\zeta)$ on Γ, the two integrals may be added to produce

$$\frac{1}{2\pi i} \int_{C} \frac{F(\zeta)}{\zeta - z}\, d\zeta = F(z)$$

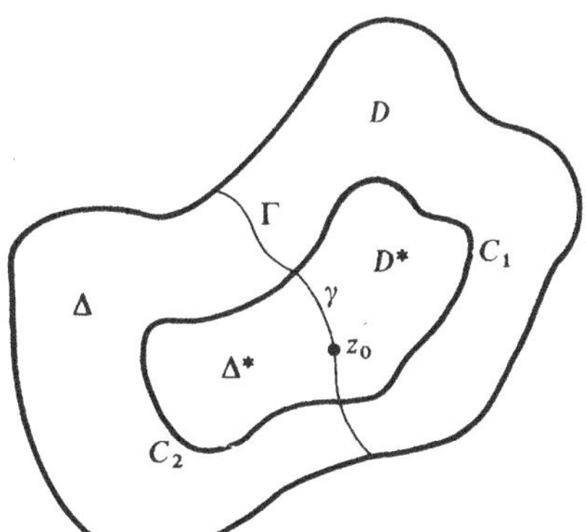

Figure 1.1

for $z \in D^* \cup \Delta^*$. But this last integral is analytic throughout the domain $D^* \cup \gamma \cup \Delta^*$ bounded by C. In particular, F is analytic at z_0. Since z_0 was an arbitrary point of Γ, this proves the theorem.

One useful consequence is the *Schwarz reflection principle.* It not only asserts the existence of an analytic continuation across a "circular" boundary arc which is mapped onto a "circular" arc, but it provides an actual formula for the continued function. The following special case will suffice to illustrate the idea. It is the case most frequently encountered in this book.

Theorem. *Let f be analytic in the unit disk $\mathbb{D}$ and have a continuous extension to an arc Γ of the boundary. Suppose further that $f(z)$ is real for every $z \in \Gamma$. Then f has an analytic continuation across Γ given by*

$$f(z) = \overline{f(1/\bar{z})}, \qquad |z| > 1.$$

Proof. For $z \in \mathbb{A} = \{z : |z| > 1\}$, define $g(z) = \overline{f(1/\bar{z})}$. Then g is analytic in $\mathbb{A}$ and continuous in $\mathbb{A} \cup \Gamma$. Furthermore, $g(z) \equiv f(z)$ on Γ, since f is real-valued there. Thus by the generalized principle of analytic continuation (as expressed in the previous theorem), the function

$$F(z) = \begin{cases} f(z), & z \in \mathbb{D} \cup \Gamma, \\ g(z), & z \in \mathbb{A} \cup \Gamma, \end{cases}$$

is analytic in $\mathbb{D} \cup \Gamma \cup \mathbb{A}$.

The Schwarz reflection principle has a local generalization to functions which map analytic boundary arcs into analytic arcs. An arc

$$z = \varphi(t), \qquad a < t < b,$$

is said to be an *analytic arc* if at each point $t_0 \in (a, b)$ the function φ has a power series expansion

$$\varphi(t) = \sum_{n=0}^{\infty} b_n(t - t_0)^n, \qquad |t - t_0| < \delta.$$

Theorem. *Let f be analytic in a domain D and have a continuous extension to an analytic boundary arc Γ. Let C be another analytic arc and suppose $f(z) \in C$ for each point $z \in \Gamma$. Then f is analytic on Γ.*

Proof. Near a point $z_0 \in \Gamma$ the analytic parametrization may be chosen to have the form

$$z = \varphi(t) = z_0 + \sum_{n=1}^{\infty} b_n(t - t_0)^n, \qquad |t - t_0| < \delta,$$

with $b_1 \neq 0$. The power series allows an extension of φ to an analytic function of the *complex* variable t in the disk $|t - t_0| < \delta$. Since $\varphi'(t_0) \neq 0$, we may assume (choosing δ smaller if necessary) that φ is univalent in this disk. Similarly, the portion of C near $w_0 = f(z_0)$ has a parametrization

$$w = \psi(s) = w_0 + \sum_{k=1}^{\infty} c_k(s - s_0)^k, \qquad |s - s_0| < \varepsilon,$$

with $c_1 \neq 0$, which may be extended to an analytic univalent function of the complex variable s in the disk $|s - s_0| < \varepsilon$. Without loss of generality, we may assume that φ maps the upper half of the disk $|t - t_0| < \delta$ into D. Then for sufficiently small δ the function $F = \psi^{-1} \circ f \circ \varphi$ is defined and analytic in this semidisk, its range is contained in the disk $|s - s_0| < \varepsilon$, and it maps the real segment $|t - t_0| < \delta$ into the real segment $|s - s_0| < \varepsilon$. By the Schwarz reflection principle, F has an analytic continuation to the full disk $|t - t_0| < \delta$ defined by $F(t) = \overline{F(\bar{t})}$ in the lower semidisk. This induces an analytic continuation of f to a full neighborhood of z_0, defined by $f = \psi \circ F \circ \varphi^{-1}$.

An examination of the proof reveals that the continued function f is locally univalent on Γ if the original function f is locally univalent near the boundary. Furthermore, it can be shown that a globally univalent function has a globally univalent extension across the boundary. In particular, a function $f \in S$ which maps the unit disk onto a Jordan domain with analytic boundary can be extended to a univalent function in a larger disk $|z| < \rho$, for some radius $\rho > 1$. The proof is left as an exercise.

§1.7. Harmonic and Subharmonic Functions

It is convenient to begin with a statement of *Green's theorem*, which will be applied repeatedly. Let D be a domain in the plane with boundary $C = \partial D$ composed of a finite number of smooth Jordan curves, oriented in the positive sense (keeping the domain D always on the left). Let $P = P(x, y)$ and $Q = Q(x, y)$ have continuous first partial derivatives in $\bar{D}$, the closure of D. Green's theorem then connects a line integral over C with an area integral over D:

$$\int_C P\,dx + Q\,dy = \iint_D \left(\frac{\partial Q}{\partial x} - \frac{\partial P}{\partial y}\right) dx\,dy.$$

An equivalent version is

$$\int_C \mathbf{F} \cdot \mathbf{n}\,ds = \iint_D \operatorname{div} \mathbf{F}\,dx\,dy,$$

where $\mathbf{F} = u\mathbf{i} + v\mathbf{j}$ is a vector-valued function with divergence

$$\operatorname{div} \mathbf{F} = \frac{\partial u}{\partial x} + \frac{\partial v}{\partial y},$$

$\mathbf{n}$ is the unit outer normal vector on C, and ds is the element of arclength. Applying this formula to the function $\mathbf{F} = \varphi \nabla\psi$, where $\nabla\psi$ is the gradient of ψ, we obtain

$$\int_C \varphi \frac{\partial \psi}{\partial n} \, ds = \iint_D \left(\frac{\partial \varphi}{\partial x} \frac{\partial \psi}{\partial x} + \frac{\partial \varphi}{\partial y} \frac{\partial \psi}{\partial y} \right) dx\, dy + \iint_D \varphi \, \Delta\psi \, dx\, dy.$$

Here $\partial\psi/\partial n$ denotes the outer normal derivative of ψ, and

$$\Delta\psi = \frac{\partial^2 \psi}{\partial x^2} + \frac{\partial^2 \psi}{\partial y^2}$$

denotes the *Laplacian* of ψ. Interchanging φ and ψ and subtracting the corresponding formula, we obtain finally

$$\int_C \left(\varphi \frac{\partial \psi}{\partial n} - \psi \frac{\partial \varphi}{\partial n} \right) ds = \iint_D (\varphi \, \Delta\psi - \psi \, \Delta\varphi) \, dx\, dy.$$

This is known as *Green's formula.*

A function $u = u(x, y)$ with continuous second partial derivatives is said to be *harmonic* in a domain D if $\Delta u = 0$ everywhere in D. In view of the Cauchy-Riemann equations

$$\frac{\partial u}{\partial x} = \frac{\partial v}{\partial y}, \qquad \frac{\partial u}{\partial y} = -\frac{\partial v}{\partial x},$$

the real and imaginary parts of an analytic function $f(z) = u(z) + iv(z)$ are both harmonic in D. A function v is called a *harmonic conjugate* of u if u and v satisfy the Cauchy–Riemann equations; that is, if $u + iv$ is analytic. When a harmonic conjugate exists, it is unique up to an arbitrary additive constant. If v is a harmonic conjugate of u, then $-u$ is a harmonic conjugate of v.

In a simply connected domain D, every harmonic function u has a harmonic conjugate v, defined by the line integral

$$v(z) = \int_\Gamma -\frac{\partial u}{\partial y} dx + \frac{\partial u}{\partial x} dy,$$

where the path of integration Γ extends from a fixed base-point $z_0 \in D$ to a general point $z \in D$. The integral is independent of path, by Green's theorem, because u is harmonic and D is simply connected.

A harmonic function of an analytic function is harmonic: if u is harmonic and g is analytic, then the composition $u \circ g$ is harmonic. This may be seen either by a calculation based on the Cauchy–Riemann equations or by the local completion of u to an analytic function $u + iv$.

Suppose now that D has a boundary C consisting of a finite number of smooth Jordan curves, and let u be harmonic in $\bar{D}$. Then by Green's theorem applied to the gradient of u,

$$\int_C \frac{\partial u}{\partial n}\, ds = \iint_D \Delta u \, dx\, dy = 0.$$

From this it is clear that the integral of u is constant over all sufficiently small circles centered at a given point $z_0 \in D$; that is,

$$\int_0^{2\pi} u(z_0 + re^{i\theta})\, d\theta$$

is independent of the radius r. Letting $r \to 0$, we obtain the *mean-value theorem*

$$u(z_0) = \frac{1}{2\pi} \int_0^{2\pi} u(z_0 + re^{i\theta})\, d\theta.$$

More generally, u can be recovered from its boundary function on a circle $|z - z_0| = R$ by the *Poisson formula*

$$u(z_0 + re^{i\theta}) = \frac{1}{2\pi} \int_0^{2\pi} P_R(r, t - \theta) u(z_0 + Re^{it})\, dt, \qquad 0 \le r < R,$$

where

$$P_R(r, t) = \frac{R^2 - r^2}{R^2 - 2Rr\cos t + r^2}$$

is the *Poisson kernel*. This is easily deduced from the mean-value theorem by considering the composition of u with a conformal self-mapping of the disk.

On the basis of the mean-value theorem, it can be shown that a function harmonic in D and continuous in $\bar{D}$ must attain its maximum and minimum values on the boundary. It can attain neither a local maximum nor a local minimum in D unless it is identically constant. This is known as the *maximum principle* for harmonic functions.

The *Dirichlet problem* is that of finding a function harmonic in D and continuous in $\bar{D}$, with prescribed boundary values on ∂D. In view of the maximum principle, there can be at most one solution. A solution exists under very general conditions (see Ahlfors [5]). In the special case where D is the disk $|z| < R$, the Poisson formula

$$u(re^{i\theta}) = \frac{1}{2\pi}\int_0^{2\pi} P_R(r, t - \theta)\varphi(Re^{it})\,dt$$

produces the solution for every choice of continuous boundary function φ.

A function u defined in a domain D is said to have the *local mean-value property* if to each point $z_0 \in D$ there corresponds a radius $\rho > 0$ such that

$$u(z_0) = \frac{1}{2\pi}\int_0^{2\pi} u(z_0 + re^{i\theta})\,d\theta$$

for all radii $r \leq \rho$. This property actually characterizes harmonic functions.

Theorem. *If a function u is continuous and has the local mean-value property in a domain D, then u is harmonic in D.*

Proof. Choose an arbitrary disk $\Delta = \{z: |z - z_0| < R\}$ with $\bar{\Delta} \subset D$, and let U be the Poisson integral of u over the boundary circle $\Gamma = \partial\Delta$. Then U is harmonic in Δ, and $U(z) = u(z)$ for all $z \in \Gamma$. But $(u - U)$ has the local mean-value property in Δ and so obeys the maximum principle there. Thus $U(z) = u(z)$ for all $z \in \Delta$, and u is harmonic.

A continuous function u is said to be *subharmonic* in a domain D if $u(z) \leq U(z)$ in each disk Δ with $\bar{\Delta} \subset D$, where U is the Poisson integral of u over the boundary of Δ. A function u has the *local sub-mean-value property* if

$$u(z_0) \leq \frac{1}{2\pi}\int_0^{2\pi} u(z_0 + re^{i\theta})\,d\theta$$

for each $z_0 \in D$ and for all $r > 0$ sufficiently small. It is clear that every subharmonic function has the local sub-mean-value property. The converse is less obvious.

Theorem. *If a function u is continuous and has the local sub-mean-value property in a domain D, then u is subharmonic in D.*

Proof. Construct the harmonic function U as in the proof of the previous theorem, and observe that $(u - U)$ has the local sub-mean-value property in Δ. It follows that $(u - U)$ attains its maximum value on $\Gamma = \partial\Delta$. But $u(z) - U(z) \equiv 0$ on Γ, so $u(z) \leq U(z)$ in Δ. Thus u is subharmonic.

As a corollary to the proof, we observe that a subharmonic function which is not identically constant cannot have a local maximum at an interior point. It is easily seen that the sum of two subharmonic functions is subharmonic. Also, the function

$$w(z) = \max\{u(z), v(z)\}$$

is subharmonic if u and v are. With the help of Jensen's inequality it can be shown that a convex nondecreasing function of a subharmonic function has the local sub-mean-value property, and is therefore subharmonic.

This last principle shows, for instance, that $|u(z)|^p$ is subharmonic for each $p \geq 1$ if u is harmonic. If f is analytic, then $\log|f(z)|$ is harmonic wherever $f(z) \neq 0$, and so $|f(z)|^p$ is subharmonic for each $p > 0$. In a region where f has zeros, $\log|f(z)|$ is subharmonic in a generalized sense which allows the value $-\infty$.

A function u is said to be *superharmonic* if $-u$ is subharmonic.

§1.8. Green's Functions

The Dirichlet problem for a general domain with smooth boundary can be solved by an integral formula analogous to the Poisson integral for the disk. The kernel is expressed in terms of *Green's function* of the domain, which is obtained by solving a special Dirichlet problem with logarithmic boundary data.

Let D be a finitely connected domain whose boundary C consists of smooth Jordan curves. Fix an arbitrary point $\zeta \in D$. Then Green's function of D is the function $g(z, \zeta)$ of the form

$$g(z, \zeta) = h(z, \zeta) - \log|z - \zeta|,$$

where $h(z, \zeta)$ is harmonic in D as a function of z, and $g(z, \zeta) = 0$ for all $z \in C$. Thus $h(z, \zeta)$ is the (unique) solution to the Dirichlet problem in D with boundary function $\log|z - \zeta|$.

Green's function has several obvious properties. For each fixed ζ, it is harmonic in D except for a logarithmic singularity at ζ. Thus it follows from the maximum principle that $g(z, \zeta) > 0$ for all $z \in D$. If the boundary curves of C are analytic, then g has a harmonic continuation across the boundary. In particular, g has a normal derivative at the boundary, and $\partial g/\partial n \leq 0$ there, where $\partial/\partial n$ denotes the derivative in the outer normal direction.

The symmetry property $g(z, \zeta) = g(\zeta, z)$ lies deeper. To prove it we fix distinct points ζ_1 and ζ_2 in D and apply Green's formula

$$\int_C \left(u\frac{\partial v}{\partial n} - v\frac{\partial u}{\partial n}\right) ds = \iint_D (u\,\Delta v - v\,\Delta u)\,dx\,dy$$

to the region D_ε obtained from D by removing small disks of radius ε centered at ζ_1 and ζ_2. Let

$$g_1(z) = g(z, \zeta_1), \qquad g_2(z) = g(z, \zeta_2),$$

and let C_ε be the boundary of D_ε, consisting of C together with circles Γ_1 and Γ_2 about ζ_1 and ζ_2. Then by Green's formula,

$$\int_{C_\varepsilon} \left(g_1 \frac{\partial g_2}{\partial n} - g_2 \frac{\partial g_1}{\partial n} \right) ds = \iint_{D_\varepsilon} (g_1 \, \Delta g_2 - g_2 \, \Delta g_1) \, dx \, dy.$$

But since g_1 and g_2 vanish on C and are harmonic in D_ε, this reduces to

$$\int_{\Gamma_1} \left(g_1 \frac{\partial g_2}{\partial n} - g_2 \frac{\partial g_1}{\partial n} \right) ds + \int_{\Gamma_2} \left(g_1 \frac{\partial g_2}{\partial n} - g_2 \frac{\partial g_1}{\partial n} \right) ds = 0.$$

Now let $\varepsilon \to 0$, observing that

$$\frac{\partial g_1}{\partial n} = \frac{\partial h_1}{\partial n} + \frac{1}{\varepsilon} \quad \text{on } \Gamma_1$$

and

$$\frac{\partial g_2}{\partial n} = \frac{\partial h_2}{\partial n} + \frac{1}{\varepsilon} \quad \text{on } \Gamma_2,$$

while the other factors grow no faster than $\log(1/\varepsilon)$. Thus we obtain

$$-2\pi g_2(\zeta_1) + 2\pi g_1(\zeta_2) = 0,$$

or $g(\zeta_1, \zeta_2) = g(\zeta_2, \zeta_1)$, the symmetry property of Green's function. As a corollary, it is clear that for each fixed $z \in D$, $g(z, \zeta)$ is a harmonic function of ζ.

We now turn to the reproducing property of Green's function. Again let D be a domain whose boundary C consists of a finite number of analytic Jordan curves, and let u be harmonic in $\bar{D}$. Fix $\zeta \in D$ and let D_ε be the domain obtained from D by removal of a small disk of radius ε centered at ζ. Let Γ_ε be the circle bounding this disk, and let $C_\varepsilon = C \cup \Gamma_\varepsilon$ be the boundary of D_ε. Then by Green's formula,

$$\int_{C_\varepsilon} \left(u \frac{\partial g}{\partial n} - g \frac{\partial u}{\partial n} \right) ds = \iint_{D_\varepsilon} (u \, \Delta g - g \, \Delta u) \, dx \, dy = 0,$$

where $g = g(z, \zeta)$ is Green's function of D. Since $g(z, \zeta) = 0$ on C, this reduces to

$$\int_C u \frac{\partial g}{\partial n} ds + \int_{\Gamma_\varepsilon} \left(u \frac{\partial g}{\partial n} - g \frac{\partial u}{\partial n} \right) ds = 0.$$

Letting $\varepsilon \to 0$ and using the local structure of g at ζ, we obtain

$$u(\zeta) = -\frac{1}{2\pi} \int_C u(z) \frac{\partial g}{\partial n}(z, \zeta)\, ds,$$

a formula which expresses $u(\zeta)$ for each point $\zeta \in D$ as an integral of its boundary values on C. If an arbitrary continuous function $\varphi(z)$ is substituted for $u(z)$ in this formula, it gives an explicit solution to the Dirichlet problem.

Unfortunately, however, the calculation of Green's function may be quite difficult. If D is simply connected, the problem is equivalent to finding the Riemann mapping function of D onto the unit disk. Indeed, if $w = f(z)$ maps D onto the unit disk $|w| < 1$, with $f(\zeta) = 0$, then

$$g(z, \zeta) = -\log|f(z)|$$

is Green's function of D. Conversely, if Green's function $g(z, \zeta)$ is known, then

$$f(z) = \exp\{-g(z, \zeta) - i\tilde{g}(z, \zeta)\}$$

is a conformal mapping of D onto the unit disk with $f(\zeta) = 0$. Here $\tilde{g}(z, \zeta)$ denotes the harmonic conjugate of $g(z, \zeta)$. It has period 2π as z winds around ζ, so f is single-valued.

For further information on Green's functions and related topics, the reader is referred to Nehari [2].

§1.9. Positive Harmonic Functions

In Chapter 2 we shall consider various subclasses of the class S of univalent functions. Several of these subclasses are defined analytically by the condition that a certain analytic function have positive real part. More specifically, they involve the class of functions f analytic in the unit disk $\mathbb{D}$, with $\mathrm{Re}\{f(z)\} > 0$ in $\mathbb{D}$ and $f(0) = 1$. Thus we are led to consider positive harmonic functions in $\mathbb{D}$, normalized by the condition $u(0) = 1$. According to the famous *Herglotz representation theorem*, found in 1911 by Herglotz [1], every positive harmonic function is the Poisson integral of a positive measure.

Recall that the Poisson kernel of $\mathbb{D}$ is

$$P(r, \theta) = \frac{1 - r^2}{1 - 2r\cos\theta + r^2}, \qquad r < 1.$$

It is helpful to bear in mind that

$$P(r, \theta) = \operatorname{Re}\left\{\frac{1 + z}{1 - z}\right\}, \qquad z = re^{i\theta}.$$

Herglotz Representation Theorem. *Let u be a positive harmonic function in $\mathbb{D}$ with $u(0) = 1$. Then there exists a unique positive unit measure $d\mu$ such that*

$$u(re^{i\theta}) = \int_0^{2\pi} P(r, \theta - t)\, d\mu(t), \qquad r < 1.$$

Corollary. *Let f be an analytic function with positive real part in $\mathbb{D}$. Then f has the form*

$$f(z) = \int_0^{2\pi} \frac{e^{it} + z}{e^{it} - z}\, d\mu(t) + i\gamma, \qquad |z| < 1,$$

where $d\mu$ is a positive measure and γ is a real constant.

The corollary is obtained by analytic completion of the Poisson kernel. The theorem is proved by appeal to the *Helly selection theorem,* also known as the *Helly–Bray theorem.* The special case for monotonic functions is as follows.

Helly Selection Theorem. *Let $\{\alpha_n\}$ be a sequence of nondecreasing functions on a bounded interval $[a, b]$, with $\alpha_n(a) = 0$ and $\alpha_n(b) = 1$. Then some subsequence $\{\alpha_{n_k}\}$ converges everywhere in $[a, b]$ to a nondecreasing function α, and for each function φ continuous on $[a, b]$,*

$$\lim_{k\to\infty} \int_a^b \varphi(t)\, d\alpha_{n_k}(t) = \int_a^b \varphi(t)\, d\alpha(t).$$

Proof. By a diagonalization process we may extract from $\{\alpha_n\}$ a subsequence $\{\beta_n\}$ such that $\beta_n(t) \to \alpha(t)$ for every rational number t in $[a, b]$. For an arbitrary t in $[a, b]$, let

$$\alpha_*(t) = \liminf_{n\to\infty} \beta_n(t)$$

and

$$\alpha^*(t) = \limsup_{n\to\infty} \beta_n(t).$$

Then $\alpha_*(t) = \alpha^*(t) = \alpha(t)$ for each rational number t. The functions α_* and α^* are nondecreasing and are therefore continuous aside from a countable set $E \subset [a, b]$. For each $t \notin E$, it is clear that $\alpha_*(t) = \alpha^*(t)$, because the rational numbers are dense in $[a, b]$. Thus the subsequence $\{\beta_n(t)\}$ converges for each $t \notin E$. Another diagonalization process applied to $\{\beta_n\}$ now produces a further subsequence $\{\alpha_{n_k}\}$ which converges everywhere on the countable set E. The function $\alpha(t) = \lim_{k\to\infty} \alpha_{n_k}(t)$ is therefore defined and nondecreasing on $[a, b]$, with $\alpha(a) = 0$ and $\alpha(b) = 1$.

To prove the statement concerning integrals, we take advantage of the *uniform* continuity of φ on $[a, b]$. Given $\varepsilon > 0$, choose a partition

$$a = t_0 < t_1 < t_2 < \cdots < t_m = b$$

such that $|\varphi(t) - \varphi(t_j)| < \varepsilon$ for $t_{j-1} \le t \le t_j$, $j = 1, 2, \ldots, m$. Let M be the maximum value of $|\varphi(t)|$ on $[a, b]$. Then

$$\begin{aligned}
&\left| \int_a^b \varphi(t)\, d\alpha_{n_k}(t) - \int_a^b \varphi(t)\, d\alpha(t) \right| \\
&\quad \le \sum_{j=1}^m \int_{t_{j-1}}^{t_j} |\varphi(t) - \varphi(t_j)|\, d\alpha_{n_k}(t) + \sum_{j=1}^m \int_{t_{j-1}}^{t_j} |\varphi(t_j) - \varphi(t)|\, d\alpha(t) \\
&\qquad + \sum_{j=1}^m |\varphi(t_j)| \left| \int_{t_{j-1}}^{t_j} d(\alpha_{n_k}(t) - \alpha(t)) \right| \\
&\quad \le 2\varepsilon + M \sum_{j=1}^m |\alpha_{n_k}(t_j) - \alpha(t_j) - \alpha_{n_k}(t_{j-1}) + \alpha(t_{j-1})| < 3\varepsilon
\end{aligned}$$

for all k sufficiently large. This completes the proof.

Proof of Herglotz Representation Theorem. For $r < 1$, define

$$\mu_r(t) = \frac{1}{2\pi} \int_0^t u(re^{i\theta})\, d\theta, \qquad r < 1.$$

Then μ_r is an increasing function with $\mu_r(0) = 0$ and $\mu_r(2\pi) = u(0) = 1$. By the Helly selection theorem, there is a sequence of radii r_n increasing to 1 for which $\mu_{r_n}(t) \to \mu(t)$, a nondecreasing function on $[0, 2\pi]$. By the Poisson formula,

$$u(r_n z) = \frac{1}{2\pi} \int_0^{2\pi} P(r, \theta - t) u(r_n e^{it})\, dt = \int_0^{2\pi} P(r, \theta - t)\, d\mu_{r_n}(t),$$

where $z = re^{i\theta} \in \mathbb{D}$. Letting $n \to \infty$ and appealing to the integration part of the Helly selection theorem, we obtain the desired representation.

To prove the uniqueness of the representing measure, suppose

$$\int_0^{2\pi} P(r, \theta - t)\, d\mu(t) \equiv 0$$

for some signed measure $d\mu$ (the difference of two positive measures). Analytic completion then gives

$$\int_0^{2\pi} \frac{e^{it} + z}{e^{it} - z}\, d\mu(t) \equiv i\gamma, \qquad |z| < 1,$$

where γ is a real constant. Expanding the kernel in power series,

$$\frac{e^{it} + z}{e^{it} - z} = 1 + 2\sum_{n=1}^{\infty} e^{-int} z^n,$$

we conclude that

$$\int_0^{2\pi} e^{int}\, d\mu(t) = 0, \qquad n = 0, \pm 1, \pm 2, \ldots .$$

It follows that the measure $d\mu$ annihilates every trigonometric polynomial. By the Weierstrass approximation theorem, it must therefore annihilate every continuous periodic function. Since the characteristic function of any interval can be approximated in L^1 norm by a continuous periodic function, this shows that the measure of each interval is zero. Thus $d\mu$ is the zero measure. This proves the uniqueness of the Herglotz representation.

It should be emphasized that although we have used the language of measure theory, the Herglotz representation may be viewed as a Riemann–Stieltjes integral. An alternate proof, based on the Krein–Milman theorem, may be found in Schober [1].

Exercises

1. Use Rouché's theorem to give a direct proof of the theorem (stated at the end of §1.1) that the uniform limit of univalent functions is either univalent or constant.

2. Use Rouché's theorem to prove that if f is analytic and univalent in a domain D, then $f'(z) \neq 0$ in D.

3. Let C be a rectifiable Jordan curve with length L, bounding a domain with area A. Prove the *isoperimetric inequality* $A \leq L^2/4\pi$, which says that among all curves of given length, the circle encloses the largest area. (*Suggestion*: Let f map the unit disk conformally onto the given domain. Express A and L as integrals involving f', and let $g = \sqrt{f'}$ to calculate these integrals in terms of power series coefficients. The idea of this proof is attributed to Carleman.)

4. Let f be analytic but not univalent in a disk $|z| < R$. Show that there are distinct points z_1 and z_2 of *equal* modulus ($|z_1| = |z_2| < R$) for which $f(z_1) = f(z_2)$.

5. Let $f(z) = \sum_{k=0}^{\infty} a_k z^k$ be analytic and univalent in $\mathbb{D}$. Show that for each fixed radius ρ, $0 < \rho < 1$, the nth partial sum $s_n(z) = \sum_{k=0}^{n} a_k z^k$ is univalent in the disk $|z| < \rho$, for each n sufficiently large.

6. Prove that the univalent polynomials are dense in S. In other words, show that each function $f \in S$ can be approximated uniformly on each compact subset of $\mathbb{D}$ by a polynomial $p \in S$.

7. Prove the *subordination principle*: If f is analytic and univalent in $\mathbb{D}$ and if g is a function analytic in $\mathbb{D}$ with $g(0) = f(0)$ and $g(\mathbb{D}) \subset f(\mathbb{D})$, then $|g'(0)| \leq |f'(0)|$ and $g(\mathbb{D}_r) \subset f(\mathbb{D}_r)$ for every $r < 1$, where $\mathbb{D}_r$ is the disk $|z| < r$.

8. Let $\mathscr{F}$ be the family of functions analytic in a domain D and omitting two fixed values: $f(z) \neq \alpha$ and $f(z) \neq \beta$ ($\alpha \neq \beta$) for each $f \in \mathscr{F}$ and for all $z \in \mathbb{D}$. Prove that $\mathscr{F}$ is a normal family. (*Hint*: Use the elliptic modular function and the monodromy theorem as in the proof of Picard's theorem. See Ahlfors [5].)

9. Let f map the unit disk conformally onto a domain bounded by an analytic Jordan curve. Show that f has an analytic and univalent continuation to a larger disk $|z| < \rho$, for some radius $\rho > 1$.

10. Calculate Green's function of the unit disk and use it to derive the Poisson formula from the general integral formula in §1.8.

11. Let $f \in S$ be a rational function which maps $\mathbb{D}$ conformally onto the complement of an arc Γ extending to infinity. Prove that Γ is a half-line. (Srebro [1].)

Chapter 2

Elementary Theory of Univalent Functions

This chapter introduces the class S of univalent functions and some of its subclasses defined by geometric conditions. A number of basic questions are answered by elementary methods. Most of the elementary results concerning the class S are direct consequences of the area theorem, which may be regarded as the cornerstone of the entire subject.

§2.1. Introduction

A single-valued function f is said to be *univalent* (or *schlicht*) in a domain $D \subset \mathbb{C}$ if it never takes the same value twice; that is, if $f(z_1) \neq f(z_2)$ for all points z_1 and z_2 in D with $z_1 \neq z_2$. The function f is said to be *locally univalent* at a point $z_0 \in D$ if it is univalent in some neighborhood of z_0. For analytic functions f, the condition $f'(z_0) \neq 0$ is equivalent to local univalence at z_0 (see §1.2). An analytic univalent function is called a *conformal mapping* because of its angle-preserving property (§1.2).

We shall be concerned primarily with the class S of functions f analytic and univalent in the unit disk $\mathbb{D} = \{z: |z| < 1\}$, normalized by the conditions $f(0) = 0$ and $f'(0) = 1$. Thus each $f \in S$ has a Taylor series expansion of the form

$$f(z) = z + a_2 z^2 + a_3 z^3 + \cdots, \qquad |z| < 1.$$

In view of the Riemann mapping theorem, most of the geometric theorems concerning functions of class S are readily translated to statements about univalent functions in arbitrary simply connected domains with more than one boundary point.

The leading example of a function of class S is the *Koebe function*

$$k(z) = z(1 - z)^{-2} = z + 2z^2 + 3z^3 + \cdots.$$

The Koebe function maps the disk $\mathbb{D}$ onto the entire plane minus the part of the negative real axis from $-\frac{1}{4}$ to infinity. This is best seen by writing

$$k(z) = \frac{1}{4}\left(\frac{1+z}{1-z}\right)^2 - \frac{1}{4}$$

and observing that the function

$$w = \frac{1+z}{1-z}$$

maps $\mathbb{D}$ conformally onto the right half-plane $\operatorname{Re}\{w\} > 0$.

Other simple examples of functions in S are:

(i) $f(z) = z$, the identity mapping;
(ii) $f(z) = z(1-z)^{-1}$, which maps $\mathbb{D}$ conformally onto the half-plane $\operatorname{Re}\{w\} > -\frac{1}{2}$;
(iii) $f(z) = z(1-z^2)^{-1}$, which maps $\mathbb{D}$ onto the entire plane minus the two half-lines $\frac{1}{2} \le x < \infty$ and $-\infty < x \le -\frac{1}{2}$;
(iv) $f(z) = \frac{1}{2}\log[(1+z)/(1-z)]$, which maps $\mathbb{D}$ onto the horizontal strip $-\pi/4 < \operatorname{Im}\{w\} < \pi/4$;
(v) $f(z) = z - \frac{1}{2}z^2 = \frac{1}{2}[1-(1-z)^2]$, which maps $\mathbb{D}$ onto the interior of a cardioid.

The sum of two functions in S need not be univalent. For example, the sum of $z(1-z)^{-1}$ and $z(1+iz)^{-1}$ has a derivative which vanishes at $\frac{1}{2}(1+i)$, as a calculation shows. However, the class S is preserved under a number of elementary transformations. Here is a partial list.

(i) *Conjugation.* If $f \in S$ and $g(z) = \overline{f(\bar{z})} = z + \bar{a}_2 z^2 + \bar{a}_3 z^3 + \cdots$, then $g \in S$.
(ii) *Rotation.* If $f \in S$ and $g(z) = e^{-i\theta} f(e^{i\theta} z)$, then $g \in S$.
(iii) *Dilation.* If $f \in S$ and $g(z) = r^{-1} f(rz)$, where $0 < r < 1$, then $g \in S$.
(iv) *Disk automorphism.* If $f \in S$ and

$$g(z) = \frac{f\left(\dfrac{z+\alpha}{1+\bar{\alpha}z}\right) - f(\alpha)}{(1-|\alpha|^2)f'(\alpha)}, \qquad |\alpha| < 1,$$

then $g \in S$.
(v) *Range transformation.* If $f \in S$ and ψ is a function analytic and univalent on the range of f, with $\psi(0) = 0$ and $\psi'(0) = 1$, then $g = \psi \circ f \in S$.
(vi) *Omitted-value transformation.* If $f \in S$ and $f(z) \ne \omega$, then $g = \omega f/(\omega - f) \in S$.
(vii) *Square-root transformation.* If $f \in S$ and $g(z) = \sqrt{f(z^2)}$, then $g \in S$.

The square-root transformation requires a word of explanation. Since $f(z) = 0$ only at the origin, a single-valued branch of the square root may be chosen by writing

$$\begin{aligned} g(z) &= \sqrt{f(z^2)} = z\{1 + a_2 z^2 + a_3 z^4 + \cdots\}^{1/2} \\ &= z + c_3 z^3 + c_5 z^5 + \cdots, \qquad |z| < 1. \end{aligned}$$

Note that g is an odd analytic function, so that $g(-z) = -g(z)$. If $g(z_1) = g(z_2)$, then $f(z_1^2) = f(z_2^2)$ and $z_1^2 = z_2^2$, which gives $z_1 = \pm z_2$. But if $z_1 = -z_2$, then $g(z_1) = g(z_2) = -g(z_1)$. Thus $g(z_1) = 0$, and $z_1 = 0$. This shows that $z_1 = z_2$ in either case, proving that g is univalent.

It should be remarked that every odd function in S is the square-root transform of some function in S. More generally, let $S^{(m)}$ denote the subclass of S consisting of all functions

$$f(z) = z + \sum_{\nu=1}^{\infty} a_{m\nu+1} z^{m\nu+1},$$

with m-fold symmetry, where $m = 2, 3, \ldots$. Then the mth-root transform $g(z) = \{f(z^m)\}^{1/m}$ of each function $f \in S$ belongs to $S^{(m)}$, and conversely every function $g \in S^{(m)}$ has this form. (See Exercise 7.)

Closely related to S is the class Σ of functions

$$g(z) = z + b_0 + b_1 z^{-1} + b_2 z^{-2} + \cdots$$

analytic and univalent in the domain $\Delta = \{z: |z| > 1\}$ exterior to $\mathbb{D}$, except for a simple pole at infinity with residue 1. Each function $g \in \Sigma$ maps Δ onto the complement of a compact connected set E. It is useful to consider the subclass Σ' of functions $g \in \Sigma$ for which $0 \in E$; that is, for which $g(z) \neq 0$ in Δ. Any function $g \in \Sigma$ will belong to Σ' after suitable adjustment of the constant term b_0. Such an adjustment will only translate the range of g and will not destroy the univalence.

For each $f \in S$, the function

$$g(z) = \{f(1/z)\}^{-1} = z - a_2 + (a_2^2 - a_3)z^{-1} + \cdots$$

belongs to Σ'. This transformation is called an *inversion*. It actually establishes a one-to-one correspondence between S and Σ'. The class Σ' is preserved under the square-root transformation

$$G(z) = \sqrt{g(z^2)} = z\{1 + b_0 z^{-2} + b_1 z^{-4} + \cdots\}^{1/2}.$$

It is important to observe that this operation cannot be applied to every function $g \in \Sigma$, but is permissible only if $g \in \Sigma'$, because the square root will introduce a branch point wherever $g(z^2) = 0$.

Sometimes it is convenient to consider the subclass Σ_0 consisting of all $g \in \Sigma$ with $b_0 = 0$. Obviously this can be achieved by suitable translation, but it may not be possible to translate a given function $g \in \Sigma$ simultaneously to both Σ_0 and Σ'.

It is also useful to distinguish the subclass $\tilde{\Sigma}$ of all functions $g \in \Sigma$ whose omitted set E has two-dimensional Lebesgue measure zero. The functions $g \in \tilde{\Sigma}$ will be called *full mappings*.

§2.2. The Area Theorem

The univalence of a function

$$g(z) = z + b_0 + \sum_{n=1}^{\infty} b_n z^{-n}, \qquad |z| > 1,$$

places a strong restriction on the size of the Laurent coefficients b_n, $n = 1, 2, \ldots$. This is expressed by the *area theorem*, which is fundamental to the theory of univalent functions. The reason for the name will be apparent from the proof. Gronwall [1] discovered the theorem in 1914.

Theorem 2.1 (Area Theorem). *If $g \in \Sigma$, then*

$$\sum_{n=1}^{\infty} n|b_n|^2 \le 1,$$

with equality if and only if $g \in \tilde{\Sigma}$.

Proof. Let E be the set omitted by g. For $r > 1$, let C_r be the image under g of the circle $|z| = r$. Since g is univalent, C_r is a simple closed curve which encloses a domain $E_r \supset E$. By Green's theorem, the area of E_r is

$$\begin{aligned} A_r &= \frac{1}{2i}\int_{C_r} \bar{w}\, dw = \frac{1}{2i}\int_{|z|=r} \overline{g(z)}g'(z)\, dz \\ &= \frac{1}{2}\int_0^{2\pi} \left\{ re^{-i\theta} + \sum_{n=0}^{\infty} \bar{b}_n r^{-n} e^{in\theta} \right\} \\ &\quad \times \left\{ 1 - \sum_{\nu=1}^{\infty} \nu b_\nu r^{-\nu-1} e^{-i(\nu+1)\theta} \right\} re^{i\theta}\, d\theta \\ &= \pi\left\{ r^2 - \sum_{n=1}^{\infty} n|b_n|^2 r^{-2n} \right\}, \qquad r > 1. \end{aligned}$$

Letting r decrease to 1, we obtain

$$m(E) = \pi\left\{1 - \sum_{n=1}^{\infty} n|b_n|^2\right\},$$

where $m(E)$ is the outer measure of E. Since $m(E) \geq 0$, this proves the theorem.

An immediate corollary is the inequality $|b_n| \leq n^{-1/2}$, $n = 1, 2, \ldots$. This inequality is not sharp if $n \geq 2$, since the function

$$g(z) = z + n^{-1/2}z^{-n}$$

is not univalent. Indeed, its derivative

$$g'(z) = 1 - n^{1/2}z^{-n-1}$$

vanishes at certain points in Δ if $n \geq 2$. However, the inequality $|b_1| \leq 1$ is sharp and has important consequences.

Corollary. *If* $g \in \Sigma$, *then* $|b_1| \leq 1$, *with equality if and only if* g *has the form*

$$g(z) = z + b_0 + b_1/z, \qquad |b_1| = 1.$$

This is a conformal mapping of Δ *onto the complement of a line segment of length* 4.

From this last result it is a short step to a theorem of Bieberbach [3] estimating the second coefficient a_2 of a function of class S. This theorem was given in 1916 and was the main basis for the famous *Bieberbach conjecture*, to be discussed in §2.4.

Theorem 2.2 (Bieberbach's Theorem). *If* $f \in S$, *then* $|a_2| \leq 2$, *with equality if and only if* f *is a rotation of the Koebe function.*

Proof. A square-root transformation and an inversion applied to $f \in S$ will produce a function

$$g(z) = \{f(1/z^2)\}^{-1/2} = z - (a_2/2)z^{-1} + \cdots$$

of class Σ. Thus $|a_2| \leq 2$, by the corollary to the area theorem. Equality occurs if and only if g has the form

$$g(z) = z - e^{i\theta}/z.$$

A simple calculation shows that this is equivalent to

$$f(\zeta) = \zeta(1 - e^{i\theta}\zeta)^{-2} = e^{-i\theta}k(e^{i\theta}\zeta),$$

a rotation of the Koebe function.

As a first application of Bieberbach's theorem, we shall now prove a famous covering theorem due to Koebe. Each function $f \in S$ is an open mapping with $f(0) = 0$, so its range contains some disk centered at the origin. As early as 1907, Koebe [1] discovered that the ranges of all functions in S contain a common disk $|w| < \rho$, where ρ is an absolute constant. The "Koebe function" shows that $\rho \leq \frac{1}{4}$, and Bieberbach [3] later established Koebe's conjecture that ρ may be taken to be $\frac{1}{4}$.

Theorem 2.3 (Koebe One-Quarter Theorem). *The range of every function of class S contains the disk $\{w: |w| < \frac{1}{4}\}$.*

Proof. If a function $f \in S$ omits the value $\omega \in \mathbb{C}$, then

$$g(z) = \frac{\omega f(z)}{\omega - f(z)} = z + \left(a_2 + \frac{1}{\omega}\right)z^2 + \cdots$$

is analytic and univalent in $\mathbb{D}$. This is the omitted-value transformation (vi) (see §2.1), which is the composition of f with a linear fractional mapping. Since $g \in S$, Bieberbach's theorem gives

$$\left|a_2 + \frac{1}{\omega}\right| \leq 2.$$

Combined with the inequality $|a_2| \leq 2$, this shows $|1/\omega| \leq 4$, or $|\omega| \geq \frac{1}{4}$. Thus every omitted value must lie outside the disk $|w| < \frac{1}{4}$.

The proof actually shows that the Koebe function and its rotations are the only functions in S which omit a value of modulus $\frac{1}{4}$. Thus the range of every other function in S covers a disk of larger radius.

It should be observed that *univalence* is the key to Koebe's theorem. For example, the analytic functions

$$f_n(z) = \frac{1}{n}(e^{nz} - 1), \qquad n = 1, 2, \ldots,$$

have the properties $f_n(0) = 0$ and $f_n'(0) = 1$, yet f_n omits the value $-1/n$, which may be chosen arbitrarily close to the origin.

§2.3. Growth and Distortion Theorems

Bieberbach's inequality $|a_2| \leq 2$ has further implications in the geometric theory of conformal mapping. One important consequence is the *Koebe distortion theorem* (German: *Verzerrungssatz*), which provides sharp upper and lower bounds for $|f'(z)|$ as f ranges over the class S. The term "distortion" arises from the geometric interpretation of $|f'(z)|$ as the infinitesimal magnification factor of arclength under the mapping f, or from that of the Jacobian $|f'(z)|^2$ as the infinitesimal magnification factor of area. The following theorem gives a basic estimate which leads to the distortion theorem and related results.

Theorem 2.4. *For each $f \in S$,*

$$\left|\frac{zf''(z)}{f'(z)} - \frac{2r^2}{1-r^2}\right| \leq \frac{4r}{1-r^2}, \qquad |z| = r < 1. \tag{1}$$

Proof. Given $f \in S$, fix $\zeta \in \mathbb{D}$ and perform a disk automorphism to construct

$$F(z) = \frac{f\left(\dfrac{z+\zeta}{1+\bar{\zeta}z}\right) - f(\zeta)}{(1-|\zeta|^2)f'(\zeta)} = z + A_2(\zeta)z^2 + \cdots. \tag{2}$$

Then $F \in S$ and a calculation gives

$$A_2(\zeta) = \frac{1}{2}\left\{(1-|\zeta|^2)\frac{f''(\zeta)}{f'(\zeta)} - 2\bar{\zeta}\right\}.$$

But by Bieberbach's theorem, $|A_2(\zeta)| \leq 2$. Simplifying this inequality and replacing ζ by z, we obtain the inequality (1). A suitable rotation of the Koebe function shows that the estimate is sharp for each $z \in \mathbb{D}$.

Theorem 2.5 (Distortion Theorem). *For each $f \in S$,*

$$\frac{1-r}{(1+r)^3} \leq |f'(z)| \leq \frac{1+r}{(1-r)^3}, \qquad |z| = r < 1. \tag{3}$$

For each $z \in \mathbb{D}$, $z \neq 0$, equality occurs if and only if f is a suitable rotation of the Koebe function.

Proof. Since an inequality $|\alpha| \leq c$ implies $-c \leq \operatorname{Re}\{\alpha\} \leq c$, it follows from (1) that

$$\frac{2r^2 - 4r}{1-r^2} \leq \operatorname{Re}\left\{\frac{zf''(z)}{f'(z)}\right\} \leq \frac{2r^2+4r}{1-r^2}.$$

Because $f'(z) \neq 0$ and $f'(0) = 1$, we can choose a single-valued branch of $\log f'(z)$ which vanishes at the origin. Now observe that

$$\mathrm{Re}\left\{\frac{zf''(z)}{f'(z)}\right\} = r\frac{\partial}{\partial r}\mathrm{Re}\{\log f'(z)\}, \qquad z = re^{i\theta}.$$

Hence

$$\frac{2r-4}{1-r^2} \leq \frac{\partial}{\partial r}\log|f'(re^{i\theta})| \leq \frac{2r+4}{1-r^2}. \tag{4}$$

Holding θ fixed, integrate with respect to r from 0 to R. A calculation then produces the inequality

$$\log\frac{1-R}{(1+R)^3} \leq \log|f'(Re^{i\theta})| \leq \log\frac{1+R}{(1-R)^3},$$

and the distortion theorem follows by exponentiation.

A suitable rotation of the Koebe function, whose derivative is

$$k'(z) = \frac{1+z}{(1-z)^3},$$

shows that both estimates of $|f'(z)|$ are best possible. Furthermore, whenever equality occurs for $z = Re^{i\theta}$ in either the upper or the lower estimate of (3), then equality must hold in the corresponding part of (4) for all r, $0 \leq r \leq R$. In particular,

$$\mathrm{Re}\left\{e^{i\theta}\frac{f''(0)}{f'(0)}\right\} = \pm 4,$$

which implies that $|a_2| = 2$. Hence by Bieberbach's theorem, f must be a rotation of the Koebe function.

The distortion theorem will now be applied to obtain the sharp upper and lower bounds for $|f(z)|$. This result is as follows.

Theorem 2.6 (Growth Theorem). *For each $f \in S$,*

$$\frac{r}{(1+r)^2} \leq |f(z)| \leq \frac{r}{(1-r)^2}, \qquad |z| = r < 1. \tag{5}$$

For each $z \in \mathbb{D}$, $z \neq 0$, equality occurs if and only if f is a suitable rotation of the Koebe function.

Proof. Let $f \in S$ and fix $z = re^{i\theta}$ with $0 < r < 1$. Observe that

$$f(z) = \int_0^r f'(\rho e^{i\theta})e^{i\theta}\, d\rho,$$

since $f(0) = 0$. Thus by the distortion theorem,

$$|f(z)| \le \int_0^r |f'(\rho e^{i\theta})|\, d\rho \le \int_0^r \frac{1+\rho}{(1-\rho)^3}\, d\rho = \frac{r}{(1-r)^2}.$$

The lower estimate is more subtle. It holds trivially if $|f(z)| \ge \frac{1}{4}$, since $r(1+r)^{-2} < \frac{1}{4}$ for $0 < r < 1$. If $|f(z)| < \frac{1}{4}$, the Koebe one-quarter theorem implies that the radial segment from 0 to $f(z)$ lies entirely in the range of f. Let C be the preimage of this segment. Then C is a simple arc from 0 to z, and

$$f(z) = \int_C f'(\zeta)\, d\zeta.$$

But $f'(\zeta)\, d\zeta$ has constant signum along C, by construction, so the distortion theorem gives

$$|f(z)| = \int_C |f'(\zeta)||d\zeta| \ge \int_0^r \frac{1-\rho}{(1+\rho)^3}\, d\rho = \frac{r}{(1+r)^2}.$$

Equality in either part of (5) implies equality in the corresponding part of (3), which implies (by Theorem 2.5) that f is a rotation of the Koebe function.

All of this information was obtained by passing to the real part in the basic inequality (1). Taking the imaginary part instead, one finds

$$-\frac{4r}{1-r^2} \le \operatorname{Im}\left\{\frac{zf''(z)}{f'(z)}\right\} \le \frac{4r}{1-r^2},$$

or

$$-\frac{4}{1-r^2} \le \frac{\partial}{\partial r} \arg f'(re^{i\theta}) \le \frac{4}{1-r^2}.$$

Radial integration now produces the inequality

$$|\arg f'(z)| \le 2 \log \frac{1+r}{1-r}, \qquad f \in S. \tag{6}$$

Here it is understood that $\arg f'(z)$ is the branch which vanishes at the origin. The quantity $\arg f'(z)$ can be interpreted geometrically as the local rotation factor under the conformal mapping f. For this reason the inequality (6) may be called a *rotation theorem*. Unfortunately, however, it is not sharp at any point $z \neq 0$ in the disk. The true rotation theorem

$$|\arg f'(z)| \leq \begin{cases} 4 \sin^{-1} r, & r \leq 1/\sqrt{2}, \\ \pi + \log \dfrac{r^2}{1 - r^2}, & r \geq 1/\sqrt{2}, \end{cases}$$

lies much deeper and will be proved in Chapter 3 by means of Loewner's method. The splitting of the sharp bound at $r = 1/\sqrt{2}$ is one of the most remarkable phenomena in univalent function theory.

One further inequality, a combined growth and distortion theorem, is sometimes useful.

Theorem 2.7. *For each $f \in S$,*

$$\frac{1 - r}{1 + r} \leq \left| \frac{zf'(z)}{f(z)} \right| \leq \frac{1 + r}{1 - r}, \qquad |z| = r < 1. \tag{7}$$

For each $z \in \mathbb{D}$, $z \neq 0$, equality occurs if and only if f is a suitable rotation of the Koebe function.

Proof. The Koebe function, for which

$$\frac{zk'(z)}{k(z)} = \frac{1 + z}{1 - z},$$

clearly offers cases of equality. Nevertheless, the inequality (7) cannot be established by direct combination of the growth and distortion theorems. Instead we must return to the function $F \in S$ given in (2), obtained from f by a disk automorphism. By the growth theorem,

$$\frac{|\zeta|}{(1 + |\zeta|)^2} \leq |F(-\zeta)| \leq \frac{|\zeta|}{(1 - |\zeta|)^2}.$$

Hence

$$\frac{1 - |\zeta|}{1 + |\zeta|} \leq \left| \frac{f(\zeta)}{\zeta f'(\zeta)} \right| \leq \frac{1 + |\zeta|}{1 - |\zeta|},$$

which proves (7).

It remains to show that rotations of the Koebe function provide the only cases of equality. Suppose first that the lower bound in (7) is attained for some $f \in S$ at some point $z = \zeta \in \mathbb{D}$. Then the related function F satisfies

$$|F(-\zeta)| = \frac{|\zeta|}{(1 - |\zeta|)^2},$$

from which we infer by the growth theorem that

$$F(z) = \frac{z}{(1 + e^{i\varphi}z)^2}, \qquad e^{i\varphi} = \bar{\zeta}/|\zeta|.$$

Let

$$w = \frac{z + \zeta}{1 + \bar{\zeta}z}, \quad \text{or} \quad z = \frac{w - \zeta}{1 - \bar{\zeta}w},$$

and let

$$G(w) = F\left(\frac{w - \zeta}{1 - \bar{\zeta}w}\right).$$

Then the relation (2) between f and F shows after a straightforward calculation that $f(w)$ is a constant multiple of

$$G(w) - G(0) = \frac{(1 + |\zeta|)^2}{1 - |\zeta|} \frac{w}{(1 + e^{-i\varphi}w)^2}.$$

Thus f is a rotation of the Koebe function. A similar argument applies to the case of equality for the upper bound in (7).

§2.4. Coefficient Estimates

We have seen that each function

$$f(z) = z + \sum_{n=2}^{\infty} a_n z^n$$

of class S has the property $|a_2| \leq 2$, with equality occurring only for rotations of the Koebe function

$$k(z) = z(1 - z)^{-2} = \sum_{n=1}^{\infty} nz^n.$$

This suggests the general problem to find

$$A_n = \sup_{f \in S} |a_n|, \qquad n = 2, 3, 4, \ldots .$$

Because the Koebe function plays the extremal role in so many problems for the class S, it is natural to suspect that it maximizes $|a_n|$ for every n. This is the famous conjecture of Bieberbach [3], first proposed in 1916.

Bieberbach Conjecture. *The coefficients of each function $f \in S$ satisfy $|a_n| \leq n$ for $n = 2, 3, \ldots$. Strict inequality holds for all n unless f is the Koebe function or one of its rotations.*

For many years this conjecture has stood as a challenge to all mathematicians and has inspired the development of important new methods in complex analysis. To this date it has been proved only for $n = 2, 3, 4, 5$, and 6. It has been proved for all n within certain subclasses of S, but the full conjecture remains open. Much of this book is devoted to partial solutions of the Bieberbach conjecture. (See Notes.)

First of all, it is easy to see that the supremum A_n is finite for each n. It follows from the compactness of S as a subset of the space $\mathscr{H}(\mathbb{D})$ of all functions analytic in $\mathbb{D}$, and from the continuity of the functional a_n, that the supremum is actually attained for some function in S. Alternatively, a crude estimation of the Cauchy integral formula for a_n, combined with the growth theorem (Theorem 2.6), gives the bound $A_n \leq Cn^2$ for some constant C. The following theorem of Littlewood [1] improves this estimate and shows that the Bieberbach conjecture has the correct order of magnitude.

Theorem 2.8 (Littlewood's Theorem). *The coefficients of each function $f \in S$ satisfy $|a_n| < en$ for $n = 2, 3, \ldots$.*

Littlewood's theorem appeared in 1925. Since that time the constant e has been replaced by a succession of smaller constants, at the sacrifice of simplicity and elegance. (Some of these refinements are discussed in Chapter 5.) Littlewood's proof is entirely elementary. It rests upon an estimate for integral means which has independent interest. With the notation

$$M_p(r, f) = \left\{ \frac{1}{2\pi} \int_0^{2\pi} |f(re^{i\theta})|^p \, d\theta \right\}^{1/p}, \qquad 0 < p < \infty,$$

this may be stated as follows.

Lemma. *For each function $f \in S$,*

$$M_1(r, f) \leq \frac{r}{1 - r}, \qquad 0 \leq r < 1. \tag{8}$$

Littlewood's theorem is an easy consequence of this lemma. The Cauchy representation for a_n gives

$$|a_n| \leq r^{-n} M_1(r, f) \leq r^{-n+1}(1 - r)^{-1},$$

an expression which attains a minimum at $r = 1 - 1/n$. Choosing this value of r, one finds

$$|a_n| \leq n\left(1 + \frac{1}{n-1}\right)^{n-1} < en,$$

which proves the theorem.

The crude estimate $|a_n| \leq r^{-n} M_1(r, f)$ can lead to nothing better than $|a_n| < (e/2)n$, because the Koebe function has integral mean

$$M_1(r, k) = \frac{1}{2\pi} \int_0^{2\pi} \frac{r}{1 - 2r\cos\theta + r^2}\, d\theta = \frac{r}{1 - r^2}.$$

The sharp bound for $M_1(r,f)$ remained undetermined until 1973, when Baernstein [1] was able to prove that $M_p(r, f) \leq M_p(r, k)$ for all $f \in S$ and for $0 < p < \infty$. In particular, this proves $|a_n| < (e/2)n$. (Baernstein's theory is described in Chapter 7. See Exercise 17.)

Littlewood proved the inequality (8) by an elegant geometric argument.

Proof of Lemma. Given $f \in S$, let

$$h(z) = \sqrt{f(z^2)} = \sum_{n=1}^{\infty} c_n z^n$$

be its square-root transform. The growth theorem gives $|f(z)| \leq r(1 - r)^{-2}$, so it follows that

$$|h(z)| \leq \frac{r}{1 - r^2}, \qquad |z| = r < 1.$$

In other words, h maps the disk $|z| < r$ conformally onto a domain D_r which lies in the disk $|w| < r(1 - r^2)^{-1}$. The area A_r of D_r is therefore no greater than the area of this disk:

$$A_r \leq \pi r^2 (1 - r^2)^{-2}.$$

But a calculation gives

$$A_r = \int_0^{2\pi} \int_0^r |h'(\rho e^{i\theta})|^2 \rho\, d\rho\, d\theta = \pi \sum_{n=1}^{\infty} n|c_n|^2 r^{2n}.$$

Consequently,

$$\sum_{n=1}^{\infty} n|c_n|^2 r^{2n-1} \le \frac{r}{(1-r^2)^2}, \qquad 0 \le r < 1.$$

Integration from 0 to r gives

$$\sum_{n=1}^{\infty} |c_n|^2 r^{2n} \le \frac{r^2}{1-r^2},$$

or

$$\frac{1}{2\pi}\int_0^{2\pi} |h(re^{i\theta})|^2 \, d\theta \le \frac{r^2}{1-r^2},$$

which is equivalent to (8).

The inequality (8) has another interesting application, also due to Littlewood [1, 3]. Let

$$L_r(f) = r\int_0^{2\pi} |f'(re^{i\theta})| \, d\theta$$

be the arclength of the image of the circle $|z| = r$ under the mapping $f \in S$. Then we have the following estimate.

Theorem 2.9. *For each $f \in S$,*

$$L_r(f) \le \frac{2\pi r(1+r)}{(1-r)^2}, \qquad 0 \le r < 1.$$

Proof. Applying the inequalities (7) and (8), we obtain

$$L_r(f) = \int_0^{2\pi} \left|\frac{zf'(z)}{f(z)}\right| |f(z)| \, d\theta \le \frac{2\pi r(1+r)}{(1-r)^2},$$

where $z = re^{i\theta}$.

It is a surprising fact that this inequality, which has been derived by such crude methods, is very nearly sharp. Indeed, it can be shown by estimation of elliptic integrals (see Chapter 7, Exercise 3) that the Koebe function has the property

$$L_r(k) > \frac{\pi r(1+r)}{2(1-r)^2}, \qquad 0 < r < 1.$$

The sharp upper bound for $L_r(f)$ is not known. If Baernstein's inequality

$$M_1(r, f) \le M_1(r, k) = \frac{r}{1 - r^2}$$

is applied instead of (8), the preceding argument gives the improved estimate

$$L_r(f) \le \frac{2\pi r}{(1 - r)^2}, \qquad 0 \le r < 1.$$

Clunie and Duren [1] have shown that $L_r(f) \le L_r(k)$ for all *close-to-convex* functions $f \in S$ (see §2.6).

§2.5. Convex and Starlike Functions

In the next few sections we shall investigate certain subclasses of S which are defined by natural geometric conditions. Among other things, we shall prove the Bieberbach conjecture for each of these subclasses.

A set $E \subset \mathbb{C}$ is said to be *starlike* with respect to a point $w_0 \in E$ if the linear segment joining w_0 to every other point $w \in E$ lies entirely in E. In more picturesque language, the requirement is that every point of E be "visible" from w_0. The set E is said to be *convex* if it is starlike with respect to each of its points; that is, if the linear segment joining any two points of E lies entirely in E. A *convex function* is one which maps the unit disk conformally onto a convex domain. A *starlike function is* a conformal mapping of the unit disk onto a domain starlike with respect to the origin. The subclass of S consisting of the convex functions is denoted by C, and S^* denotes the subclass of starlike functions. Thus $C \subset S^* \subset S$. Note that the Koebe function is starlike but not convex.

Closely related to the classes C and S^* is the class P of all functions φ analytic and having positive real part in $\mathbb{D}$, with $\varphi(0) = 1$. According to the Herglotz formula (§1.9), every $\varphi \in P$ can be represented as a Poisson–Stieltjes integral

$$\varphi(z) = \int_0^{2\pi} \frac{e^{it} + z}{e^{it} - z}\, d\mu(t), \tag{9}$$

where $d\mu(t) \ge 0$ and $\int d\mu(t) = 1$. The following lemma, first proved by Carathéodory [1], is often useful.

Carathéodory's Lemma. *If $\varphi \in P$ and*

$$\varphi(z) = 1 + \sum_{n=1}^{\infty} c_n z^n,$$

then $|c_n| \leq 2$, $n = 1, 2, \ldots$. This inequality is sharp for each n.

Proof. Since

$$\frac{e^{it} + z}{e^{it} - z} = 1 + 2\sum_{n=1}^{\infty} e^{-int} z^n,$$

the representation (9) gives

$$c_n = 2\int_0^{2\pi} e^{-int}\, d\mu(t), \qquad n = 1, 2, \ldots.$$

Thus $|c_n| \leq 2$, with equality if and only if e^{-int} has constant signum on the support of the measure $d\mu$. In particular, equality holds for all n for the function

$$\varphi(z) = \frac{1 + z}{1 - z} = 1 + 2\sum_{n=1}^{\infty} z^n.$$

The following theorem gives an analytic description of starlike functions.

Theorem 2.10. *Let f be analytic in $\mathbb{D}$, with $f(0) = 0$ and $f'(0) = 1$. Then $f \in S^*$ if and only if $zf'(z)/f(z) \in P$.*

Proof. Suppose first that $f \in S^*$. Then we claim that f maps each subdisk $|z| < \rho < 1$ onto a starlike domain. An equivalent assertion is that $g(z) = f(\rho z)$ is starlike in $\mathbb{D}$. In other words, we must show that for each fixed t ($0 < t < 1$) and for each $z \in \mathbb{D}$, the point $tg(z)$ is in the range of g. But since $f \in S^*$, an application of the Schwarz lemma gives $tf(z) = f(\omega(z))$ for some function ω analytic in $\mathbb{D}$ and satisfying $|\omega(z)| \leq |z|$. Thus

$$tg(z) = tf(\rho z) = f(\omega(\rho z)) = g(\omega_1(z)),$$

where $\omega_1(z) = \omega(\rho z)/\rho$ and $|\omega_1(z)| \leq |z|$.

This proves that f maps each circle $|z| = \rho < 1$ onto a curve C_ρ that bounds a starlike domain. It follows that $\arg f(z)$ increases as z moves around the circle $|z| = \rho$ in the positive direction. In other words,

$$\frac{\partial}{\partial\theta}\{\arg f(\rho e^{i\theta})\} \geq 0.$$

But

$$\frac{\partial}{\partial\theta}\{\arg f(\rho e^{i\theta})\} = \operatorname{Im}\left\{\frac{\partial}{\partial\theta}\log f(\rho e^{i\theta})\right\}$$
$$= \operatorname{Im}\left\{\frac{izf'(z)}{f(z)}\right\} = \operatorname{Re}\left\{\frac{zf'(z)}{f(z)}\right\}, \qquad z = \rho e^{i\theta}.$$

Thus $zf'(z)/f(z) \in P$, by the maximum principle for harmonic functions.

Conversely, suppose f is a normalized analytic function with $zf'(z)/f(z) \in P$. Then f has a simple zero at the origin and no zeros elsewhere in the disk. Retracing the above calculation, we see that for each $\rho < 1$,

$$\frac{\partial}{\partial\theta}\{\arg f(\rho e^{i\theta})\} > 0, \qquad 0 \le \theta \le 2\pi.$$

Thus as z runs around the circle $|z| = \rho$ in the counter-clockwise direction, the point $f(z)$ traverses a closed curve C_ρ with increasing argument. Because f has exactly one zero inside the circle $|z| = \rho$, the argument principle tells us that C_ρ surrounds the origin exactly once. But if C_ρ winds about the origin only once with increasing argument, it can have no self-intersections. Thus C_ρ is a simple closed curve which bounds a starlike domain D_ρ, and f assumes each value $w \in D_\rho$ exactly once in the disk $|z| < \rho$. Since this is true for every $\rho < 1$, it follows that f is univalent and starlike in $\mathbb{D}$. This concludes the proof.

The convex functions can be described in a similar way.

Theorem 2.11. *Let f be analytic in $\mathbb{D}$, with $f(0) = 0$ and $f'(0) = 1$. Then $f \in C$ if and only if $[1 + zf''(z)/f'(z)] \in P$.*

Proof. Suppose first that $f \in C$. We claim that f must map each subdisk $|z| < r$ onto a convex domain. To show this, choose points z_1 and z_2 with $|z_1| \le |z_2| < r$. Let $w_1 = f(z_1)$ and $w_2 = f(z_2)$. Let

$$w_0 = tw_1 + (1-t)w_2, \qquad 0 < t < 1.$$

Then since f is a convex mapping, there is a unique point $z_0 \in \mathbb{D}$ for which $f(z_0) = w_0$. We have to show that $|z_0| < r$. But the function

$$g(z) = tf(z_1 z/z_2) + (1-t)f(z)$$

is analytic in $\mathbb{D}$, with $g(0) = 0$ and $g(z_2) = w_0$. Since $f \in C$, the function

$$h(z) = f^{-1}(g(z))$$

is well-defined. Since $h(0) = 0$ and $|h(z)| \leq 1$, the Schwarz lemma tells us that $|h(z)| \leq |z|$. Thus

$$|z_0| = |h(z_2)| \leq |z_2| < r,$$

which was to be shown. Hence f maps each circle $|z| = r < 1$ onto a curve C_r which bounds a convex domain. The convexity implies that the slope of the tangent to C_r is nondecreasing as the curve is traversed in the positive direction. Analytically, this condition is

$$\frac{\partial}{\partial\theta}\left(\arg\left\{\frac{\partial}{\partial\theta} f(re^{i\theta})\right\}\right) \geq 0,$$

or

$$\operatorname{Im}\left\{\frac{\partial}{\partial\theta}\log[ire^{i\theta}f'(re^{i\theta})]\right\} \geq 0,$$

which reduces to the condition

$$\operatorname{Re}\left\{1 + \frac{zf''(z)}{f'(z)}\right\} \geq 0, \qquad |z| = r.$$

Thus by the maximum principle for harmonic functions, $[1 + zf''(z)/f'(z)] \in P$.

Conversely, suppose f is a normalized analytic function with $[1 + zf''(z)/f'(z)] \in P$. The above calculation shows that the slope of the tangent to the curve C_r increases monotonically. But as a point makes a complete circuit of C_r, the argument of the tangent vector has a net change

$$\int_0^{2\pi} \frac{\partial}{\partial\theta}\left(\arg\left\{\frac{\partial}{\partial\theta} f(re^{i\theta})\right\}\right) d\theta = \int_0^{2\pi} \operatorname{Re}\left\{1 + \frac{zf''(z)}{f'(z)}\right\} d\theta$$

$$= \operatorname{Re}\left\{\int_{|z|=r}\left[1 + \frac{zf''(z)}{f'(z)}\right]\frac{dz}{iz}\right\} = 2\pi, \qquad z = re^{i\theta}.$$

This shows that C_r is a simple closed curve bounding a convex domain. This for arbitrary $r < 1$ implies that f is a univalent function with convex range.

The two preceding theorems reveal a surprisingly close analytic connection between convex and starlike mappings. This was first observed by Alexander [1] in 1915.

Theorem 2.12 (Alexander's Theorem). *Let f be analytic in $\mathbb{D}$, with $f(0) = 0$ and $f'(0) = 1$. Then $f \in C$ if and only if $zf'(z) \in S^*$.*

Proof. If $g(z) = zf'(z)$, then

$$\frac{zg'(z)}{g(z)} = 1 + \frac{zf''(z)}{f'(z)}.$$

Thus the left-hand function is analytic and has positive real part in $\mathbb{D}$ if and only if the same is true of the right-hand function. The assertion now follows from Theorems 2.10 and 2.11.

Near the origin each function $f \in S$ is close to the identity mapping. It is to be expected that f will map small circles $|z| = \rho$ onto curves which bound convex domains. The following theorem expresses this notion in quantitative terms.

Theorem 2.13. *For every positive number $\rho \leq 2 - \sqrt{3}$, each function $f \in S$ maps the disk $|z| < \rho$ onto a convex domain. This is false for every $\rho > 2 - \sqrt{3}$.*

Proof. In view of the inequality (1), we have the estimate

$$\operatorname{Re}\left\{1 + \frac{zf''(z)}{f'(z)}\right\} \geq \frac{1 - 4r + r^2}{1 - r^2}, \qquad |z| = r < 1.$$

But $1 - 4r + r^2 > 0$ for $r < 2 - \sqrt{3}$, so f must map such a disk $|z| < r$ onto a convex domain. The Koebe function, for which

$$1 + \frac{zk''(z)}{k'(z)} = \frac{1 + 4z + z^2}{1 - z^2},$$

shows that the bound $2 - \sqrt{3}$ is sharp. This number $2 - \sqrt{3} = 0.267\ldots$ is called the *radius of convexity* for the class S. The *radius of starlikeness* is known to be $\tanh \pi/4 = 0.655\ldots$, but this result lies deeper. A proof using Loewner's method is given in Chapter 3.

We now turn to the Bieberbach conjecture for the class S^* of starlike functions. This was first proved by R. Nevanlinna [1] in 1920.

Theorem 2.14. *The coefficients of each function $f \in S^*$ satisfy $|a_n| \leq n$ for $n = 2, 3, \ldots$. Strict inequality holds for all n unless f is a rotation of the Koebe function.*

Proof. Given $f \in S^*$, define

$$\varphi(z) = \frac{zf'(z)}{f(z)} = 1 + \sum_{n=1}^{\infty} c_n z^n.$$

Then $\varphi \in P$ by Theorem 2.10, and Carathéodory's lemma gives $|c_n| \leq 2$. Now write

$$zf'(z) = \varphi(z)f(z)$$

and compare coefficients of z^n to obtain

$$na_n = a_n + \sum_{k=1}^{n-1} c_{n-k}a_k, \qquad n = 1, 2, 3, \ldots,$$

where $a_1 = 1$. The proof now proceeds by induction. Suppose we have proved $|a_k| \leq k$ for $k = 1, 2, \ldots, n-1$, where $n \geq 2$. Then

$$(n-1)|a_n| \leq \sum_{k=1}^{n-1} |c_{n-k}||a_k| \leq 2\sum_{k=1}^{n-1} k = n(n-1), \tag{10}$$

which proves $|a_n| \leq n$. According to Bieberbach's theorem, $|a_2| < 2$ if f is not a rotation of the Koebe function. It then follows from the inequality (10) that $|a_n| < n$ for all $n \geq 2$ if $f \in S^*$ and f is not a rotation of the Koebe function.

Corollary. *If $f \in C$, then $|a_n| \leq 1$ for $n = 2, 3, \ldots$. Strict inequality holds for all n unless f is a rotation of the function ℓ defined by $\ell(z) = z(1-z)^{-1}$.*

Proof. This follows easily from Alexander's theorem. If $f \in C$, then $zf'(z) \in S^*$, so $n|a_n| \leq n$. The function

$$\ell(z) = \frac{z}{1-z} = \sum_{n=1}^{\infty} z^n$$

satisfies $z\ell'(z) = k(z)$ and maps $\mathbb{D}$ onto the half-plane $\text{Re}\{w\} > -\frac{1}{2}$. The corollary was first proved (directly) by Loewner [1].

Various inequalities for the class S, such as the growth and distortion theorems, remain sharp in S^* because the Koebe function is starlike and is extremal in the full class S. However, these estimates can be improved for the class C, which excludes the Koebe function. As may be expected, the half-plane mapping ℓ is the typical extremal function in C. For instance, the following theorem improves upon the Koebe one-quarter theorem.

Theorem 2.15. *The range of every function $f \in C$ contains the disk $|w| < \frac{1}{2}$.*

Proof. It is geometrically obvious, and easy to prove analytically, that if $f \in C$ and $f(z) \neq \omega$, then $g(z) = [f(z) - \omega]^2$ is univalent. Indeed, if $g(z_1) = g(z_2)$, then either $f(z_1) = f(z_2)$ or $\frac{1}{2}[f(z_1) + f(z_2)] = \omega$. The latter is impossible for a convex function f which omits the value ω. Thus

$$h(z) = \frac{\omega^2 - g(z)}{2\omega} \in S.$$

But $h(z) \neq \frac{1}{2}\omega$ because $g(z) \neq 0$, so it follows from the Koebe one-quarter theorem that $|\frac{1}{2}\omega| \geq \frac{1}{4}$, or $|\omega| \geq \frac{1}{2}$. This proves the theorem. The function ℓ shows that the radius $\frac{1}{2}$ is best possible.

This proof is due to MacGregor [1]. Another proof is given by Nehari [2], p. 223.

§2.6. Close-to-Convex Functions

We now turn to an interesting subclass of S which contains S^* and has a simple geometric description. This is the class of close-to-convex functions, introduced by Kaplan [1] in 1952.

A function f analytic in the unit disk is said to be *close-to-convex* if there is a convex function g such that

$$\operatorname{Re}\left\{\frac{f'(z)}{g'(z)}\right\} > 0 \quad \text{for all } z \in \mathbb{D}.$$

We shall denote by K the class of close-to-convex functions f normalized by the usual conditions $f(0) = 0$ and $f'(0) = 1$. Note that f is not required *a priori* to be univalent. Note also that the associated function g need not be normalized. The additional condition that $g \in C$ defines a proper subclass of K which will be denoted by K_0.

Every convex function is obviously close-to-convex. More generally, every starlike function is close-to-convex. Indeed, each $f \in S^*$ has the form $f(z) = zg'(z)$ for some $g \in C$, and

$$\operatorname{Re}\left\{\frac{f'(z)}{g'(z)}\right\} = \operatorname{Re}\left\{\frac{zf'(z)}{f(z)}\right\} > 0.$$

These remarks are summarized by the chain of proper inclusions

$$C \subset S^* \subset K_0 \subset K.$$

Every close-to-convex function is univalent. This can be inferred from the following simple but important criterion for univalence. (In fact, it is an equivalent statement.) The criterion is due to Noshiro [1] and Warschawski [1].

Theorem 2.16 (Noshiro–Warschawski Theorem). *If f is analytic in a convex domain D and* $\operatorname{Re}\{f'(z)\} > 0$ *there, then f is univalent in D.*

Proof. Let z_1 and z_2 be distinct points in D. Then f is defined on the linear segment joining z_1 to z_2, and

$$\begin{aligned} f(z_2) - f(z_1) &= \int_{z_1}^{z_2} f'(z)\,dz \\ &= (z_2 - z_1) \int_0^1 f'(tz_2 + (1-t)z_1)\,dt \neq 0, \end{aligned}$$

since $\operatorname{Re}\{f'(z)\} > 0$.

Theorem 2.17. *Every close-to-convex function is univalent.*

Proof. If f is close-to-convex, then $\operatorname{Re}\{f'(z)/g'(z)\} > 0$ for some convex function g. Let D be the range of g and consider the function

$$h(w) = f(g^{-1}(w)), \qquad w \in D.$$

Then

$$h'(w) = \frac{f'(g^{-1}(w))}{g'(g^{-1}(w))} = \frac{f'(z)}{g'(z)},$$

so $\operatorname{Re}\{h'(w)\} > 0$ in D. Thus h is univalent, by Theorem 2.16, and so f is univalent.

Close-to-convex functions can be characterized by a geometric condition somewhat similar to the defining properties of convex and starlike functions. Let f be analytic in $\mathbb{D}$ and let C_r be the image under f of the circle $|z| = r$, where $0 < r < 1$. Roughly speaking, f is close-to-convex if and only if none of the curves C_r makes a "reverse hairpin turn." More precisely, the requirement is that as θ increases, the tangent direction $\arg\{(\partial/\partial\theta) f(re^{i\theta})\}$ should never decrease by as much as π from any previous value. Because

$$\frac{\partial}{\partial\theta}\left(\arg\left\{\frac{\partial}{\partial\theta} f(re^{i\theta})\right\}\right) = \operatorname{Re}\left\{1 + \frac{zf''(z)}{f'(z)}\right\}, \qquad z = re^{i\theta},$$

this theorem, which is due to Kaplan [1], can be stated as follows.

Theorem 2.18 (Kaplan's Theorem). *Let f be analytic and locally univalent in $\mathbb{D}$. Then f is close-to-convex if and only if*

$$\int_{\theta_1}^{\theta_2} \operatorname{Re}\left\{1 + \frac{zf''(z)}{f'(z)}\right\} d\theta > -\pi, \qquad z = re^{i\theta}, \tag{11}$$

for each r $(0 < r < 1)$ and for each pair of real numbers θ_1 and θ_2 with $\theta_1 < \theta_2$.

The proof makes use of an elementary lemma.

Lemma. *Let φ be a real-valued continuous function on $(-\infty, \infty)$, with the properties $\varphi(t + 2\pi) = \varphi(t) + 2\pi$ and*

$$\varphi(t_2) - \varphi(t_1) > -\pi, \qquad t_1 < t_2.$$

Then there is a continuous nondecreasing function ψ such that $\psi(t + 2\pi) = \psi(t) + 2\pi$ and $|\varphi(t) - \psi(t)| \le \pi/2$.

Proof of Lemma. Consider the function

$$\psi(t) = \max_{s \le t} \varphi(s) - \frac{\pi}{2}.$$

Clearly, ψ is continuous and nondecreasing. In view of the properties of φ,

$$\psi(t + 2\pi) = \max_{s \le t} \varphi(s + 2\pi) - \frac{\pi}{2} = \psi(t) + 2\pi$$

and

$$\varphi(t) - \frac{\pi}{2} \le \psi(t) \le [\varphi(t) + \pi] - \frac{\pi}{2} = \varphi(t) + \frac{\pi}{2}.$$

Proof of Theorem. First suppose f is close-to-convex, and let g be an associated convex function. Then for a suitable choice of arguments,

$$|\arg f'(z) - \arg g'(z)| < \frac{\pi}{2}.$$

Let

$$F(r, \theta) = \arg\left\{\frac{\partial}{\partial\theta} f(re^{i\theta})\right\} = \arg f'(re^{i\theta}) + \frac{\pi}{2} + \theta$$

and

$$G(r, \theta) = \arg\left\{\frac{\partial}{\partial\theta} g(re^{i\theta})\right\} = \arg g'(re^{i\theta}) + \frac{\pi}{2} + \theta.$$

Since g is convex, $G(r, \theta)$ is an increasing function of θ. The close-to-convexity condition takes the form

$$|F(r, \theta) - G(r, \theta)| < \frac{\pi}{2}.$$

Thus for $\theta_1 < \theta_2$,

$$\begin{aligned} F(r, \theta_2) - F(r, \theta_1) &= [F(r, \theta_2) - G(r, \theta_2)] \\ &\quad + [G(r, \theta_2) - G(r, \theta_1)] + [G(r, \theta_1) - F(r, \theta_1)] \\ &> -\frac{\pi}{2} + 0 - \frac{\pi}{2} = -\pi, \end{aligned}$$

which is equivalent to the condition (11).

Conversely, suppose f is a locally univalent function with the property (11), and let

$$\varphi_r(t) = \int_0^t \mathrm{Re}\left\{1 + \frac{zf''(z)}{f'(z)}\right\} d\theta, \qquad z = re^{i\theta}.$$

Since $f'(z) \neq 0$ in the disk, $\arg f'(re^{i\theta})$ is a periodic function of θ, and so

$$\varphi_r(t + 2\pi) - \varphi_r(t) = F(r, t + 2\pi) - F(r, t) = 2\pi.$$

The condition (11) takes the form

$$\varphi_r(t_2) - \varphi_r(t_1) > -\pi, \qquad t_1 < t_2.$$

An appeal to the lemma now produces a continuous nondecreasing function ψ_r with $\psi_r(t + 2\pi) = \psi_r(t) + 2\pi$ and

$$|\varphi_r(t) - \psi_r(t)| \leq \frac{\pi}{2}.$$

For fixed $\rho < 1$, we next define a function h_ρ analytic in $|z| < \rho$ by the Poisson integral

$$h_\rho(z) = \frac{1}{2\pi}\int_0^{2\pi} \frac{\zeta + z}{\zeta - z}[\psi_\rho(t) - t]\, dt, \qquad \zeta = \rho e^{it}.$$

Then

$$\operatorname{Re}\{h_\rho(re^{i\theta})\} = \frac{1}{2\pi}\int_0^{2\pi} P_\rho(r, \theta - t)[\psi_\rho(t) - t]\, dt,$$

where

$$P_\rho(r, \theta) = \frac{\rho^2 - r^2}{\rho^2 - 2r\rho\cos\theta + r^2}$$

is the Poisson kernel for the disk $|z| < \rho$. Since the function $[\psi_\rho(t) - t]$ is periodic and ψ_ρ is nondecreasing, an integration by parts gives

$$\frac{\partial}{\partial\theta}\operatorname{Re}\{h_\rho(re^{i\theta})\} = \frac{1}{2\pi}\int_0^{2\pi} P_\rho(r, \theta - t)[d\psi_\rho(t) - dt] > -1.$$

Applying this to the analytic function

$$g_\rho(z) = e^{i\alpha_\rho}\int_0^z e^{ih_\rho(w)}\, dw, \qquad |z| < \rho,$$

where α_ρ is a real number to be chosen later, we find

$$\begin{aligned}\operatorname{Re}\left\{1 + \frac{zg_\rho''(z)}{g_\rho'(z)}\right\} &= 1 + \operatorname{Re}\{izh_\rho'(z)\}\\ &= 1 + \operatorname{Re}\left\{\frac{\partial}{\partial\theta}h_\rho(re^{i\theta})\right\} = 1 + \frac{\partial}{\partial\theta}\operatorname{Re}\{h_\rho(re^{i\theta})\} > 0.\end{aligned}$$

Thus $g_\rho(z)$ is convex in the disk $|z| < \rho$. Furthermore,

$$\arg g_\rho'(z) = \operatorname{Re}\{h_\rho(z)\} + \alpha_\rho$$

and

$$\begin{aligned}\arg f'(re^{i\theta}) &= F(r, \theta) - \frac{\pi}{2} - \theta\\ &= \varphi_r(\theta) + F(r, 0) - \frac{\pi}{2} - \theta.\end{aligned}$$

But $\arg f'(z)$ is a harmonic function, so it can be expressed as a Poisson integral:

$$\arg f'(re^{i\theta}) = \frac{1}{2\pi}\int_0^{2\pi} P_\rho(r, \theta - t) \arg f'(\rho e^{it})\, dt, \qquad r < \rho.$$

We now choose $\alpha_\rho = F(\rho, 0) - \pi/2$ and obtain for $|z| < \rho$

$$\arg f'(z) - \arg g_\rho'(z) = \frac{1}{2\pi}\int_0^{2\pi} P_\rho(r, \theta - t)[\varphi_\rho(t) - \psi_\rho(t)]\, dt, \qquad z = re^{i\theta}.$$

Since $|\varphi_\rho(t) - \psi_\rho(t)| \leq \pi/2$, it follows that

$$|\arg f'(z) - \arg g_\rho'(z)| \leq \frac{\pi}{2}, \qquad |z| < \rho.$$

Finally, we observe that $g_\rho(0) = 0$ and $|g_\rho'(0)| = 1$ for all $\rho < 1$. From this we may infer that the functions g_ρ constitute a normal family. More precisely, for each fixed $\rho_0 < 1$, the functions g_ρ with $\rho > \rho_0$ constitute a normal family on the disk $|z| < \rho_0$. Therefore, by the familiar diagonal process, we may extract a sequence $\{g_{\rho_j}\}$ with $\rho_j \to 1$ which converges uniformly on each compact subset of $\mathbb{D}$ to an analytic function g. Because of the properties of the functions g_{ρ_j}, the limit function g is convex and

$$\mathrm{Re}\left\{\frac{f'(z)}{g'(z)}\right\} \geq 0, \qquad |z| < 1.$$

Hence f is close-to-convex. This concludes the proof of Kaplan's theorem.

Lewandowski [2] has observed that the class of close-to-convex functions is the same as the class of linearly accessible functions introduced by Biernacki [3] in 1936. A domain is said to be *linearly accessible* if its complement is the union of a family of nonintersecting half-lines. Such a domain is clearly simply connected. A univalent function in the unit disk whose range is linearly accessible is called a linearly accessible function. The preceding geometric description of the close-to-convex functions (Kaplan's theorem) makes it seem plausible that the two classes coincide, but the proof is not easy. Bielecki and Lewandowski [1] have simplified the part of the argument showing that every close-to-convex function is linearly accessible.

The Bieberbach conjecture is valid for the class K of close-to-convex functions. This result is due to Reade [1]. The proof for starlike functions, as given in §2.5, can be adapted to this more general situation.

The radius of close-to-convexity within the class S is 0.80 . . . , as Krzyż [2] has shown. His calculation uses Loewner's method, which will be developed in Chapter 3.

§2.7. Spirallike Functions

There is another natural generalization of starlikeness which again leads to a useful criterion for univalence. We refer to the class of spirallike functions, introduced by Špaček [1] in 1933.

A *logarithmic spiral* is a curve in the complex plane of the form

$$w = w_0 e^{-\lambda t}, \qquad -\infty < t < \infty,$$

where w_0 and λ are complex constants with $w_0 \neq 0$ and $\operatorname{Re}\{\lambda\} \neq 0$. There is no loss of generality in assuming that $\lambda = e^{i\alpha}$ with $-\pi/2 < \alpha < \pi/2$. The curve is then called an *α-spiral*. Observe that 0-spirals are radial half-lines. For each α ($|\alpha| < \pi/2$) there is a unique α-spiral which joins a given point $w_0 \neq 0$ to the origin.

A domain D containing the origin is said to be *α-spirallike* if for each point $w_0 \neq 0$ in D the arc of the α-spiral from w_0 to the origin lies entirely in D. This obviously implies that D is simply connected. A function f analytic and univalent in the unit disk, with $f(0) = 0$, is said to be *α-spirallike* if its range is α-spirallike. Finally, such a function is *spirallike* if it is α-spirallike for some α. The 0-spirallike functions are simply the starlike functions.

Spirallike functions can be characterized by an analytic condition which is a slight generalization of the condition for starlikeness (*cf.* Theorem 2.10).

Theorem 2.19. *Let f be analytic in $\mathbb{D}$, with $f(0) = 0$, $f'(0) \neq 0$, and $f(z) \neq 0$ for $0 < |z| < 1$. Let α lie in the interval $-\pi/2 < \alpha < \pi/2$. Then f is α-spirallike if and only if*

$$\operatorname{Re}\left\{e^{-i\alpha}\frac{zf'(z)}{f(z)}\right\} > 0, \qquad |z| < 1. \tag{12}$$

The theorem asserts, in particular, that (12) is a sufficient condition for univalence. We shall give an elegant proof due to Brickman [2], who obtained a more general result. This approach involves differential equations and provides a new proof of Theorem 2.10 when specialized to starlike functions (the case $\alpha = 0$). We begin with a lemma.

Lemma. *Let φ be an analytic function with $\operatorname{Re}\{\varphi(z)\} > 0$ in $\mathbb{D}$. Then for each $\zeta \in \mathbb{D}$ the initial-value problem*

$$\frac{dz}{dt} = -z\varphi(z), \qquad z(0) = \zeta, \tag{13}$$

defines a function $z = z(t;\zeta)$ with strictly decreasing modulus on the interval $0 \leq t < \infty$, tending to zero as $t \to \infty$.

Proof. Taking the real part of (13), we find

$$\frac{d}{dt}\log|z| = -\operatorname{Re}\{\varphi(z)\} < 0,$$

so $|z|$ decreases as t increases. The solution $z = z(t;\zeta)$ is therefore defined for all $t \geq 0$, and $|z(t;\zeta)| \leq |\zeta|$. Thus by the maximum principle for harmonic functions,

$$\operatorname{Re}\{\varphi(z(t;\zeta))\} \geq \delta, \qquad 0 \leq t < \infty,$$

for some $\delta > 0$. It follows that

$$\frac{d}{dt}\log|z(t;\zeta)| < -\delta,$$

so that

$$|z(t;\zeta)| < |\zeta|e^{-\delta t} \to 0 \quad \text{as } t \to \infty.$$

Proof of Theorem. Suppose first that f has the property (12), and let φ be the function with positive real part given by

$$\varphi(z) = \frac{\lambda f(z)}{zf'(z)}, \qquad \lambda = e^{i\alpha}.$$

For each $\zeta \in \mathbb{D}$ let $z = z(t;\zeta)$ be the corresponding function defined by the lemma, and let

$$w = w(t;\zeta) = f(z(t;\zeta)).$$

Then (13) gives $w(0;\zeta) = f(\zeta)$ and

$$\frac{dw}{dt} = f'(z)\frac{dz}{dt} = -z\varphi(z)f'(z) = -\lambda w,$$

which implies that

$$w(t;\zeta) = f(\zeta)e^{-\lambda t}, \qquad 0 \leq t < \infty.$$

This shows for each $\zeta \in \mathbb{D}$ that f maps the curve $z = z(t;\zeta)$ onto the arc of the α-spiral from ζ to 0. Thus the range of f is α-spirallike.

To show that f is univalent, suppose $f(\zeta_1) = f(\zeta_2)$ for some points ζ_1 and ζ_2 in $\mathbb{D}$. Then $w(0;\zeta_1) = w(0;\zeta_2)$, and so $w(t;\zeta_1) = w(t;\zeta_2)$ for all $t \geq 0$. Now the hypothesis $f'(0) \neq 0$ implies that f is univalent in some disk $|z| < \varepsilon$.

But by the lemma, $|z(t;\zeta_1)|$ and $|z(t;\zeta_2)|$ are both less than ε for all $t > t_0$, for some t_0. It follows that $z(t;\zeta_1) = z(t;\zeta_2)$ for all $t > t_0$, and so by uniqueness that $z(t;\zeta_1) = z(t;\zeta_2)$ for $t \ge 0$. In particular, $z(0;\zeta_1) = z(0;\zeta_2)$, which means that $\zeta_1 = \zeta_2$. This proves the univalence of f in $\mathbb{D}$.

Conversely, let f be (univalent and) α-spirallike for some α, $-\pi/2 < \alpha < \pi/2$. Then for each $\zeta \in \mathbb{D}$ the range of f contains the full arc of the α-spiral $w = f(\zeta)e^{-\lambda t}$, $t \ge 0$, where again $\lambda = e^{i\alpha}$. It is therefore possible to define the curve

$$z = z(t;\zeta) = f^{-1}(f(\zeta)e^{-\lambda t}), \qquad 0 \le t < \infty. \tag{14}$$

Clearly, $z(0;\zeta) = \zeta$. For each fixed t, the function $z(t;\zeta)$ is analytic in ζ, and it has the properties $|z(t;\zeta)| < 1$ and $z(t;0) = 0$. Thus $|z(t;\zeta)| \le |\zeta|$, by the Schwarz lemma. On the other hand, it is clear from (14) that

$$f'(z(t;\zeta))\frac{\partial z}{\partial t}(t;\zeta) = -\lambda e^{-\lambda t}f(\zeta),$$

so the proof of (12) reduces to showing that

$$\mathrm{Re}\left\{\frac{1}{\zeta}\frac{\partial z}{\partial t}(0;\zeta)\right\} \le 0,$$

or

$$\lim_{t\to 0}\frac{1}{t}\,\mathrm{Re}\left\{\frac{z(t;\zeta)}{\zeta} - 1\right\} \le 0,$$

an obvious consequence of the inequality $|z(t;\zeta)| \le |\zeta|$. This shows that f has the property (12) and completes the proof of the theorem.

The condition (12) has a nice geometric interpretation. The *radial angle* of a curve $w = w(t)$ is $\arg\{w'/w\}$, the angle between the radius and tangent vectors at a point where $w(t) \ne 0$. The image under f of the circle $|z| = r$ is the curve C_r given by $w = f(re^{i\theta})$, and so its radial angle is $\arg\{izf'(z)/f(z)\}$. The condition (12) simply requires that this radial angle lie between α and $\alpha + \pi$. On the other hand, α-spirals are the curves with constant radial angle α, if oriented with increasing modulus. Thus a univalent function f has the property (12) if and only if each of its level curves C_r intersects all α-spirals at angles between 0 and π. From this point of view, Theorem 2.19 is geometrically obvious.

Simple geometric considerations show that unless $\alpha = 0$, an α-spirallike function need not be close-to-convex. If $\alpha \ne 0$, a function may be α-spirallike and yet have level curves which exhibit the "reverse hairpin turns" denied to all close-to-convex functions. An explicit example is the function

$$f(z) = z(1 - z)^{-2e^{i\alpha}\cos\alpha},$$

which maps the disk onto the complement of an arc of an α-spiral. (See Exercise 19.) This function plays the role of the Koebe function in extremal problems for α-spirallike functions.

It is also easy to see that a close-to-convex function need not be spirallike. One example is the function

$$f(z) = \frac{z - z^2\cos\psi}{(1 - e^{i\psi}z)^2}, \qquad \cos\psi \neq 0,$$

which maps the disk onto the complement of a nonradial half-line. (See Exercise 21.)

§2.8. Typically Real Functions

Our next objective is to prove the Bieberbach conjecture for functions in S with real coefficients. Let S_R denote the class of univalent functions

$$f(z) = z + a_2 z^2 + a_3 z^3 + \cdots, \qquad |z| < 1,$$

whose coefficients a_n are all real. If $f \in S_R$, then $f(z)$ is real on the real axis and is not real anywhere else in the disk. Indeed, $f(z) = f(\bar{z})$ wherever $f(z)$ is real, and this violates the univalence of f unless $z = \bar{z}$. It is also clear that the image of each function $f \in S_R$ is symmetric with respect to the real axis. Furthermore, f must map the upper semidisk into the upper half-plane and the lower semidisk into the lower half-plane, since $f'(0) > 0$.

Many of the theorems concerning the class S_R depend only on these latter properties and have little to do with univalence. In fact, the proofs are easier if applied to a wider class of functions. A function f analytic in $\mathbb{D}$ is said to be *typically real* if it has real values on the real axis and nonreal values elsewhere. Thus the sign of $\text{Im}\{f(z)\}$ is constant in each of the semidisks above and below the real axis. It is clear that every typically real function has real coefficients, because $a_n = f^{(n)}(0)/n!$. However, a typically real function need not be univalent.

Let T denote the class of all typically real functions f such that $f(0) = 0$ and $f'(0) = 1$. If $f \in T$, then $\text{Im}\{f(z)\} > 0$ when $\text{Im}\{z\} > 0$ and $\text{Im}\{f(z)\} < 0$ when $\text{Im}\{z\} < 0$, because the normalization gives f this property near the origin. We have already observed that $S_R \subset T$. The importance of the class T

is that unlike S_R it can be described in a purely analytic manner, similar to the descriptions of convex and starlike functions. This representation is due to Rogosinski [2], who introduced the concept of a typically real function. Recall that P is the class of analytic functions φ with $\operatorname{Re}\{\varphi(z)\} > 0$ in $\mathbb{D}$ and $\varphi(0) = 1$. Let P_R denote the set of functions $\varphi \in P$ whose coefficients are all real: $\varphi(\bar{z}) = \overline{\varphi(z)}$.

Theorem 2.20 (Rogosinski's Theorem). *If $f \in T$, then*

$$\varphi(z) = \frac{1 - z^2}{z} f(z) \in P_R.$$

Conversely, if $\varphi \in P_R$, then

$$f(z) = \frac{z}{1 - z^2} \varphi(z) \in T.$$

Proof. The function $h(z) = (1 - z^2)/z$ is the reciprocal of the square-root transform of the Koebe function, which maps $\mathbb{D}$ onto the complement of two radial arcs on the imaginary axis from $\pm i/2$ to ∞. Thus $h(z)$ is imaginary on the unit circle, and

$$h(e^{i\theta}) = -2i \sin \theta,$$

so $\operatorname{Im}\{h(e^{i\theta})\}$ is negative on the upper semicircle and positive on the lower semicircle. Choosing $f \in T$, define

$$\varphi_\rho(z) = h(z) f(\rho z), \qquad 0 < \rho < 1.$$

Then φ_ρ is analytic on $\overline{\mathbb{D}}$ and

$$\operatorname{Re}\{\varphi_\rho(e^{i\theta})\} = 2 \sin \theta \operatorname{Im}\{f(\rho e^{i\theta})\} \geq 0$$

on the entire unit circle, since $f \in T$. It follows from the maximum principle for harmonic functions that $\operatorname{Re}\{\varphi_\rho(z)\} > 0$ in $\mathbb{D}$. Letting ρ tend to 1, we conclude that $\varphi \in P$. Clearly, then, $\varphi \in P_R$.

The converse is more subtle. For $0 < \rho < 1$, the function $f_\rho(z) = \varphi(\rho z)/h(z)$ is analytic in $\overline{\mathbb{D}}$ except for simple poles at ± 1. The hypothesis that φ has real coefficients implies that $f_\rho(z)$ is real on the real axis. Since $f'_\rho(0) = 1$, it is clear that f_ρ is univalent in some neighborhood of the origin, where $\operatorname{Im}\{f_\rho(z)\}$ is positive in the upper half-plane and negative in the lower half-plane. Consider now the reciprocal

$$g_\rho(z) = 1/f_\rho(z) = h(z)\psi(\rho z),$$

where $\psi = 1/\varphi \in P_R$. Observe that g_ρ is analytic in $\overline{\mathbb{D}}$ except for a simple pole at the origin. On the unit circle,

$$\operatorname{Im}\{g_\rho(e^{i\theta})\} = -2\sin\theta \operatorname{Re}\{\psi(\rho e^{i\theta})\},$$

and so $\operatorname{Im}\{g_\rho(e^{i\theta})\}$ is negative on the upper semicircle and positive on the lower semicircle. For $0 < \varepsilon < 1$, let $\mathbb{D}_\varepsilon^+$ and $\mathbb{D}_\varepsilon^-$ denote the parts of the annulus $\varepsilon < |z| < 1$ which lie in the upper and lower half-plane, respectively. We have found that $g_\rho(z)$ is analytic on the closure of $\mathbb{D}_\varepsilon^+$, and that $\operatorname{Im}\{g_\rho(z)\} \le 0$ on the entire boundary $\partial\mathbb{D}_\varepsilon^+$, for each ε sufficiently small. Thus by the maximum principle, $\operatorname{Im}\{g_\rho(z)\} < 0$ in $\mathbb{D}_\varepsilon^+$. Similarly, $\operatorname{Im}\{g_\rho(z)\} > 0$ in $\mathbb{D}_\varepsilon^-$. This for arbitrary $\varepsilon > 0$ implies that $\operatorname{Im}\{f_\rho(z)\}$ is positive in the upper semidisk and negative in the lower semidisk. Letting ρ tend to 1, we conclude that f has the same property. This shows that $f \in T$, which completes the proof.

On the basis of Rogosinki's theorem, it is easy to give explicit examples of typically real functions which are not univalent. For instance, let $\varphi(z) = 1 - z^4$. Then $\varphi \in P_R$ and so

$$f(z) = z + z^3 = \frac{z}{1 - z^2}\varphi(z) \in T.$$

But $f \notin S$, since $f'(z) = 1 + 3z^2 = 0$ for $z = \pm i/\sqrt{3}$.

It is obvious from the Rogosinski representation that T is a *convex* subset of $\mathscr{H}(\mathbb{D})$, the space of all analytic functions on the disk. In fact, it can be shown (see Chapter 9, Exercise 2) that T is the closure in $\mathscr{H}(\mathbb{D})$ of the convex hull of S_R. This may explain why so many properties of univalent functions with real coefficients are shared by typically real functions. (This result is due to Brickman, MacGregor, and Wilken [1]. See also Schober [1].) It is an interesting unsolved problem to characterize the functions $\varphi \in P_R$ which generate univalent functions through the Rogosinski representation.

The Bieberbach conjecture for functions with real coefficients was proved independently around 1931 by Dieudonné [2], Rogosinski [2], and Szász [1]. It is an immediate consequence of the following inequality, proved by Dieudonné for $f \in S_R$ and by Rogosinski for $f \in T$.

Theorem 2.21. *If* $f(z) = \sum_{n=0}^{\infty} a_n z^n \in T$, *then* $|a_{n+2} - a_n| \le 2$, $n = 0, 1, 2, \ldots$.

Proof. By Rogosinski's theorem, the function

$$\varphi(z) = \frac{1 - z^2}{z} f(z) = 1 + \sum_{n=0}^{\infty} (a_{n+2} - a_n) z^{n+1}$$

is of class P_R. It therefore follows from Carathéodory's lemma (see §2.5) that

$$|a_{n+2} - a_n| \leq 2, \qquad n = 0, 1, 2, \ldots,$$

where $a_0 = 0$ and $a_1 = 1$.

Corollary. *If $f \in T$, then $|a_n| \leq n$ for $n = 2, 3, \ldots$. Strict inequality occurs for all even n unless f is the Koebe function or its real rotation $-k(-z)$. Strict inequality occurs for all odd n unless f is a convex combination of these two functions.*

Here the conditions for equality follow from a close examination of the proof of Theorem 2.21, including that of Carathéodory's lemma. (See Exercise 28.) In particular, this proves the Bieberbach conjecture for functions $f \in S_R$, with $|a_n| < n$ for all $n \geq 2$ unless f is the Koebe function or its real rotation.

Stronger results have been obtained by more powerful methods. Schiffer [14] used a variational method to show that $\text{Re}\{a_n\} \leq n$ for all functions $f \in S$ whose coefficients $a_2, \ldots, a_{n-1}$ are real. G. S. Goodman [1] obtained the same result from a theorem of Teichmüller, and Obrock [2] relaxed the hypothesis to assume that a_m is real only for $m = 2, \ldots, [n/2]$.

The Koebe function has real coefficients, but most of its rotations do not. Thus the problem of finding sharp bounds for $|f(\zeta)|$ or $|f'(\zeta)|$ at a fixed point $\zeta \in \mathbb{D}$ assumes new interest in the class S_R. Goluzin [19] solved these problems for typically real functions, and Jenkins [10] determined the full region of values of $f(\zeta)$ as f ranges through S_R. More recently, A. W. Goodman [7] found the region of values in the class T. Some related results were obtained by Reich [1].

Leeman [1, 2] found the sharp upper and lower bounds on a_n among all functions $f \in T$ with fixed second coefficient a_2. Curiously, the bounds take two different forms depending on the parity of n. The corresponding problem for the class S_R has not been solved.

§2.9. A Primitive Variational Method

One general approach to the proof of an inequality within a given class of functions is to formulate an associated extremal problem and to look for properties of the extremal functions. For example, the nth coefficient problem may be viewed as a search for the set of functions in S for which $|a_n|$ is largest. The Bieberbach conjecture asserts that these extremal functions consist precisely of the Koebe function and its rotations. One line of attack is to determine analytic or geometric properties of the extremal functions, in the hope of excluding all but the Koebe function and its rotations.

Because S is a compact normal family, there exists a function $f \in S$ whose nth coefficient has maximum modulus. But every rotation of f is again an extremal function, so there is an extremal function whose nth coefficient is real and positive. This shows that the problem of maximizing $|a_n|$ is equivalent to that of maximizing $\operatorname{Re}\{a_n\}$.

The strategy of a variational method is to obtain information about an extremal function by comparing it with its "neighbors" in the class S. The problem is then to construct a sufficiently general perturbation which will preserve univalence. We shall now describe an elementary variational technique introduced by Marty [2] in 1934.

Let $f(z) = z + a_2 z^2 + \cdots$ be a function in S whose nth coefficient has maximum real part. Consider again the function

$$F(z) = \frac{f\left(\dfrac{z+\zeta}{1+\bar{\zeta}z}\right) - f(\zeta)}{(1-|\zeta|^2)f'(\zeta)} = z + A_2(\zeta)z^2 + \cdots$$

obtained from f by a disk automorphism, where $\zeta \in \mathbb{D}$. If ζ is near the origin, F may be viewed as a slight perturbation of f. Since $F \in S$ and f is extremal, it is clear that

$$\operatorname{Re}\{A_n(\zeta)\} \le \operatorname{Re}\{a_n\}, \qquad |\zeta| < 1. \tag{15}$$

The effective use of this inequality depends upon an asymptotic analysis of $A_n(\zeta)$ as $\zeta \to 0$. Observe first that

$$\frac{z+\zeta}{1+\bar{\zeta}z} = z + (\zeta - \bar{\zeta}z^2) + O(|\zeta|^2).$$

The binomial theorem therefore gives

$$\left(\frac{z+\zeta}{1+\bar{\zeta}z}\right)^m = z^m + m(\zeta - \bar{\zeta}z^2)z^{m-1} + O(|\zeta|^2)$$

for $m = 1, 2, \ldots$, and it follows that

$$f\left(\frac{z+\zeta}{1+\bar{\zeta}z}\right) = \sum_{m=1}^{\infty} a_m[z^m + m(\zeta - \bar{\zeta}z^2)z^{m-1}] + O(|\zeta|^2),$$

where $a_1 = 1$. On the other hand,

$$(1-|\zeta|^2)f'(\zeta) = 1 + 2a_2\zeta + O(|\zeta|^2).$$

These last two expansions lead to the asymptotic formula

$$A_n(\zeta) = a_n + \zeta[(n+1)a_{n+1} - 2a_2 a_n] - \bar{\zeta}(n-1)a_{n-1} + O(|\zeta|^2).$$

Putting this into the inequality (15), we find

$$\operatorname{Re}\{\zeta[(n+1)a_{n+1} - 2a_2 a_n - (n-1)\overline{a_{n-1}}] + O(|\zeta|^2)\} \le 0.$$

Now divide by $|\zeta|$ and let ζ tend to zero along a ray to obtain (since $\arg \zeta$ is arbitrary)

$$(n+1)a_{n+1} - 2a_2 a_n - (n-1)\overline{a_{n-1}} = 0.$$

This is known as the *Marty relation.* It must be satisfied by the coefficients of each function in S whose nth coefficient has maximum real part.

Observe that the Koebe function does satisfy the Marty relation:

$$(n+1)^2 - 4n - (n-1)^2 = 0.$$

(Otherwise, the Bieberbach conjecture would have been disproved!) But unfortunately, there are many other functions in S, certainly not extremal, which also satisfy the Marty relation. For instance, every odd function trivially satisfies it for every odd integer n. It is hardly surprising to find so many extraneous solutions, because the Marty relation was derived by comparing the extremal function only with neighbors of a very special type. More sophisticated variational methods, with more general devices for producing comparison functions, are developed in Chapter 10. Applied to the nth coefficient problem, these methods give results much stronger than the Marty relation.

§2.10. Growth of Integral Means

We now turn to a remarkable inequality discovered by Prawitz [1] in 1927. It gives an estimate for the integral means

$$M_p(r, f) = \left\{\frac{1}{2\pi}\int_0^{2\pi} |f(re^{i\theta})|^p \, d\theta\right\}^{1/p}, \qquad 0 < p < \infty,$$

of a function $f \in S$ in terms of the maximum modulus

$$M_\infty(r, f) = \max_{|z|=r} |f(z)|.$$

This important result is best understood when viewed in a more general context. We shall therefore begin with a brief discussion of H^p spaces. (See Duren [6] for more information.)

Let f be a function analytic in $\mathbb{D}$, univalent or not. For $0 < p \leq \infty$, each of the means $M_p(r, f)$ is a nondecreasing function of r. The function f is said to belong to the *Hardy space* H^p if $M_p(r, f)$ remains bounded as $r \to 1$. Because $M_p(r, f) \leq M_q(r, f)$ for $p < q$, the H^p spaces contract as p increases: $H^p \supset H^q$ for $p < q$. If $f \in H^p$, then it has a *radial limit*

$$f(e^{i\theta}) = \lim_{r \to 1} f(re^{i\theta})$$

in almost every direction. The boundary function $f(e^{i\theta})$ cannot vanish on any set of positive measure unless f is the zero function. Although a function $f \in H^p$ need not belong to H^q for any $q > p$, it is easy to show that

$$M_q(r, f) = O((1 - r)^{1/q - 1/p}), \qquad 0 < p < q \leq \infty.$$

Hardy and Littlewood strengthened this result by showing that if $f \in H^p$, then for each $\lambda \geq p$,

$$\int_0^1 (1 - r)^{\lambda(1/p - 1/q) - 1} M_q^\lambda(r, f)\, dr < \infty, \qquad 0 < p < q \leq \infty.$$

In particular, each $f \in H^p$ has the property

$$\int_0^1 M_\infty^p(r, f)\, dr < \infty.$$

The converse of this last statement is totally false. It is possible to construct an analytic function f whose maximum modulus $M_\infty(r, f)$ increases arbitrarily slowly to infinity, but which fails to have a radial limit on any set of positive measure. In particular, such a function f is not in any H^p space. For *univalent* functions, however, the converse is true and can be expressed in quantitative form as follows.

Theorem 2.22 (Prawitz' Theorem). *If $f \in S$, then for $0 < p < \infty$,*

$$M_p^p(r, f) \leq p \int_0^r \frac{1}{t} M_\infty^p(t, f)\, dt, \qquad 0 < r < 1.$$

Corollary. *$S \subset H^p$ for all $p < \frac{1}{2}$.*

Proof of Corollary. If $f \in S$, then by the growth theorem $M_\infty(r, f) \leq r(1-r)^{-2}$, so the integral

$$\int_0^1 M_\infty^p(r, f)\, dr$$

converges for every $p < \frac{1}{2}$. The result is best possible because the Koebe function $k \notin H^{1/2}$.

It follows from the corollary that every function of class S has a radial limit in almost every direction, even if its range has inaccessible boundary points. Beurling [1] sharpened this result by showing that each function of class S has a radial limit at every point on the unit circle except for a set of *capacity* zero. (See also Collingwood and Lohwater [1].)

The following lemma is the key to the proof of Prawitz' theorem. A Jordan curve will be called *smooth* if it has a parametric representation with two continuous derivatives. Line integrals over the boundary of a Jordan domain are understood to be traversed in the positive direction, keeping the domain on the left.

Lemma. *Let C_1 and C_2 be smooth Jordan curves surrounding the origin, with C_1 inside C_2. Then*

$$\int_{C_1} r^p\, d\theta \leq \int_{C_2} r^p\, d\theta, \qquad 0 < p < \infty,$$

where (r, θ) are polar coordinates.

Proof of Lemma. Let D be the annular domain between C_1 and C_2, and let C be its boundary. In view of the relation

$$d\theta = \frac{\partial \theta}{\partial x}\, dx + \frac{\partial \theta}{\partial y}\, dy,$$

Green's theorem gives

$$\int_{C_2} r^p\, d\theta - \int_{C_1} r^p\, d\theta = \int_C r^p\, d\theta$$

$$= p \iint_D r^{p-1} \left\{ \frac{\partial r}{\partial x} \frac{\partial \theta}{\partial y} - \frac{\partial r}{\partial y} \frac{\partial \theta}{\partial x} \right\} dx\, dy.$$

But the Cauchy–Riemann equations for the function $\log z = \log r + i\theta$ are

$$\frac{1}{r}\frac{\partial r}{\partial x} = \frac{\partial \theta}{\partial y}; \qquad \frac{1}{r}\frac{\partial r}{\partial y} = -\frac{\partial \theta}{\partial x}.$$

Thus

$$\int_C r^p \, d\theta = p \iint_D r^p \left\{ \left(\frac{\partial \theta}{\partial x}\right)^2 + \left(\frac{\partial \theta}{\partial y}\right)^2 \right\} dx\, dy \geq 0,$$

which proves the lemma.

An inspection of the proof reveals that r^p can be replaced by any non-decreasing function with a continuous derivative.

Proof of Theorem. Write $w = f(re^{i\theta}) = Re^{i\Phi}$. One of the Cauchy–Riemann equations for $\log f$ then takes the form

$$\frac{1}{R}\frac{\partial R}{\partial r} = \frac{1}{r}\frac{\partial \Phi}{\partial \theta}.$$

Let C_r be the image under f of the circle $|z| = r$. Then

$$\begin{aligned} 2\pi \frac{d}{dr} M_p^p(r, f) &= \int_0^{2\pi} \frac{\partial}{\partial r} R^p \, d\theta \\ &= \frac{p}{r} \int_0^{2\pi} R^p \frac{\partial \Phi}{\partial \theta} \, d\theta = \frac{p}{r} \int_{C_r} R^p \, d\Phi. \end{aligned}$$

Observe now that for each $\varepsilon > 0$, the curve C_r lies inside the circle Γ_r defined by $|w| = R = M_\infty(r, f) + \varepsilon$. Thus by the lemma,

$$\int_{C_r} R^p \, d\Phi \leq \int_{\Gamma_r} R^p \, d\Phi = 2\pi[M_\infty(r, f) + \varepsilon]^p.$$

Letting ε tend to zero, we obtain

$$\frac{d}{dr} M_p^p(r, f) \leq \frac{p}{r} M_\infty^p(r, f),$$

and Prawitz' theorem follows by integration.

§2.11. Odd Univalent Functions

For each function $f \in S$, the square-root transform

$$h(z) = \sqrt{f(z^2)} = z + c_3 z^3 + c_5 z^5 + \cdots$$

is an odd univalent function. Conversely, it is easy to see that every odd function $h \in S$ is the square-root transform of some $f \in S$. The set of all odd functions in S is denoted by $S^{(2)}$. More generally, for each integer $m \geq 2$, the class of all mth-root transforms

$$h(z) = \{f(z^m)\}^{1/m} = z + c_{m+1} z^{m+1} + c_{2m+1} z^{2m+1} + \cdots$$

of functions $f \in S$ is denoted by $S^{(m)}$. This is precisely the set of all functions $h \in S$ with m-fold symmetry.

The square-root transform of the Koebe function is

$$\frac{z}{1 - z^2} = z + z^3 + z^5 + \cdots.$$

It is to be expected that this function will play the role of the Koebe function in the class $S^{(2)}$. It does provide the maximum and minimum values of $|h(z)|$ and $|h'(z)|$, for instance (see Exercise 4). Thus it is reasonable to expect the coefficients of odd univalent functions to be bounded, and Littlewood and Paley [1] proved this in 1932. Their proof is ingenious and is still of interest although it leads to a relatively poor numerical value for the bound.

Theorem 2.23 (Littlewood–Paley Theorem). *The coefficients of every function $h \in S^{(2)}$ satisfy $|c_n| \leq A$, $n = 3, 5, 7, \ldots$, where A is an absolute constant.*

Proof. The idea is to show that the derivative of every function $h \in S^{(2)}$ has an integral mean with the property

$$M_1(r, h') \leq B(1 - r)^{-1}, \tag{16}$$

where B is an absolute constant. The result will then follow from the Cauchy representation of c_n, as in the proof of Littlewood's theorem (§2.4).

Each function $h \in S^{(2)}$ has the form $h(z) = \sqrt{f(z^2)}$ for some $f \in S$. Two more square-root transformations produce the univalent function

$$g(z) = \{h(z^4)\}^{1/4} = \{f(z^8)\}^{1/8}.$$

Since

$$h'(z^4) = z^{-3}\{g(z)\}^3 g'(z),$$

the Schwarz inequality gives

$$M_1(r^4, h') \le r^{-3} M_6^3(r, g) M_2(r, g').$$

By Prawitz' theorem (§2.10),

$$\begin{aligned} M_6^6(r, g) &\le 6 \int_0^r M_\infty^6(\rho, g)\rho^{-1}\, d\rho \\ &= 6 \int_0^r M_\infty^{3/4}(\rho^8, f)\rho^{-1}\, d\rho \le 6 \int_0^r (1-\rho^8)^{-3/2}\rho^5\, d\rho \\ &\le C(1-r)^{-1/2}. \end{aligned}$$

On the other hand, by the nondecreasing property of the integral means,

$$\begin{aligned} M_2^2(r, g') &\le \frac{1}{\pi r(1-r)} \int_r^{\sqrt{r}} \rho\, d\rho \int_0^{2\pi} |g'(\rho e^{i\theta})|^2\, d\theta \\ &\le [r(1-r)]^{-1} M_\infty^2(\sqrt{r}, g) < (1-r)^{-3/2}, \end{aligned}$$

since the double integral represents the area of a region which lies inside the circle with center at the origin and radius $M_\infty(\sqrt{r}, g)$. Combining these estimates, we obtain (16).

Finally, the Cauchy integral formula gives

$$n|c_n| \le r^{1-n} M_1(r, h').$$

Applying the estimate (16) and choosing $r = 1 - 1/n$, we conclude that $|c_n| < eB$.

The proof actually shows that $|c_n| \le A < 14$. Littlewood and Paley remarked in a footnote, "No doubt the true bound is given by $A = 1$." This has become known as the *Littlewood–Paley conjecture.* It was promptly disproved by Fekete and Szegö [1], who applied the Loewner method to obtain the *sharp* inequality

$$|c_5| \le \tfrac{1}{2} + e^{-2/3} = 1.013\ldots.$$

The more general Fekete–Szegö inequality, which contains this result, is presented in §3.8. It may be remarked that the Littlewood–Paley conjecture is true for odd starlike functions. Privalov [1] gave a proof in 1924, and Goluzin [3] later showed that $|a_n| \le 1$ for all functions $f \in S^*$ with $a_2 = 0$. On the other hand, the Littlewood–Paley conjecture is false even for functions with real coefficients. For each given odd index $n \ge 5$, an elementary construction by Schaeffer and Spencer [1] produces a function $h \in S^{(2)}$ with all coefficients real and with $|c_n| > 1$. (See §3.9.)

The Littlewood–Paley conjecture easily implies the Bieberbach conjecture. Indeed, if $f \in S$ and $h \in S^{(2)}$ is its square-root transform, then $\{h(z)\}^2 = f(z^2)$. Now compare the coefficients of z^{2n} to obtain the relation

$$a_n = c_1c_{2n-1} + c_3c_{2n-3} + \cdots + c_{2n-1}c_1, \qquad c_1 = 1.$$

Since there are n terms on the right-hand side, the uniform bound $|c_k| \leq 1$ would imply $|a_n| \leq n$. In fact, a simple application of the Cauchy–Schwarz inequality shows that the following conjecture, proposed by Robertson [4] in 1936, also implies the Bieberbach conjecture.

Robertson Conjecture. *For each function $h(z) = z + c_3z^3 + c_5z^5 + \cdots$ of class $S^{(2)}$,*

$$1 + |c_3|^2 + \cdots + |c_{2n-1}|^2 \leq n.$$

This conjecture has been proved up to $n = 4$ (see §3.10), but for larger n it remains open, even for functions with real coefficients. V. I. Milin [1] has obtained the general bound $1.21\,n$.

Landau [5] found an alternate proof of the Littlewood–Paley theorem which makes no appeal to the theorem of Prawitz.

§2.12. Asymptotic Bieberbach Conjecture

We shall now consider two other coefficient conjectures, both weaker than the Bieberbach conjecture, and establish a surprising connection between them. For each integer $n \geq 2$, let A_n denote the maximum of $|a_n|$ for all $f \in S$. Hayman [3] has proved the existence of the limit

$$\lambda = \lim_{n\to\infty} A_n/n.$$

The Bieberbach conjecture asserts that $A_n = n$ for all n, while the *asymptotic Bieberbach conjecture* is the weaker assertion that $\lambda = 1$.

Asymptotic Bieberbach Conjecture. *Let $A_n = \max_{f\in S}|a_n|$. Then $\lim_{n\to\infty} A_n/n = 1$.*

Another conjecture was proposed by Littlewood [1] in 1925.

Littlewood Conjecture. *Let $f \in S$ and suppose $f(z) \neq \omega$ for all $z \in \mathbb{D}$. Then $|a_n| \leq 4|\omega|n$ for all n.*

Both of these conjectures remain unsettled. In view of the Koebe one-quarter theorem, Littlewood's conjecture is weaker than Bieberbach's. It was the discovery of Nehari [5] that Littlewood's conjecture is also implied by the asymptotic Bieberbach conjecture. In fact, he proved the following theorem.

Theorem 2.24 (Nehari's Theorem). *Let $f \in S$ and suppose $f(z) \neq \omega$ for all $z \in \mathbb{D}$. Then*

$$|a_n| \leq 4|\omega|\lambda n, \qquad n = 2, 3, \ldots,$$

where $\lambda = \lim_{n\to\infty} A_n/n$.

The proof employs a curious lemma.

Lemma. *Let g be analytic and univalent in $\mathbb{D}$, with $g(0) = 0$ and $g(z) \neq 1$. Then the function G defined by*

$$G(z) = 2g(z^2) - 2\{g(z^2)[g(z^2) - 1]\}^{1/2}$$

has the same properties.

Proof. Let $g(z) = c_1 z + c_2 z^2 + \cdots$. Note first that

$$h(z) = \{g(z^2)[g(z^2) - 1]\}^{1/2} = ic_1^{1/2}z + \cdots$$

is an odd analytic function which vanishes only at the origin. Now suppose that $G(z) = G(\zeta)$ for some points z and ζ in $\mathbb{D}$. Then

$$g(z^2) - g(\zeta^2) = h(z) - h(\zeta).$$

Square both sides to obtain

$$g(z^2) + g(\zeta^2) - 2g(z^2)g(\zeta^2) = -2h(z)h(\zeta).$$

Now square again and reduce the result to

$$[g(z^2) - g(\zeta^2)]^2 = 0.$$

Since g is univalent, this implies that $z^2 = \zeta^2$, or $z = \pm\zeta$. But h is an odd function with $h(z) \neq 0$ for $z \neq 0$, and so

$$G(z) - G(-z) = -4h(z) \neq 0$$

unless $z = 0$. Thus $z = \zeta$, which proves that G is univalent. If $G(z) = 1$ for some z, then

$$w - \tfrac{1}{2} = \{w(w-1)\}^{1/2}, \quad \text{where } w = g(z^2).$$

Squaring both sides, we find $\frac{1}{4} = 0$. This contradiction shows that $G(z) \neq 1$ in $\mathbb{D}$, and the proof is complete.

Proof of Theorem. If $f \in S$ and $f(z) \neq \omega$, then

$$g(z) = \frac{1}{\omega} f(z) = c_1 z + c_2 z^2 + \cdots$$

satisfies the hypotheses of the lemma. The operation of the lemma may be iterated to produce a sequence of functions

$$g_k(z) = c_1^{(k)} z + c_2^{(k)} z^2 + \cdots, \qquad k = 0, 1, 2, \ldots,$$

where $g_0 = g$ and

$$g_{k+1}(z) = 2g_k(z^2) - 2\{g_k(z^2)[g_k(z^2) - 1]\}^{1/2}, \qquad k = 0, 1, 2, \ldots.$$

A comparison of coefficients gives

$$c_{2n}^{(k+1)} = 2c_n^{(k)}, \qquad n = 1, 2, \ldots, \tag{17}$$

and

$$|c_1^{(k+1)}| = 2|c_1^{(k)}|^{1/2}, \qquad k = 0, 1, 2, \ldots, \tag{18}$$

where $c_n^{(0)} = c_n = a_n/\omega$. Now $|c_1^{(0)}| = |\omega^{-1}| \leq 4$, by the Koebe one-quarter theorem, and it follows inductively from (18) that $|c_1^{(k)}| \leq 4$ for all k. Since $g_k/c_1^{(k)} \in S$, it is therefore clear that

$$|c_m^{(k)}| \leq |c_1^{(k)}| A_m \leq 4A_m, \qquad m = 2, 3, \ldots.$$

For arbitrary $n \geq 2$, the iteration of (17) now yields

$$2^k |c_n| = |c_{m_k}^{(k)}| \leq 4A_{m_k}, \qquad k = 1, 2, \ldots,$$

where $m_k = 2^k n$. Consequently,

$$|a_n| = |\omega||c_n| \leq 4|\omega| n m_k^{-1} A_{m_k},$$

and the desired result follows as $k \to \infty$.

Another proof of Nehari's theorem was found by Bombieri [2], using a result of Hayman [3] relating the class S to a certain class of functions univalent in a half-plane. Along similar lines, Hamilton [3] has recently given a proof that Littlewood's conjecture implies the asymptotic Bieberbach conjecture.

Notes

The theory of univalent functions began to take shape around the turn of the century. In the year 1907, Koebe [1] proved the existence of an absolute constant $\kappa > 0$ for which the disk $|w| < \kappa$ is contained in the range of every function $f \in S$. The value $\kappa = \frac{1}{4}$ was determined by Bieberbach [2, 3] a few years later. Koebe [2] also gave a primitive form of the distortion theorem by proving the existence of positive upper and lower bounds for $|f'(z)|$, depending only on $|z|$. Pick [1] obtained related results. Gronwall [1] established the area theorem in 1914, and Bieberbach [2, 3] applied it to prove $|a_2| \leq 2$ together with the sharp forms of Koebe's covering and distortion theorems. At about the same time, Gronwall [2] stated the sharp growth and distortion theorems without proof. Faber [1, 2] gave another proof that $\kappa = \frac{1}{4}$. Bieberbach [4] obtained the primitive form of the rotation theorem (the inequality (6) in §2.3), but the sharp bound remained undetermined until the work of Goluzin [1] in 1936. Landau [1] gave an elegant exposition of these early results in his well-known little book on function theory. Another early account appears in the book of Bieberbach [5].

The Bieberbach conjecture has a rich and continuing history. Bieberbach [3] used the notation k_n instead of A_n for the maximum value of $|a_n|$ in S. Having proved that $k_2 = 2$, Bieberbach [3, *p.* 946] suggested tentatively in a footnote, "Dass $k_n \geq n$ zeigt das Beispiel $\Sigma\, nz^n$. Vielleicht ist überhaupt $k_n = n$." This is the source of his famous conjecture. A few years later, Loewner [3] developed the parametric representation of slit mappings (see Chapter 3) and applied it to prove $|a_3| \leq 3$. The fourth coefficient eluded many efforts until 1955, when Garabedian and Schiffer [1] finally proved $|a_4| \leq 4$ by a variational method. Their proof was quite difficult, but 5 years later Charzyński and Schiffer [1, 2] found two relatively simple proofs. One proof was based upon the Grunsky inequalities (see Chapter 4), a direct extension of the area principle, and was entirely elementary. This discovery focused new attention on the Grunsky inequalities, which Grunsky [3] had introduced in 1939. Since 1960, the Grunsky inequalities have been the basis for a number of important advances, including the theory of I. M. Milin [6] and the proof by FitzGerald [1] that $|a_n| < \sqrt{\frac{7}{6}}n$ for all n (see Chapter 5). In 1968, Pederson [2] and Ozawa [1, 2] used the Grunsky inequalities to prove $|a_6| \leq 6$. Several years later, Pederson and Schiffer [1] applied the Garabedian–Schiffer inequalities, a generalization of the Grunsky inequalities, to prove $|a_5| \leq 5$. (See the notes at the end of Chapter 4 for further information.)

EXERCISES

1. Show that $|a_2^2 - a_3| \le 1$ for every function $f \in S$, and that the rotations of the Koebe function provide the only extremal functions. (*Hint*: Invert.)

2. Show that $|a_2^2 - a_3| \le \frac{1}{3}(1 - |a_2|^2)$ for every function $f \in C$. (Trimble [1].)

3. Show that $f(z) = \frac{1}{2}[z(1 - z)^{-2} + z(1 + z)^{-2}]$ is the average of two functions in S but is not univalent.

4. For odd functions $h \in S^{(2)}$, prove the sharp inequalities

$$\frac{r}{1 + r^2} \le |h(z)| \le \frac{r}{1 - r^2}$$

and

$$\frac{1 - r^2}{(1 + r^2)^2} \le |h'(z)| \le \frac{1 + r^2}{(1 - r^2)^2}, \qquad r = |z| < 1.$$

5. For convex functions $f \in C$, prove the sharp inequalities

$$\frac{r}{1 + r} \le |f(z)| \le \frac{r}{1 - r}, \qquad r = |z| < 1,$$

with equality occurring only for functions of the form $f(z) = z(1 - e^{i\varphi}z)^{-1}$, $0 \le \varphi \le 2\pi$.

6. Assuming the Bieberbach conjecture, prove the sharp inequality

$$|f^{(n)}(z)| \le n!\frac{n + r}{(1 - r)^{n+2}}, \qquad r = |z| < 1, \quad n = 1, 2, \ldots,$$

for every function $f \in S$. (Marty [1].)

7. Show that for each integer $m \ge 2$, the mth-root transform

$$g(z) = \{f(z^m)\}^{1/m} = z + \sum_{\nu=1}^{\infty} c_{m\nu+1} z^{m\nu+1}$$

is univalent and so belongs to the subclass $S^{(m)}$ of all functions in S with m-fold symmetry. Conversely, show that every $g \in S^{(m)}$ is the mth-root transform of some $f \in S$.

8. Show that the disk automorphism transformation is invertible. More specifically, show that for each fixed $\alpha \in \mathbb{D}$, every function $g \in S$ arises as the image of a unique function $f \in S$ through the disk automorphism formula given in §2.1.

9. Prove the Littlewood–Paley conjecture $|a_n| \leq 1$ for odd starlike functions $f \in S^*$. (Privalov [1]. More generally, Goluzin [3] proved this for functions $f \in S^*$ with $a_2 = 0$.)

10. Let $k(z) = z(1 - z)^{-2}$ be the Koebe function.

(a) Show that $\log[k(z)/z]$ is a convex function.
(b) Show that $\log k'(z)$ is a starlike function. (Duren and McLaughlin [1].)

(In particular, both functions are univalent.)

11. Use the Herglotz formula to show that each $f \in S^*$ has a unique integral representation of the form

$$f(z) = z \exp\left\{\int_0^{2\pi} \log[k_\varphi(z)/z]\, d\mu(\varphi)\right\},$$

where $d\mu$ is a unit measure and $k_\varphi(z) = z(1 - e^{i\varphi}z)^{-2}$ is a rotation of the Koebe function. Observe further that this formula represents a starlike function f for arbitrary choice of the unit measure $d\mu$.

12. Let n be an arbitrary positive integer, let $\varphi_1, \ldots, \varphi_n$ be distinct points in the interval $[0, 2\pi)$, and let α_j be positive numbers with sum $\alpha_1 + \cdots + \alpha_n = 2$. Show that the function

$$f(z) = z\prod_{j=1}^{n}(1 - ze^{i\varphi_j})^{-\alpha_j}$$

is starlike, and describe the range of a function of this form.

13. Describe the region of values of $\log[f(z)/z]$ for fixed $z \in \mathbb{D}$ as f ranges over the class S^*. Establish the sharp estimate

$$\left|\arg\frac{f(z)}{z}\right| \leq 2\sin^{-1}r, \qquad r = |z| < 1,$$

for $f \in S^*$, and identify the extremal functions. Deduce the sharp rotation theorem $|\arg f'(z)| \leq 2\sin^{-1} r$ for $f \in C$.

14. Show that for each fixed $z \in \mathbb{D}$, some rotation of the Koebe function maximizes $|\arg f'(z)|$ within the class S^*. (Stroganoff [1], A. W. Goodman [1].)

15. Establish the inequality $M_1(r, f) \leq M_1(r, k) = r(1 - r^2)^{-1}$ for the integral means of functions $f \in S^*$. (*Note*: Baernstein [1] has extended this result to the full class S. The proof is developed in Chapter 7.)

16. Use Prawitz' theorem to prove Littlewood's theorem that $|a_n| < en$.

17. Given the result of Baernstein [1] that $M_1(r, f) \leq r(1 - r^2)^{-1}$ for all $f \in S$, deduce that $|a_n| < (e/2)n$ for $n = 2, 3, \ldots$. *Caveat*: This depends upon the inequality

$$\left(1 + \frac{2}{n}\right)^{n/2} < \frac{n+1}{n+2}\, e, \qquad n = 1, 2, \ldots.$$

18. Prove the Bieberbach conjecture for close-to-convex functions. In other words, show that $|a_n| \leq n$ for $f \in K$, with equality only for rotations of the Koebe function. (Reade [1].)

19. Let α be a real number with $0 < |\alpha| < \pi/2$. Show that the function

$$f(z) = z(1 - z)^{-2e^{i\alpha}\cos\alpha}$$

is α-spirallike but not close-to-convex. (*Hint*: Show that the condition (11) of Theorem 2.18 is violated.)

20. Suppose $0 < \alpha < \pi/2$. Show that the function $f(z) = z(1 - z)^{-\rho e^{i\alpha}}$ is α-spirallike if $0 < \rho \leq 2\cos\alpha$, but is not spirallike if $\rho > 2\cos\alpha$.

21. Show that if $\cos\varphi \neq 0$, the function

$$f(z) = (z - z^2\cos\varphi)(1 - e^{i\varphi}z)^{-2}$$

maps $\mathbb{D}$ onto the complement of a nonradial half-line. Verify that f is close-to-convex but not spirallike.

22. Let $f(z) = z + a_2 z^2 + \cdots$ be a function in S.

(a) Prove that $|a_n| \leq n$ for all n if $\operatorname{Re}\{\sqrt{f(z)/z}\} > \frac{1}{2}$ in $\mathbb{D}$.
(b) Prove that $|a_n| \leq 1$ for all n if f is odd and $\operatorname{Re}\{f(z)/z\} > \frac{1}{2}$ in $\mathbb{D}$.

(Dvořák [1].)

23. For each convex function $f \in C$, show that

$$F(z, \zeta) = \begin{cases} \dfrac{2zf'(z)}{f(z) - f(\zeta)} - \dfrac{z + \zeta}{z - \zeta}, & z \neq \zeta, \\[2ex] 1 + \dfrac{zf''(z)}{f'(z)}, & z = \zeta, \end{cases}$$

has positive real part for all points $(z, \zeta) \in \mathbb{D}^2$. (*Hint*: A domain is convex if and only if it is starlike with respect to each of its points.) Deduce that each $f \in C$

has the properties

$$\mathrm{Re}\left\{\frac{zf'(z)}{f(z)}\right\} > \frac{1}{2} \quad \text{and} \quad \mathrm{Re}\left\{\frac{f(z)}{z}\right\} > \frac{1}{2}$$

for all $z \in \mathbb{D}$, and that the constant $\frac{1}{2}$ is best possible in each instance. (Sheil-Small [1], Suffridge [3], Strohhäcker [1], Marx [1].)

24. Let $f(z) = z + \sum_{n=2}^{\infty} a_n z^n$ be analytic in $\mathbb{D}$.

(a) Show that if $\sum_{n=2}^{\infty} n|a_n| \leq 1$, then $f \in S$. Show also that the bound is best possible: for each $\varepsilon > 0$ there exist nonunivalent functions f with $\sum_{n=2}^{\infty} n|a_n| \leq 1 + \varepsilon$.

(b) Under the same hypothesis $\sum_{n=2}^{\infty} n|a_n| \leq 1$, show further that $f \in S^*$ and that f maps the disk $|z| < \frac{1}{2}$ onto a convex domain. Give an example to show that the radius $\frac{1}{2}$ is best possible. (Alexander [1], Remak [1], Lewandowski [1], A. W. Goodman [3], Clunie and Keogh [1].)

25. Prove that if $f \in S$ is a polynomial of degree n, then $|a_n| \leq 1/n$. Give an example to show that the bound $1/n$ is best possible.

26. As in §2.8, let P_R be the class of functions $\varphi \in P$ with real coefficients. Appealing to the uniqueness of the Herglotz representation, show that each $\varphi \in P_R$ has the unique representation

$$\varphi(z) = \int_0^{\pi} \frac{1 - z^2}{1 - 2z\cos t + z^2}\, d\mu(t),$$

where $d\mu$ is a unit measure on $[0, \pi]$. Conclude from Rogosinski's theorem that each typically real function $f \in T$ has a unique representation

$$f(z) = \int_0^{\pi} \frac{z}{1 - 2z\cos t + z^2}\, d\mu(t).$$

Finally, show that the coefficients of each function $f \in T$ have the form

$$a_n = \int_0^{\pi} \frac{\sin nt}{\sin t}\, d\mu(t), \qquad n = 1, 2, \ldots,$$

where $d\mu$ is a unit measure on $[0, \pi]$. (Robertson [2].)

27. Show that the coefficients of each function $f \in T$ satisfy the inequality

$$a_n^2 \leq 1 + a_3 + \cdots + a_{2n-1}.$$

(*Hint*: Use the representation of Exercise 26 and the Schwarz inequality. This proof was suggested by M. S. Robertson.) (FitzGerald [1].)

28. Prove the corollary to Theorem 2.21, and deduce that $|a_n| \le n$ for each $f \in S_R$, with strict inequality for all n unless $f(z) = z(1 - z)^{-2}$ or $f(z) = z(1 + z)^{-2}$. (*Suggestion*: Show that if $\varphi(z) = 1 + c_1 z + \cdots$ belongs to P_R, then $c_2 < 2$ unless φ has the form

$$\varphi(z) = a\frac{1+z}{1-z} + (1-a)\frac{1-z}{1+z}, \qquad 0 \le a \le 1.)$$

29. Let C_r be the image of the circle $|z| = r < 1$ under a mapping $f \in S$.

(a) Show that the curvature of C_r at a point $w = f(z)$ is given by

$$K_z = \frac{1}{r|f'(z)|}\operatorname{Re}\left\{1 + \frac{zf''(z)}{f'(z)}\right\}, \qquad |z| = r < 1.$$

(b) Obtain the (crude) estimates

$$\left(\frac{1-r}{1+r}\right)^2 \frac{1 - 4r + r^2}{r} \le K_z \le \left(\frac{1+r}{1-r}\right)^2 \frac{1 + 4r + r^2}{r}, \qquad |z| = r < 1.$$

30. Show that for each function $f \in S$ which is not the identity mapping $f(z) \equiv z$, there are points z_1 and z_2 in $\mathbb{D}$ such that $f(z_1)f(z_2) = 1$. (A. W. Goodman [2], Crum [1].)

31. Sharpen the Koebe one-quarter theorem by showing that each omitted value ω of a function $f \in S$ satisfies $|\omega| \ge (2 + |a_2|)^{-1}$, where a_2 is the second coefficient of f.

32. Let a function $f \in S$ omit two values α and β which lie on opposite sides of the origin: $\arg\{\beta\} = \arg\{\alpha\} + \pi$.

(a) Show that $\max\{|\alpha|, |\beta|\} \ge \frac{1}{2}$, with equality only if $|\alpha| = |\beta| = \frac{1}{2}$. Give an example to show that the lower bound $\frac{1}{2}$ is best possible. (Szegö [1].)

(b) Show that $|\alpha| + |\beta| \ge 1$, and verify that equality is possible. Thus the segment joining α and β must have length at least 1.

33. Let a function $f \in S$ be bounded by a number $M > 1$, so that $|f(z)| < M$ for all $z \in \mathbb{D}$. Improve Bieberbach's bound $|a_2| \le 2$ to $|a_2| \le 2(1 - 1/M)$, and show that equality occurs if and only if f maps $\mathbb{D}$ onto the disk $|w| < M$ minus a radial segment. (*Hint*: Consider $k(e^{i\theta}f(z)/M)$, where k is the Koebe function.) (Pick [2].)

34. Show that a polynomial $p(z) = z + a_2 z^2 + \cdots + a_n z^n$ is univalent in $\mathbb{D}$ if and only if its associated polynomials

$$q(z;\theta) = 1 + a_2 \frac{\sin 2\theta}{\sin\theta} z + \cdots + a_n \frac{\sin n\theta}{\sin\theta} z^{n-1}$$

have no zeros in $\mathbb{D}$ for any choice of the parameter θ, $0 \le \theta \le \pi/2$. (*Hint*: Refer to Chapter 1, Exercise 4. It is equivalent to require that $q(z;\theta) \ne 0$ in $\mathbb{D}$ for $0 \le \theta \le \pi$.) (Dieudonné [1].)

35. Show that the polynomials $z + z^2/2 + \cdots + z^n/n$ and $z + z^3/3 + \cdots + z^{2n+1}/(2n+1)$ are univalent in $\mathbb{D}$. (Alexander [1].)

Chapter 3

Parametric Representation of Slit Mappings

In the preceding chapter we applied elementary methods to obtain a wide variety of basic results in the theory of univalent functions. Although these methods are clever and elegant, they are aptly described as a "mixed bag of tricks." More general and more powerful techniques are needed both to obtain deeper results and to unify the theory.

In 1923, Charles Loewner [3] (Karl Löwner) developed and applied the first nonelementary method. To this day it remains one of the most effective approaches to extremal problems for univalent functions. Loewner's method focuses upon the single-slit mappings, the functions which map the disk onto the complement of an arc. Because these functions are dense in the class S, the sharp estimation of any continuous functional over S reduces to its estimation over the subclass of single-slit mappings. Loewner derived a differential equation which may be viewed as giving a parametric representation of a dense subclass of S, containing all of the single-slit mappings. This is an important analytic device which generates a number of sharp inequalities not easily accessible by other methods.

This chapter opens with a discussion of the Carathéodory convergence theorem, a fundamental result of independent interest which is used to prove the density of the slit mappings. Loewner's differential equation is then derived, together with a proof that it generates only univalent functions. The chapter concludes with a variety of applications: the Bieberbach conjecture for the third coefficient, the radius of starlikeness, the sharp rotation theorem, the disproof of the Littlewood–Paley conjecture for odd functions, and several other results.

§3.1. Carathéodory Convergence Theorem

We now turn to a classical theorem of Carathéodory [2] which plays a central role in geometric function theory. Let $D_1, D_2, \ldots$ be a sequence of simply connected domains in the complex plane $\mathbb{C}$, all containing the origin and none coinciding with $\mathbb{C}$. Let $w = f_n(z)$ be the conformal mapping of the unit disk $\mathbb{D} = \{z: |z| < 1\}$ onto D_n, normalized by the conditions $f_n(0) = 0$ and $f'_n(0) > 0$. Carathéodory's theorem connects the analytic behavior of the sequence of mapping functions f_n with the geometric behavior of their ranges D_n.

The theorem involves an unusual notion of convergence of the sequence of domains D_n. Two cases are distinguished. First suppose that the origin is an interior point of the intersection of the domains D_n, or equivalently that some disk $\{w: |w| < \rho\}$ lies in D_n for all n. Then the *kernel* of the sequence $\{D_n\}$ is defined as the largest domain D containing the origin and having the property that each compact subset of D lies in all but a finite number of the domains D_n. A straightforward application of the Heine–Borel theorem shows that an arbitrary union of domains with this property must inherit the same property. This proves the existence of a largest domain D and shows that the kernel is well-defined in this case. If the origin is not an interior point of the intersection, the kernel is defined as $D = \{0\}$. In either case, the sequence $\{D_n\}$ is said to *converge* to its kernel D if every subsequence of $\{D_n\}$ has the same kernel. This is indicated by the notation $D_n \to D$.

It is instructive to compare this with a more common definition of domain convergence. The sets

$$D_* = \liminf_{n\to\infty} D_n = \bigcup_{k=1}^{\infty} \bigcap_{n=k}^{\infty} D_n$$

and

$$D^* = \limsup_{n\to\infty} D_n = \bigcap_{k=1}^{\infty} \bigcup_{n=k}^{\infty} D_n$$

consist respectively of the points belonging to all but a finite number, and to infinitely many of the domains D_n. If $D_* = D^*$, the sequence $\{D_n\}$ is said to *converge topologically* to this common set.

If the sequence $\{D_n\}$ is increasing, so that $D_1 \subset D_2 \subset \cdots$, then clearly the kernel is $D = \bigcup_{n=1}^{\infty} D_n$ and $D_n \to D$. It is equally clear that $D_* = D^* = D$, so the two notions of convergence agree in this case.

On the other hand, the two notions need not agree even if the sequence $\{D_n\}$ is decreasing. Here is a typical example. Let D_n consist of the entire

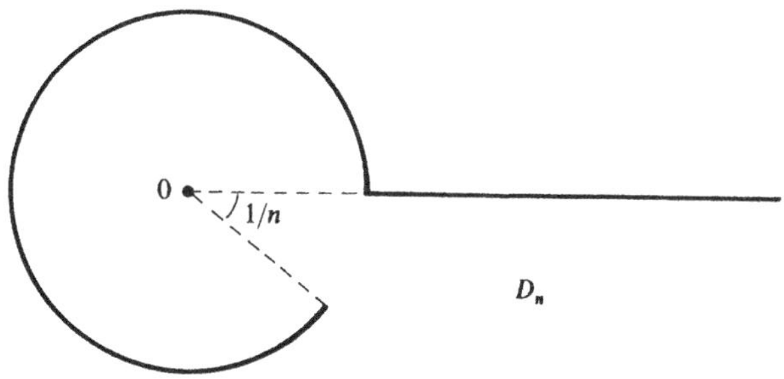

Figure 3.1

plane $\mathbb{C}$ minus the segment $1 \leq z < \infty$ of the real axis and the arc $0 \leq \theta \leq 2\pi - 1/n$ of the unit circle. (See Figure 3.1.) Then the kernel of $\{D_n\}$ is simply the unit disk $\mathbb{D}$, and $D_n \to \mathbb{D}$. However,

$$D_* = D^* = \mathbb{D} \cup \Delta - \{z: 1 \leq z < \infty\},$$

the entire plane $\mathbb{C}$ minus the segment $\{z: 1 \leq z < \infty\}$ and the full unit circle. This is a disconnected set.

We are now prepared to state the basic theorem.

Theorem 3.1 (Carathéodory Convergence Theorem). *Let $\{D_n\}$ be a sequence of simply connected domains with $0 \in D_n \subsetneq \mathbb{C}$, $n = 1, 2, \ldots$. Let f_n map the unit disk $\mathbb{D}$ conformally onto D_n and satisfy $f_n(0) = 0$ and $f_n'(0) > 0$. Let D be the kernel of $\{D_n\}$. Then $f_n \to f$ uniformly on each compact subset of $\mathbb{D}$ if and only if $D_n \to D \neq \mathbb{C}$. In the case of convergence there are two possibilities. If $D = \{0\}$, then $f = 0$. If $D \neq \{0\}$, then D is a simply connected domain, f maps $\mathbb{D}$ conformally onto D, and the inverse functions $f_n^{-1} \to f^{-1}$ uniformly on each compact subset of D.*

Proof. Suppose first that $f_n(z) \to f(z)$ uniformly on compact subsets of $\mathbb{D}$. Then f is analytic in $\mathbb{D}$, and by a well-known theorem (see §1.1) f is either constant or univalent in $\mathbb{D}$. If f is constant, then $f(z) \equiv 0$ because $f_n(0) = 0$ for all n.

Case I: $f(z) \equiv 0$. Then we must show that $D = \{0\}$. Otherwise, some disk $\{w: |w| < \rho\}$ is contained in D_n for every n. The inverse functions φ_n are then defined in this disk and have the properties $\varphi_n(0) = 0$ and $|\varphi_n(w)| < 1$ there. It follows from the Schwarz lemma that $|\varphi_n'(0)| \leq 1/\rho$, or $|f_n'(0)| \geq \rho > 0$, which contradicts the assumption that $f_n(z) \to 0$ uniformly on compact sets. The same argument shows that every subsequence of $\{D_n\}$ has kernel $\{0\}$, so $D_n \to D = \{0\}$.

Case II: $f(z) \not\equiv 0$. Then f is univalent and maps $\mathbb{D}$ onto some domain Δ, with $f(0) = 0$ and $f'(0) > 0$. We must show that $\Delta = D$ and that $D_n \to D$.

We show first that $\Delta \subset D$. For this purpose, choose an arbitrary compact subset $E \subset \Delta$ and surround E by a rectifiable Jordan curve Γ in $\Delta - E$. Let δ be the (positive) distance from E to Γ, and let $\gamma = f^{-1}(\Gamma)$. We shall now prove that $E \subset D_n$ for all n sufficiently large. Fix a point $w_0 \in E$ and observe that $|f(z) - w_0| \geq \delta$ for all $z \in \gamma$. By the uniform convergence, $|f_n(z) - f(z)| < \delta$ on γ for all $n \geq N$. Hence by Rouché's theorem (see §1.1),

$$f_n(z) - w_0 = [f(z) - w_0] + [f_n(z) - f(z)]$$

has the same number of zeros inside γ as does $[f(z) - w_0]$; namely, one zero. This shows that $w_0 \in D_n$ for all $n \geq N$, where N depends on E but not on w_0. In other words, $E \subset D_n$ for all $n \geq N$. In view of the definition of the kernel D, this proves that $\Delta \subset D$.

It follows, in particular, that the inverse functions $\varphi_n = f_n^{-1}$ are eventually defined (for all $n \geq N$) on E and are uniformly bounded there: $|\varphi_n(w)| \leq 1$. Now choose an expanding sequence of compact sets $E_m \subset \Delta$ and apply a diagonal argument to extract a subsequence $\{\varphi_{n_k}\}$ which converges uniformly on each compact subset of Δ to a function φ analytic in Δ with $\varphi(0) = 0$ and $\varphi'(0) \geq 0$. In fact,

$$0 < \frac{1}{f'(0)} = \lim_{n\to\infty} \frac{1}{f_n'(0)} = \lim_{n\to\infty} \varphi_n'(0) = \varphi'(0),$$

so φ is univalent in Δ.

The next step is to show that $\varphi = f^{-1}$. Fix $z_0 \in \mathbb{D}$ and let $w_0 = f(z_0)$. Choose $\varepsilon > 0$ so small that the circle C defined by $|z - z_0| = \varepsilon$ lies in $\mathbb{D}$, let $\Gamma = f(C)$, and let δ be the distance from w_0 to Γ. Then $|f(z) - w_0| \geq \delta$ on C, while $|f_{n_k}(z) - f(z)| < \delta$ on C for all $k \geq k_0$. Consequently, for each $k \geq k_0$ it follows from Rouché's theorem that $f_{n_k}(z_k) = w_0$ for some z_k inside C. Thus $|z_k - z_0| < \varepsilon$ and $z_k = \varphi_{n_k}(w_0)$. Therefore,

$$|\varphi(w_0) - z_0| \leq |\varphi(w_0) - \varphi_{n_k}(w_0)| + |z_k - z_0| < 2\varepsilon$$

for k sufficiently large. Letting $\varepsilon \to 0$, we conclude that $\varphi(w_0) = z_0$. Because z_0 was chosen arbitrarily in $\mathbb{D}$, this proves that $\varphi = f^{-1}$.

The preceding argument applies equally well to every subsequence of $\{\varphi_n\}$ and shows that some further subsequence converges to f^{-1} uniformly on compact subsets. It follows that $\varphi_n \to f^{-1}$ uniformly on each compact subset of Δ. In fact, the same argument shows that $\{\varphi_n\}$ converges uniformly on compact subsets of D to a univalent function ψ which satisfies $|\psi(w)| < 1$ there. This function ψ is an analytic continuation of f^{-1} from Δ to D. However, f^{-1} already maps Δ conformally *onto* $\mathbb{D}$, so this is impossible unless $\Delta = D$.

It remains to show that $D_n \to D$. But the entire argument can be repeated for any subsequence $\{D_{n_k}\}$ to conclude that f maps $\mathbb{D}$ onto the kernel of $\{D_{n_k}\}$, which must therefore coincide with the kernel of $\{D_n\}$. Hence $D_n \to D$. This completes the discussion of Case II.

Conversely, suppose $D_n \to D \neq \mathbb{C}$. Again there are two cases.

Case I: $D = \{0\}$. Then we claim that $f_n'(0) \to 0$. If not, there exist $\varepsilon > 0$ and a subsequence $\{f_{n_k}\}$ such that $f_{n_k}'(0) \geq \varepsilon$. In view of the Koebe one-quarter theorem, each domain D_{n_k} must then contain the disk $|w| < \varepsilon/4$, contradicting the assumption that each subsequence of $\{D_n\}$ has kernel $\{0\}$. Thus $f_n'(0) \to 0$. On the other hand, the growth theorem (Theorem 2.6) gives

$$|f_n(z)| \leq f_n'(0) \frac{|z|}{(1 - |z|)^2}, \qquad |z| < 1.$$

It follows that $f_n(z) \to 0$ uniformly on each compact subset of $\mathbb{D}$.

Case II: $D \neq \{0\}$, $D \neq \mathbb{C}$. Then we claim that $\{f_n'(0)\}$ is a bounded sequence. Indeed, if $f_{n_k}'(0) \to \infty$ for some subsequence, the Koebe one-quarter theorem would imply that $\{D_{n_k}\}$ has kernel $\mathbb{C}$. This contradiction shows that $\{f_n'(0)\}$ is bounded. We now infer from the growth theorem (as given above) that the functions f_n are uniformly bounded on each compact subset of $\mathbb{D}$ and therefore constitute a normal family. In order to conclude that $\{f_n\}$ converges uniformly on compact subsets of $\mathbb{D}$, it will now suffice to show that it converges pointwise. (Apply Vitali's theorem, stated in §1.3.) Because the functions f_n form a normal family, however, two subsequences with different limits at some point $z_0 \in \mathbb{D}$ would have further subsequences $\{f_{n_k}\}$ and $\{f_{m_k}\}$ converging uniformly on compact sets to different functions f and $\tilde{f}$, with $f(z_0) \neq \tilde{f}(z_0)$. In view of what we have already proved, the corresponding sequences $\{D_{n_k}\}$ and $\{D_{m_k}\}$ would then have different kernels, the images of $\mathbb{D}$ under f and $\tilde{f}$, respectively. But this contradicts the hypothesis that $D_n \to D$. Thus we have shown that $f_n \to f$ uniformly on each compact subset of $\mathbb{D}$. This completes the proof.

§3.2. Density of Slit Mappings

A *slit mapping* is a function which maps a domain conformally onto the complex plane minus a set of Jordan arcs. A *single-slit mapping* is a slit mapping whose range is the complement of a single Jordan arc. We shall be concerned primarily with single-slit mappings defined on the unit disk $\mathbb{D}$. The omitted Jordan arc must then extend to infinity.

The point of departure for Loewner's method is the observation that the single-slit mappings are dense in the class S. In other words, each function in S can be approximated uniformly on compact subsets of $\mathbb{D}$ by single-slit mappings. The following theorem states this more precisely.

Theorem 3.2. *To each function $f \in S$ there corresponds a sequence of single-slit mappings $f_n \in S$ such that $f_n \to f$ uniformly on each compact subset of $\mathbb{D}$.*

This theorem has a simple but very important consequence. Let $\mathscr{A} = \mathscr{H}(\mathbb{D})$ be the space of all functions analytic in $\mathbb{D}$, endowed with the topology of uniform convergence on compact subsets. Let ϕ be a complex-valued functional defined and continuous on some open subset of $\mathscr{A}$ containing the compact set S. Then the theorem reduces the problem of finding the maximum of $\operatorname{Re}\{\phi\}$ over S to the problem of finding the supremum of $\operatorname{Re}\{\phi\}$ over the subset of single-slit mappings in S. The latter problem can often be solved by appeal to Loewner's representation of the single-slit mappings, to be developed in this chapter. This is a very effective method which applies to many extremal problems. One inherent weakness, however, is its failure to identify the extremal functions.

Proof of Theorem. It is sufficient to produce a single-slit mapping $g \in S$ such that

$$|f(z) - g(z)| < \varepsilon, \qquad |z| \le \rho < 1,$$

where ε and $\rho < 1$ are given positive numbers and f is a given function in S. (The theorem then follows by choosing sequences $\{\varepsilon_n\}$ and $\{\rho_n\}$ with $\varepsilon_n \to 0$ and $\rho_n \to 1$.) Observe first that each $f \in S$ can be approximated uniformly on compact sets by a function in S which maps $\mathbb{D}$ onto the interior of an analytic Jordan curve. For example, the dilations $f(rz)/r$ provide such an approximation, where $0 < r < 1$.

Consider now a function $f \in S$ which maps $\mathbb{D}$ onto a domain D bounded by an analytic Jordan curve C. Let Γ_n be a Jordan arc which runs from infinity to a point w_0 on C, then part of the way around C to a point w_n, as shown in Figure 3.2. Let D_n be the complement of Γ_n and let g_n map $\mathbb{D}$ conformally onto D_n, with $g_n(0) = 0$ and $g_n'(0) > 0$. Let the endpoints w_n

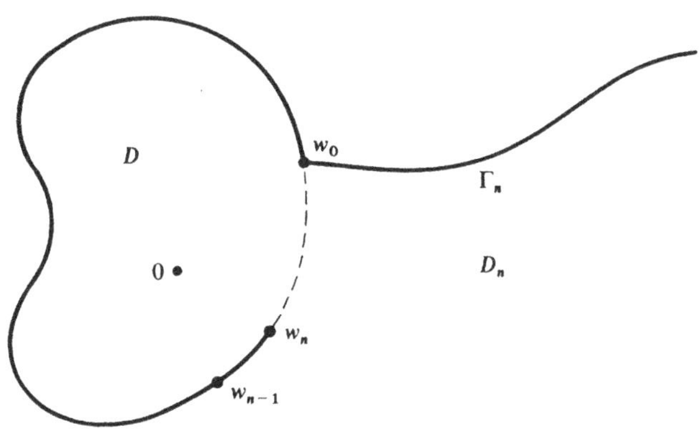

Figure 3.2

be chosen so that $\Gamma_n \subset \Gamma_{n+1}$ and $w_n \to w_0$. It is then clear that D is the kernel of the sequence $\{D_n\}$, and that $D_n \to D$. Therefore, in view of the Carathéodory convergence theorem, $g_n \to f$ uniformly on compact subsets of $\mathbb{D}$. Because this implies (by the Cauchy formula) that $g_n'(0) \to f'(0) = 1$, it follows that

$$h_n = g_n/g_n'(0) \in S$$

are single-slit mappings which converge to f uniformly on compact subsets of $\mathbb{D}$. This proves the theorem.

§3.3. Loewner's Differential Equation

We now proceed to develop Loewner's representation of the single-slit mappings. Let $f \in S$ map the unit disk $\mathbb{D}$ onto a domain D which is the complement of a Jordan arc Γ extending from a finite point w_0 to infinity. Let $w = \psi(t)$, $0 \le t < T$, be a continuous parametric representation of Γ, with $\psi(0) = w_0$ and $\psi(s) \ne \psi(t)$ if $s \ne t$. Let Γ_t denote the portion of Γ from $\psi(t)$ to ∞, and let D_t be the complement of Γ_t. Then $D_0 = D$ and $D_s \subset D_t$ if $s < t$. Let

$$g(z, t) = \beta(t)\{z + b_2(t)z^2 + b_3(t)z^3 + \cdots\}$$

be the conformal mapping of $\mathbb{D}$ onto D_t for which $g(0, t) = 0$ and $g'(0, t) = \beta(t) > 0$. All of the Taylor coefficients of g are continuous functions of t, as is shown by the Carathéodory convergence theorem and the Cauchy formula. In particular, $\beta(t)$ is continuous. It is clear that $\beta(0) = 1$, because $g(z, 0) = f(z)$. It follows from the subordination principle (Chapter 1, Exercise 7), a variant of the Schwarz lemma, that $\beta(t)$ is strictly increasing. Thus the parametric representation of Γ may be rechosen so that $\beta(t) = e^t$, $0 \le t < T$. To see this, let $w = \tilde{\psi}(s) = \psi(\sigma(s))$ be a new parametrization of Γ, giving a leading coefficient $\tilde{\beta}(s) = \beta(\sigma(s))$. Then $\tilde{\beta}(s) = e^s$ if we choose $\sigma(s) = \beta^{-1}(e^s)$.

With this choice of parametric representation, the endpoint T must be ∞. This may be seen as follows. Let M be a fixed positive number. Then Γ_t lies entirely outside the circle $|w| = M$ for all t sufficiently close to T. It follows from the maximum modulus theorem that

$$\left|\frac{z}{g(z, t)}\right| \le \frac{1}{M}, \qquad |z| < 1.$$

In particular, $M \le |g'(0, t)| = e^t$ for all t sufficiently close to T. Since M is arbitrary, this shows that $e^t \to \infty$ as $t \to T$. Thus $T = \infty$.

In summary, we have chosen the parametrization $w = \psi(t)$ of the omitted arc Γ in such a way that

$$g(z, t) = e^t\left\{z + \sum_{n=2}^{\infty} b_n(t)z^n\right\}, \qquad 0 \le t < \infty. \tag{1}$$

This will be called the *standard parametrization* of Γ. Each coefficient $b_n(t)$ is a continuous function of t.

Now consider the function

$$f(z, t) = g^{-1}(f(z), t) = e^{-t}\left\{z + \sum_{n=2}^{\infty} a_n(t)z^n\right\}, \qquad 0 \le t < \infty, \tag{2}$$

which maps $\mathbb{D}$ conformally onto $\mathbb{D}$ minus an arc extending in from the boundary. Of course, $f(z, 0) = z$, the identity mapping. Each coefficient $a_n(t)$ is a polynomial function of $b_2(t), \ldots, b_n(t)$; hence $a_n(t)$ is also continuous. We shall see that

$$e^t f(z, t) \to f(z) = z + a_2 z^2 + \cdots$$

as $t \to \infty$, uniformly on compact subsets of $\mathbb{D}$, which will imply that $a_n(t) \to a_n$. This is contained in the following theorem, basic to Loewner's method, which may be viewed as providing a structural formula for the single-slit mappings in S.

Theorem 3.3. *Let $f \in S$ be a single-slit mapping with omitted arc Γ. Let $w = \psi(t)$, $0 \le t < \infty$, be the standard parametrization of Γ, and let $f(z, t)$ be defined as in equation (2). Then $f(z, t)$ satisfies the differential equation*

$$\frac{\partial f}{\partial t} = -f \frac{1 + \kappa f}{1 - \kappa f}, \tag{3}$$

where $\kappa = \kappa(t)$ is a continuous complex-valued function with $|\kappa(t)| = 1$, $0 \le t < \infty$. Furthermore,

$$\lim_{t \to \infty} e^t f(z, t) = f(z), \qquad |z| < 1, \tag{4}$$

and the convergence is uniform on each compact subset of $\mathbb{D}$.

Intuitively speaking, the theorem says that $e^t f(z, t)$ represents a "flow" from z to $f(z)$ whose dynamics is governed by the differential equation (3). This equation is known as *Loewner's differential equation.*

Proof. Consider first the limit relation (4). A slightly stronger statement is that $e^t g^{-1}(w, t) \to w$ uniformly on compact subsets of $\mathbb{C}$. By the growth theorem (Theorem 2.6),

$$\frac{e^t |z|}{(1 + |z|)^2} \le |g(z, t)| \le \frac{e^t |z|}{(1 - |z|)^2}, \qquad |z| < 1.$$

Given $w \in \mathbb{C}$, set $z = g^{-1}(w, t)$ for sufficiently large t and rearrange these inequalities to obtain

$$\{1 - |g^{-1}(w, t)|\}^2 \le e^t \left| \frac{g^{-1}(w, t)}{w} \right| \le \{1 + |g^{-1}(w, t)|\}^2. \tag{5}$$

In particular, $|g^{-1}(w, t)| \le 4|w|e^{-t}$, so $g^{-1}(w, t) \to 0$ as $t \to \infty$, uniformly on each compact set. Thus the inequalities (5) show that $e^t|g^{-1}(w, t)/w| \to 1$ uniformly on compact sets. Therefore, the functions $e^t g^{-1}(w, t)/w$ constitute a normal family, and so converge uniformly on compact sets to an analytic function $G(w)$ as t tends to infinity through a suitable sequence of values. But it is clear that $G(w) \equiv 1$, since $|G(w)| \equiv 1$ and $G(0) = 1$. Because the limit is independent of the choice of sequence, it follows that $e^t g^{-1}(w, t)/w \to 1$ as $t \to \infty$, or that $e^t g^{-1}(w, t) \to w$, uniformly on compact sets. This proves (4).

We now turn to the derivation of Loewner's differential equation (3). For $0 \le s < t < \infty$, consider the function

$$\zeta = h(z, s, t) = g^{-1}(g(z, s), t) = e^{s-t}z + \cdots,$$

which maps the disk $\mathbb{D}$ in the z-plane onto the disk $\mathbb{D}$ in the ζ-plane minus a Jordan arc J_{st} extending in from the boundary. Let B_{st} be the arc of the circle $|z| = 1$ which corresponds to J_{st}. Let $\lambda(t) = g^{-1}(\psi(t), t)$ be the point on the unit circle which the function $g(z, t)$ maps onto the tip of Γ_t. Then $\lambda(s)$ is an interior point of the arc B_{st}, while $\lambda(t)$ is the point where J_{st} meets the unit circle. The situation is illustrated in Figure 3.3.

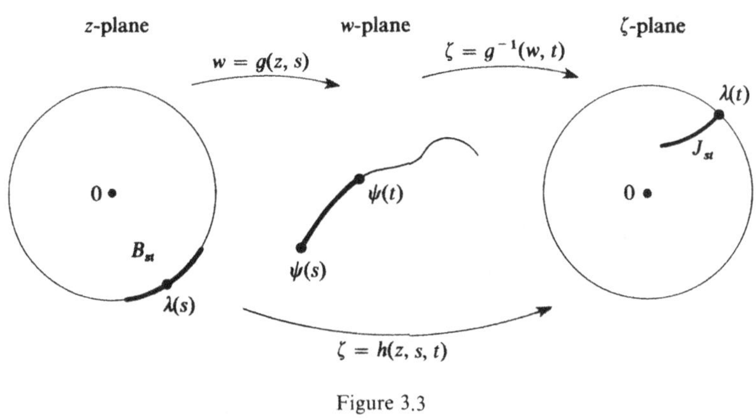

Figure 3.3

By the Carathéodory extension theorem (see §1.5), the function $g^{-1}(w, s)$ is continuous up to the (two-sided) slit Γ_s. Thus as t decreases to the fixed valued s, the arc B_{st} shrinks to the point $\lambda(s)$. Similarly, if t is held fixed and s increases to t, the arc J_{st} contracts to $\lambda(t)$.

Our next aim is to show that λ is a continuous function. This is intuitively obvious but rather difficult to prove. The first step is to continue $h(z, s, t)$ analytically, by Schwarz reflection, over the circular arc complementary to B_{st}. The extended function then maps the complement of B_{st} conformally onto the complement of $J_{st} \cup J_{st}^*$, where J_{st}^* denotes the reflection of J_{st}

across the unit circle. It is clear from the Koebe one-quarter theorem that the arc J_{st} lies outside the disk

$$\{\zeta : |\zeta| < \tfrac{1}{4}e^{s-t}\},$$

so its reflection J_{st}^* lies inside the disk

$$\{\zeta : |\zeta| < 4e^{t-s}\}.$$

Furthermore, by the reflection property,

$$\lim_{z\to\infty} \frac{h(z, s, t)}{z} = \lim_{z\to 0} \frac{z}{h(z, s, t)} = e^{t-s}.$$

It follows from the maximum modulus theorem that

$$\left|\frac{h(z, s, t)}{z}\right| \le 4e^{t-s}$$

throughout the complement of B_{st}.

Recall now that the arc B_{st} contracts to the point $\lambda(s)$ as t decreases to s. A normal family argument therefore shows that as t decreases to s through some sequence, the functions $h(z, s, t)/z$ converge uniformly on compact sets to a function $\varphi(z)$ analytic and bounded on the extended complex plane punctured at $\lambda(s)$, with $\varphi(0) = 1$. By Liouville's theorem, $\varphi(z) \equiv 1$. Because the limit is independent of the choice of sequence, it follows that $h(z, s, t) \to z$ as t decreases to s, and the convergence is uniform on each compact set not containing $\lambda(s)$.

We can now show that λ is continuous. Let $s \ge 0$ be fixed. Given $\varepsilon > 0$, choose $\delta > 0$ so small that for all t satisfying $0 < t - s < \delta$, the arc B_{st} lies inside the circle C with center $\lambda(s)$ and radius ε. Let $\tilde{C}$ be the image of C under the extended mapping $\zeta = h(z, s, t)$. Then $\tilde{C}$ is a Jordan curve which surrounds $J_{st} \cup J_{st}^*$. (See Figure 3.4.) In particular, the point $\lambda(t)$ lies inside $\tilde{C}$. Because $h(z, s, t) \to z$ uniformly on C as $t \to s$, the curve $\tilde{C}$ has diameter less than 3ε for all t sufficiently close to s. Thus for any point $z_0 \in C$,

$$\begin{aligned} |\lambda(s) - \lambda(t)| &\le |\lambda(s) - z_0| + |z_0 - h(z_0)| + |h(z_0) - \lambda(t)| \\ &\le \varepsilon + \varepsilon + 3\varepsilon = 5\varepsilon \end{aligned}$$

for all $t > s$ sufficiently near s. (Here $h(z) = h(z, s, t)$.) This proves that λ is right-continuous. A similar argument shows that λ is also left-continuous. Thus λ is a continuous function.

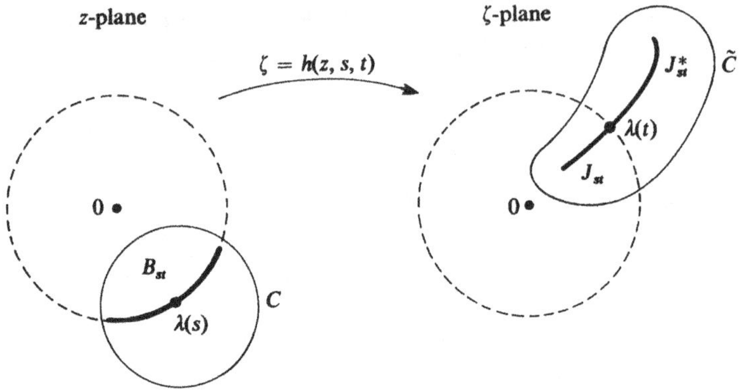

Figure 3.4

We can now derive the Loewner differential equation. Let

$$\Phi(z) = \Phi(z, s, t) = \log\left\{\frac{h(z, s, t)}{z}\right\}$$

denote the branch of the logarithm for which $\Phi(0) = s - t$. Observe that Φ is analytic in $\mathbb{D}$ and continuous in $\overline{\mathbb{D}}$. It is clear from the mapping property of h that $\operatorname{Re}\{\Phi(z)\} = 0$ everywhere on the unit circle except inside the arc B_{st}, where $\operatorname{Re}\{\Phi(z)\} < 0$. Analytic completion of the Poisson formula therefore gives

$$\Phi(z) = \frac{1}{2\pi}\int_{\alpha}^{\beta} \operatorname{Re}\{\Phi(e^{i\theta})\}\, \frac{e^{i\theta} + z}{e^{i\theta} - z}\, d\theta, \tag{6}$$

where $e^{i\alpha}$ and $e^{i\beta}$ are the endpoints of B_{st}. In particular,

$$s - t = \Phi(0) = \frac{1}{2\pi}\int_{\alpha}^{\beta} \operatorname{Re}\{\Phi(e^{i\theta})\}\, d\theta. \tag{7}$$

In view of the identity

$$h(f(z, s), s, t) = f(z, t),$$

the substitution of $f(z, s)$ for z in (6) gives

$$\log\frac{f(z, t)}{f(z, s)} = \frac{1}{2\pi}\int_{\alpha}^{\beta} \operatorname{Re}\{\Phi(e^{i\theta})\}\, \frac{e^{i\theta} + f(z, s)}{e^{i\theta} - f(z, s)}\, d\theta. \tag{8}$$

The mean-value theorem, applied separately to the real and imaginary parts of the integral in (8), now gives

$$\log \frac{f(z,t)}{f(z,s)} = \frac{1}{2\pi}\left[\operatorname{Re}\left\{\frac{e^{i\sigma}+f(z,s)}{e^{i\sigma}-f(z,s)}\right\} + i\operatorname{Im}\left\{\frac{e^{i\tau}+f(z,s)}{e^{i\tau}-f(z,s)}\right\}\right]\int_\alpha^\beta \operatorname{Re}\{\Phi(e^{i\theta})\}\,d\theta,$$

where $e^{i\sigma}$ and $e^{i\tau}$ are points on the arc B_{st}. Dividing this result by $(t - s)$, using (7), and letting t decrease to s, we find

$$\frac{\partial}{\partial s}\{\log f(z,s)\} = -\frac{\lambda(s)+f(z,s)}{\lambda(s)-f(z,s)}, \tag{9}$$

since B_{st} contracts to the point $\lambda(s)$. The derivative in (9) was calculated as a right-hand derivative, but the same argument shows that the left-hand derivative also exists and satisfies (9); for this we need only observe that B_{st} contracts to $\lambda(t)$ as s increases to t. With the definition $\kappa(t) = 1/\lambda(t)$, the equation (9) is Loewner's differential equation (3). Because λ is continuous and $|\lambda(t)| \equiv 1$, it is clear that κ has the same properties. This completes the proof of Theorem 3.3.

§3.4. Univalence of Solutions

We saw in the preceding section that every single-slit mapping $f \in S$ is the uniform limit of $e^t f(z,t)$ as $t \to \infty$, where $f(z,t)$ satisfies the initial condition $f(z,0) = z$ and is generated through the Loewner differential equation by some continuous function $\kappa(t)$ of unit modulus. Furthermore, each function $f(z,t)$ maps the disk $\mathbb{D}$ onto a slit disk consisting of $\mathbb{D}$ with a certain Jordan arc removed. The function κ is determined in a canonical way by the geometry of the original slit domain, the image of $\mathbb{D}$ under f.

It is natural to ask whether every continuous function κ of unit modulus arises in this fashion from some slit domain. This is a delicate question whose answer is negative. A counterexample is cited at the end of this section.

Fortunately, the more important question from the viewpoint of applications is whether every such function κ generates a *univalent* function f, a slit mapping or not. The next theorem asserts that this is true under very general conditions. In particular, it is true if κ has unit modulus and is *piecewise continuous*; that is, if it is continuous apart from a finite number of jump discontinuities. Thus in solving extremal problems over the class S we need consider only those functions f which are generated through the Loewner differential equation by some piecewise continuous unimodular function κ.

The next theorem is stated for the more general *Loewner–Kufarev equation*

$$\frac{\partial w}{\partial t} = -wp(w, t), \tag{10}$$

where p is an analytic function with positive real part. Recall that P is the class of functions φ analytic in $\mathbb{D}$ with $\operatorname{Re}\{\varphi(w)\} > 0$ and $\varphi(0) = 1$. It should be observed that for each fixed t the function

$$p(w, t) = \frac{1 + \kappa(t)w}{1 - \kappa(t)w}$$

belongs to the class P if $|\kappa(t)| = 1$. The more general equation was introduced by Kufarev [1, 3].

Theorem 3.4. *Let $p(w, t)$ be defined for $w \in \mathbb{D}$ and $0 \le t < \infty$. For each fixed $w \in \mathbb{D}$, let $p(w, t)$ be integrable over each interval $0 \le t \le T < \infty$. For each fixed $t \in [0, \infty)$, let $p(w, t) \in P$. Then the differential equation* (10) *has a unique solution $w = f(z, t)$ for $0 \le t < \infty$ which satisfies the initial condition $f(z, 0) = z$. For each fixed t, $f(z, t)$ is analytic and univalent in $\mathbb{D}$, and $e^t f(z, t) \in S$. As $t \to \infty$, $e^t f(z, t)$ converges uniformly on each compact subset of $\mathbb{D}$ to a function $f(z) \in S$.*

This theorem can be deduced from more general results on differential equations, but we shall give a self-contained proof based on the standard method of successive approximations. The estimates in this proof will take advantage of the special properties of equation (10). Two lemmas will be needed.

Lemma 1. *For all complex numbers α and β with positive real part,*

$$|e^{-\alpha} - e^{-\beta}| \le |\alpha - \beta|.$$

Lemma 2. *For each $\varphi \in P$ and for each pair of numbers $\alpha, \beta \in \mathbb{D}$, the inequality*

$$|\varphi(\alpha) - \varphi(\beta)| \le 2(1 - r)^{-2}|\alpha - \beta|$$

holds, where $|\alpha| \le r$ and $|\beta| \le r$.

Proof of Lemma 1. Fix β with $\operatorname{Re}\{\beta\} > 0$ and consider the function

$$F(z) = \frac{e^{-z} - e^{-\beta}}{z - \beta}, \qquad \operatorname{Re}\{z\} \ge 0,$$

where $F(\beta) = -e^{-\beta}$. Observe that F is analytic in the right half-plane, and $F(z) \to 0$ as $z \to \infty$ in this region. By the maximum modulus theorem, it suffices to prove that $|F(iy)| \leq 1$, $-\infty < y < \infty$. But this is equivalent to proving that $|G(w)| \leq 1$ for $\text{Re}\{w\} \geq 0$, where $G(w) = (1 - e^{-w})/w$ and $G(0) = 1$. This again reduces to showing that $|G(iv)| \leq 1$, $-\infty < v < \infty$. But this is the inequality $|1 - e^{i\theta}| \leq |\theta|$, which for $|\theta| \leq \pi$ is simply the geometric statement that the length of a circular arc majorizes the length of the chord.

Proof of Lemma 2. It follows from the Herglotz representation (see §2.5) that

$$\varphi'(z) = 2\int_0^{2\pi} e^{it}(e^{it} - z)^{-2}\, d\mu(t), \qquad |z| < 1,$$

where $d\mu(t) \geq 0$ and $\int d\mu(t) = 1$. Thus a trivial estimate gives

$$|\varphi'(z)| \leq 2(1 - |z|)^{-2}.$$

The desired result now follows from the formula

$$\varphi(\beta) - \varphi(\alpha) = \int_\alpha^\beta \varphi'(z)\, dz,$$

where the path of integration is a line segment.

Proof of Theorem. The differential equation (10) and the initial condition are equivalent to the integral equation

$$w = z\exp\left\{-\int_0^t p(w, t)\, dt\right\}. \tag{11}$$

We shall use the method of successive approximations to prove the existence and uniqueness of a solution to (11). Let the sequence of functions $w_n = w_n(z, t)$ be defined by $w_0(z, t) \equiv 0$ and

$$w_{n+1} = z\exp\left\{-\int_0^t p(w_n, t)\, dt\right\}, \qquad n = 0, 1, \ldots. \tag{12}$$

Since $\text{Re}\{p(w, t)\} > 0$, an inductive argument easily shows that $|w_n(z, t)| \leq |z|$ for all $t \geq 0$. For each fixed t, the functions w_n are analytic in $\mathbb{D}$ and have the properties

$$w_n(0, t) = 0, \qquad w_n'(0, t) = e^{-t}, \qquad w_n(z, 0) = z, \qquad n = 1, 2, \ldots.$$

Using the recursive definition (12) and appealing to Lemmas 1 and 2, we find

$$|w_{n+2} - w_{n+1}| \leq |z| \left| \int_0^t [p(w_{n+1}, t) - p(w_n, t)]\, dt \right|$$
$$\leq 2|z|(1 - |z|)^{-2} \int_0^t |w_{n+1} - w_n|\, dt.$$

It follows by induction that

$$|w_{n+1}(z, t) - w_n(z, t)| \leq \frac{2^n |z|^n t^n}{(1 - |z|)^{2n} n!}, \qquad n = 0, 1, \ldots. \tag{13}$$

The formula

$$w_n = \sum_{k=1}^{n} [w_k - w_{k-1}]$$

now shows that the functions $w_n(z, t)$ converge uniformly in $|z| \leq r$ and $0 \leq t \leq T$ for each $r < 1$ and $T < \infty$. The limit function

$$f(z, t) = \lim_{n \to \infty} w_n(z, t)$$

is analytic in $\mathbb{D}$ for each t and continuous in $[0, \infty)$ for each z. It satisfies the integral equation (11) and therefore has the properties $f(0, t) = 0$, $f'(0, t) = e^{-t}$, $f(z, 0) = z$, and $|f(z, t)| \leq |z|$.

We now turn to the question of uniqueness. Suppose the equation (11) has two solutions $w = f(z, t)$ and $v = g(z, t)$. Estimating the difference as in the proof of (13), we conclude that

$$|w - v| \leq 2|z|(1 - |z|)^{-2} \int_0^t |w - v|\, dt,$$

which gives inductively

$$|f(z, t) - g(z, t)| \leq \frac{2^{n+1} |z|^{n+1} t^n}{(1 - |z|)^{2n} n!}, \qquad n = 0, 1, \ldots.$$

As $n \to \infty$, this implies that $f(z, t) \equiv g(z, t)$. Thus the solution is unique.

The next step is to prove that $f(z, t)$ is univalent in $\mathbb{D}$ for each fixed t. Suppose that $f(z_1, \tau) = f(z_2, \tau)$ for some τ and for some pair of points z_1

and z_2 with $|z_1| \le r$ and $|z_2| \le r$. Let $w = f(z_1, t)$ and $v = f(z_2, t)$. Then the differential equation (10) gives

$$\begin{aligned}\frac{\partial}{\partial t}(w - v) &= vp(v, t) - wp(w, t) \\ &= v[p(v, t) - p(w, t)] + p(w, t)(v - w).\end{aligned}$$

From this we obtain, using Lemma 2 and the Herglotz representation of p,

$$\left|\frac{\partial}{\partial t}|w - v|\right| \le \left|\frac{\partial}{\partial t}(w - v)\right| \le A|w - v|,$$

where

$$A = \frac{2r}{(1 - r)^2} + \frac{1 + r}{1 - r}.$$

In particular,

$$\frac{\partial}{\partial t}|w - v| \ge -A|w - v|,$$

which implies

$$\frac{\partial}{\partial t}\{e^{At}|w - v|\} \ge 0.$$

Integrating from 0 to τ and recalling that $f(z, 0) = z$, we conclude that $-|z_1 - z_2| \ge 0$, or $z_1 = z_2$. This proves the univalence of $f(z, t)$. Thus $e^t f(z, t) \in S$ for each t.

It remains to prove the uniform convergence of $e^t f(z, t)$ as $t \to \infty$. Because $w = f(z, t)$ satisfies the integral equation (11), we have the formula

$$e^t f(z, t) = z \exp\left\{\int_0^t [1 - p(f(z, t), t)]\, dt\right\}. \tag{14}$$

The problem therefore reduces to showing that the integral in (14) converges uniformly on each compact subset of $\mathbb{D}$ as $t \to \infty$. In fact, we shall see that it converges absolutely. Bearing in mind that $p(0, t) = 1$ and $|f(z, t)| \le |z|$, we first appeal to Lemma 2 to obtain the estimate

$$|1 - p(f(z, t), t)| \le 2(1 - r)^{-2}|f(z, t)|, \qquad |z| \le r < 1.$$

Next we recall that $e^t f(z, t) \in S$ and apply the growth theorem to obtain

$$|f(z, t)| \leq r(1 - r)^{-2} e^{-t}, \qquad |z| \leq r.$$

These two estimates show that the integral

$$\int_0^\infty |1 - p(f(z, t), t)| \, dt$$

converges uniformly for $|z| \leq r$. It follows from (14) that $e^t f(z, t)$ converges uniformly on compact subsets of $\mathbb{D}$ as $t \to \infty$. Thus the limit function $f(z)$ belongs to the class S. This finishes the proof of Theorem 3.4.

A corollary is that if $\kappa(t)$ is a continuous (or piecewise continuous) function of unit modulus, the Loewner equation (3) has a unique solution $f(z, t)$ satisfying the initial condition $f(z, 0) = z$. This function $f(z, t)$ is univalent on $\mathbb{D}$ and maps it into $\mathbb{D}$. However, its range need not be a slit disk ($\mathbb{D}$ with a Jordan arc removed). Kufarev [2] has given the following counterexample. If

$$\kappa(t) = [e^{it} + i\sqrt{1 - e^{-2t}}]^3,$$

then the corresponding function $f(z, t)$ maps $\mathbb{D}$ onto a region consisting of $\mathbb{D}$ minus part of the disk bounded by the circle which intersects the unit circle orthogonally at the points $\kappa(t)$ and $\sqrt[3]{\kappa(t)}$, as shown in Figure 3.5. It also follows from Kufarev's analysis that the function $f(z) = \lim_{t \to \infty} e^t f(z, t)$ is not a slit mapping.

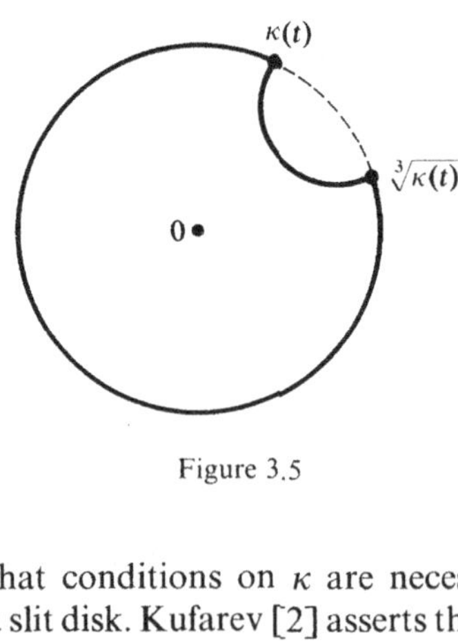

Figure 3.5

It is not known what conditions on κ are necessary and sufficient for $f(z, t)$ to map $\mathbb{D}$ onto a slit disk. Kufarev [2] asserts that a sufficient condition is that κ have a derivative which is continuous on $[0, \infty)$. (In his counterexample, $\kappa'(t)$ is unbounded near the origin.)

§3.5. The Third Coefficient

As a first application of Loewner's method, we now propose to prove the Bieberbach conjecture for the third coefficient. Loewner [3] gave the same application in his original paper. The problem is to show that $|a_3| \leq 3$ for all functions

$$f(z) = z + a_2 z^2 + a_3 z^3 + \cdots$$

of class S. Because S is preserved under rotation, it is equivalent to show that $\operatorname{Re}\{a_3\} \leq 3$. The Loewner theory reduces the problem to functions of the form

$$f(z) = \lim_{t \to \infty} e^t f(z, t),$$

where $f(z, t)$ is the solution to some Loewner differential equation

$$\frac{\partial f}{\partial t} = -f \frac{1 + \kappa f}{1 - \kappa f}, \qquad f(z, 0) = z,$$

corresponding to a continuous function κ of unit modulus. As before, let

$$f(z, t) = e^{-t}[z + a_2(t)z^2 + a_3(t)z^3 + \cdots].$$

Then it is clear that $a_n(0) = 0$ and

$$\lim_{t \to \infty} a_n(t) = a_n, \qquad n = 2, 3, \ldots.$$

Equating the coefficients of z^2 and z^3 in the Loewner differential equation, we find

$$a_2'(t) = -2e^{-t}\kappa(t), \tag{15}$$

$$a_3'(t) = -2e^{-2t}[\kappa(t)]^2 - 4e^{-t}\kappa(t)a_2(t). \tag{16}$$

Integration of (15) gives

$$a_2 = \int_0^\infty a_2'(t)\, dt = -2 \int_0^\infty e^{-t}\kappa(t)\, dt.$$

Since $|\kappa(t)| \equiv 1$, it follows that

$$|a_2| \leq 2 \int_0^\infty e^{-t}\, dt = 2,$$

with equality only if κ is constant. But if $\kappa(t) \equiv \alpha$, the Loewner equation leads to $f(z) = z(1 + \alpha z)^{-2}$, a rotation of the Koebe function (see Exercise 2).

In view of (15), the equation (16) has the form

$$a_3'(t) = -2e^{-2t}[\kappa(t)]^2 + 2a_2(t)a_2'(t),$$

so that integration gives

$$a_3 = -2\int_0^\infty e^{-2t}[\kappa(t)]^2\,dt + 4\left\{\int_0^\infty e^{-t}\kappa(t)\,dt\right\}^2.$$

Setting $\kappa(t) = e^{i\theta(t)}$, we obtain

$$\mathrm{Re}\{a_3\} \le 2\int_0^\infty e^{-2t}[1 - 2\cos^2\theta(t)]\,dt + 4\left\{\int_0^\infty e^{-t}\cos\theta(t)\,dt\right\}^2.$$

Applying the Schwarz inequality, we conclude that

$$\begin{aligned}\mathrm{Re}\{a_3\} &\le 1 - 4\int_0^\infty e^{-2t}\cos^2\theta(t)\,dt \\ &\quad + 4\left\{\int_0^\infty e^{-t}\,dt\right\}\left\{\int_0^\infty e^{-t}\cos^2\theta(t)\,dt\right\} \\ &= 1 + 4\int_0^\infty (e^{-t} - e^{-2t})\cos^2\theta(t)\,dt \\ &\le 1 + 4\int_0^\infty (e^{-t} - e^{-2t})\,dt = 3.\end{aligned}$$

Equality occurs if and only if $\cos^2\theta(t) \equiv 1$. Since κ is required to be continuous, this implies that either $\kappa(t) \equiv 1$ or $\kappa(t) \equiv -1$. The corresponding functions $f(z)$ are $z(1 + z)^{-2}$ and $z(1 - z)^{-2}$.

One disadvantage of the Loewner method is its possible failure to identify all of the extremal functions. The above calculations show that among all single-slit mappings $f \in S$, the coefficients a_2 and a_3 attain their maximum moduli only for rotations of the Koebe function. However, the proof does not exclude the possibility that other functions in S may also maximize $|a_2|$ or $|a_3|$. For a_2 we already know (Theorem 2.2) that this does not happen, but for a_3 the possibility appears to remain. Fortunately, it is known in general that for arbitrary n the only functions in S which maximize $|a_n|$ are single-slit mappings. This and much more will be proved in Chapter 10 by a variational method.

In principle, the Loewner method offers a general method of attack on the Bieberbach conjecture. An explicit formula for a_n can be calculated in terms of κ alone, but it contains multiple integrals and becomes rather formidable as n increases. Other methods were developed to prove the Bieberbach conjecture for $n = 4$, 5, and 6. Only later did Nehari [8] succeed in proving $|a_4| \leq 4$ by the Loewner method.

§3.6. Radius of Starlikeness

We have already found (Theorem 2.13) that the radius of convexity in the class S is $2 - \sqrt{3} = 0.267\ldots$. This means that every function $f \in S$ maps the disk of radius $2 - \sqrt{3}$ onto a convex domain, and this is false for all larger radii. The Koebe function is extremal because it maps each disk of larger radius onto a nonconvex domain.

Using the Loewner theory, we shall now calculate the radius of starlikeness in the class S. This is a much deeper result. The Koebe function is certainly not extremal for this problem because it is starlike in the full unit disk.

The radius of starlikeness will be derived from a theorem providing the sharp upper and lower bounds for $\arg\{f(z)/z\}$ in S, where the branch is chosen which vanishes at the origin. Observe first that these bounds will depend only on $|z|$, because S is preserved under rotation. Furthermore, the upper and lower bounds are symmetric because S is preserved under conjugation. The theorem is as follows.

Theorem 3.5. *For each $f \in S$,*

$$\left|\arg \frac{f(z)}{z}\right| \leq \log \frac{1 + |z|}{1 - |z|}, \qquad |z| < 1.$$

This bound is sharp for each $z \in \mathbb{D}$.

Proof. Again it is enough to prove the inequality for functions generated by solutions $f(z, t)$ to the Loewner differential equation for arbitrary continuous functions κ of unit modulus, with $f(z, 0) = z$. The Loewner equation may be expressed in the form

$$\frac{\partial}{\partial t}\{\log f(z, t)\} = -\frac{1 + \kappa(t)f(z, t)}{1 - \kappa(t)f(z, t)}.$$

Separation into real and imaginary parts gives the equivalent pair of equations

$$\frac{\partial}{\partial t}|f(z, t)| = -|f(z, t)|\frac{1 - |f(z, t)|^2}{|1 - \kappa(t)f(z, t)|^2} \tag{17}$$

and

$$\frac{\partial}{\partial t}\{\arg f(z,t)\} = -2\frac{\operatorname{Im}\{\kappa(t)f(z,t)\}}{|1-\kappa(t)f(z,t)|^2}. \tag{18}$$

The first equation (17) shows that $|f(z,t)|$ decreases from $|z|$ to 0 as t increases from 0 to ∞. Hence there is a one-to-one correspondence between t and $|f|$, and it is possible to introduce $|f|$ as the independent variable. Equation (17) gives

$$dt = -\frac{d|f|}{|f|}\frac{|1-\kappa f|^2}{1-|f|^2}, \tag{19}$$

so it follows from (18) that

$$|d\arg f(z,t)| \le \frac{2|f|\,dt}{|1-\kappa f|^2} = -\frac{2d|f|}{1-|f|^2}.$$

This inequality can be integrated to produce

$$\left|\arg\frac{f(z,t)}{z}\right| = \left|\int_0^t \frac{\partial}{\partial t}\arg f(z,t)\,dt\right| \le 2\int_0^\infty \frac{|f|}{|1-\kappa f|^2}\,dt$$

$$= 2\int_0^{|z|}\frac{d|f|}{1-|f|^2} = \log\frac{1+|z|}{1-|z|}, \qquad |z|<1.$$

Since every single-slit mapping $f \in S$ is the limit of $e^t f(z,t)$ for some choice of κ, this gives the desired bound.

The proof of sharpness is rather delicate. A reexamination of the estimates reveals that equality will occur for fixed $z \in \mathbb{D}$ if

$$\operatorname{Im}\{\kappa(t)f(z,t)\} \equiv -|f(z,t)|, \qquad 0 \le t < \infty. \tag{20}$$

But since $\kappa(t)$ and $f(z,t)$ must be related through the Loewner differential equation, it is not obvious that any choice of κ will generate a function $f(z,t)$ satisfying (20). To show that such a function κ does exist, we argue as follows. If the equation (20) were satisfied, then $\operatorname{Re}\{\kappa f\} = 0$ and the original pair of differential equations (17) and (18) would reduce to

$$\frac{\partial|f|}{\partial t} = -|f|\frac{1-|f|^2}{1+|f|^2} \tag{21}$$

and

$$\frac{\partial}{\partial t}\{\arg f\} = \frac{2|f|}{1+|f|^2}, \tag{22}$$

where $f = f(z, t)$ and $f(z, 0) = z$. Integration of (21) gives

$$\frac{|f(z,t)|}{1 - |f(z,t)|^2} = \frac{|z|e^{-t}}{1 - |z|^2}.$$

This equation uniquely determines $|f(z, t)|$ because the function $x/(1 - x^2)$ increases from 0 to ∞ in the interval $0 \le x < 1$. (Note that $|f(z, 0)| = |z|$.) These values of $|f(z, t)|$ can now be used to determine $\arg\{f(z, t)/z\}$ by integrating equation (22) with the initial condition $f(z, 0) = z$. Having specified both the modulus and the argument of $f(z, t)$, we have therefore constructed $f(z, t)$. This function can now be combined with the relation (20) to *define* $\kappa(t)$. The resulting function κ is continuous and has unit modulus. Through its associated Loewner differential equation it must generate precisely the function $f(z, t)$ we have just constructed. Since the relation (20) holds by definition, this shows that the bound given in the theorem is sharp for each $z \in \mathbb{D}$.

It is now a short step to a closely related theorem which can be used to compute the radius of starlikeness.

Theorem 3.6. *For each* $f \in S$,

$$\left|\arg \frac{zf'(z)}{f(z)}\right| \le \log \frac{1 + |z|}{1 - |z|}, \qquad |z| < 1.$$

This bound is sharp for each $z \in \mathbb{D}$.

Proof. Given $f \in S$ and $\zeta \in \mathbb{D}$, consider the function $g \in S$ obtained from f through a disk automorphism:

$$g(z) = \frac{f\left(\dfrac{z + \zeta}{1 + \bar{\zeta}z}\right) - f(\zeta)}{(1 - |\zeta|^2)f'(\zeta)}, \qquad |z| < 1.$$

The choice $z = -\zeta$ gives the identity

$$\frac{g(-\zeta)}{-\zeta} = \frac{1}{1 - |\zeta|^2} \cdot \frac{f(\zeta)}{\zeta f'(\zeta)}.$$

In particular,

$$\arg\left\{\frac{\zeta f'(\zeta)}{f(\zeta)}\right\} = -\arg\left\{\frac{g(-\zeta)}{-\zeta}\right\}.$$

Thus the desired bound is obtained by appeal to Theorem 3.5. The bound is sharp because for each fixed $\zeta \in \mathbb{D}$, every function $g \in S$ is the image of some $f \in S$ under the disk automorphism induced by ζ. (See Chapter 2, Exercise 8.)

Corollary. *For every radius $r \leq \rho = \tanh(\pi/4)$, each function $f \in S$ maps the disk $|z| < r$ onto a domain starlike with respect to the origin. This is false for every $r > \rho$.*

Proof. Recall that the starlikeness of f is described analytically by the condition $\operatorname{Re}\{zf'(z)/f(z)\} > 0$. It therefore follows from Theorem 3.6 that the disk $|z| < r$ has a starlike image under every function $f \in S$ if and only if $r \leq \rho$, where ρ is determined by the equation

$$\log \frac{1+\rho}{1-\rho} = \frac{\pi}{2},$$

or $\rho = \tanh(\pi/4) = 0.655\ldots$. This number ρ is called the *radius of starlikeness* in S.

The radius of starlikeness was first found by Grunsky [2] in 1934. Grunsky [1] also established Theorems 3.5 and 3.6 while obtaining more general results. For instance, he obtained a striking extension of Theorem 3.5 by showing that for each fixed $z \in \mathbb{D}$, the region of values of $\log[f(z)/z]$ for all $f \in S$ is precisely the closed disk

$$\left\{w \in \mathbb{C}: \left|w - \log \frac{1}{1-r^2}\right| \leq \log \frac{1+r}{1-r}\right\},$$

where $r = |z|$. We shall prove this in Chapter 10 by a variational method. A proof closer to Grunsky's original approach may be found in the survey article by Goluzin [5]. In 1936, Goluzin [1, 2] used Loewner's method to establish the results proved in this section.

Krzyż [2] has applied Loewner's method to show that the radius of close-to-convexity in S is $0.80\ldots$.

§3.7. The Rotation Theorem

We now return to the problem of finding sharp upper and lower bounds for $\arg f'(z)$ in the class S. These bounds constitute what is called the *rotation theorem* for S because of the geometric interpretation of $\arg f'(z)$ as the local rotation factor under the mapping $f \in S$. As usual, $\arg f'(z)$ is understood to be the branch which vanishes at the origin. Because S is preserved under rotation and conjugation, it is again clear that the sharp upper and lower bounds are symmetric and depend only on $|z|$. As a consequence of Bieberbach's theorem, we have already derived (§2.3) the inequality

$$|\arg f'(z)| \le 2 \log \frac{1+r}{1-r}, \qquad r = |z| < 1,$$

but this bound is not sharp for any $z \neq 0$. The sharp bound remained undetermined until 1936, when Goluzin [1] found it by skillful use of the Loewner differential equation. The result has the striking feature that the sharp bound undergoes a change of form on the circle $|z| = 1/\sqrt{2}$, at which points $\arg f'(z)$ may attain the values $\pm\pi$.

Theorem 3.7 (Rotation Theorem). *For each $f \in S$,*

$$|\arg f'(z)| \le \begin{cases} 4 \sin^{-1} r, & r \le 1/\sqrt{2}, \\ \pi + \log \dfrac{r^2}{1-r^2}, & r \ge 1/\sqrt{2}, \end{cases}$$

where $r = |z| < 1$. The bound is sharp for each $z \in \mathbb{D}$.

Proof. Again it suffices to consider functions $f \in S$ of the form

$$f(z) = \lim_{t \to \infty} e^t f(z, t),$$

where $f(z, t)$ is the solution to the Loewner differential equation (3) corresponding to some continuous function κ of unit modulus, with $f(z, 0) = z$. Here the convergence is uniform on compact subsets of $\mathbb{D}$, so it follows from the Cauchy formula that

$$f'(z) = \lim_{t \to \infty} e^t f'(z, t),$$

where $f'(z, t) = (\partial/\partial z) f(z, t)$ and the convergence is again uniform on compact subsets of $\mathbb{D}$.

Differentiation of the Loewner equation (3) with respect to z gives

$$\frac{\partial f'}{\partial t} = -f' \frac{1 + 2\kappa f - \kappa^2 f^2}{(1 - \kappa f)^2}.$$

This may be put into the form

$$\frac{\partial}{\partial t}\{\log f'\} = 1 - \frac{2}{(1 - \kappa f)^2},$$

with imaginary part

$$\frac{\partial}{\partial t}\{\arg f'\} = \frac{2\,\mathrm{Im}\{(1-\kappa f)^2\}}{|1-\kappa f|^4}. \tag{23}$$

At this point it is advantageous to introduce $|f|$ as the independent variable, as in the preceding section. Recall that $|f|$ decreases from $|z|$ to 0 as t goes from 0 to ∞. After substitution of the expression (19) for dt, the equation (23) becomes

$$d\arg f' = -\frac{2\,\mathrm{Im}\{(1-\kappa f)^2\}\,d|f|}{|1-\kappa f|^2|f|(1-|f|^2)}. \tag{24}$$

Observe that

$$\frac{\mathrm{Im}\{(1-\kappa f)^2\}}{|1-\kappa f|^2} = \sin(2\arg\{1-\kappa f\}). \tag{25}$$

But since $|\kappa(t)| = 1$, the point κf lies on the circle with radius $|f|$, so it is clear geometrically (see Figure 3.6) that

$$|\arg\{1-\kappa f\}| \le \sin^{-1}|f|. \tag{26}$$

Now if $|f| \le 1/\sqrt{2}$, so that $\sin^{-1}|f| \le \pi/4$, the inequality (26) gives

$$\begin{aligned}|\sin(2\arg\{1-\kappa f\})| &\le \sin(2\sin^{-1}|f|)\\ &= 2|f|\sqrt{1-|f|^2}.\end{aligned}$$

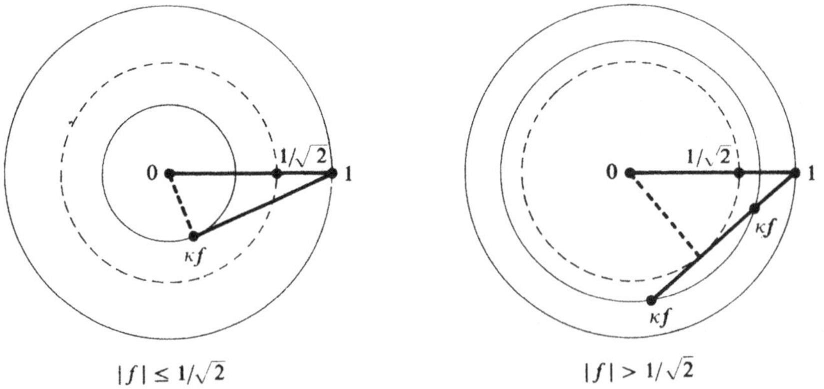

Figure 3.6

On the other hand, if $|f| > 1/\sqrt{2}$, it is clear from Figure 3.6 that there are two values of κ for which $\arg\{1 - \kappa f\} = \pi/4$, so we cannot improve upon the trivial bound

$$|\sin(2\arg\{1 - \kappa f\})| \leq 1.$$

In view of the relation (25), these estimates give

$$\frac{|\operatorname{Im}\{(1 - \kappa f)^2\}|}{|1 - \kappa f|^2} \leq \begin{cases} 2|f|\sqrt{1 - |f|^2}, & |f| \leq 1/\sqrt{2}, \\ 1, & |f| > 1/\sqrt{2}. \end{cases}$$

Combining this with equation (24), we find

$$|d\arg f'| \leq \begin{cases} \dfrac{-4\,d|f|}{\sqrt{1 - |f|^2}}, & |f| \leq 1/\sqrt{2}, \\[2ex] \dfrac{-2\,d|f|}{|f|(1 - |f|^2)}, & |f| > 1/\sqrt{2}. \end{cases}$$

Since

$$\arg f'(z) = \lim_{t \to \infty} \arg f'(z, t),$$

we conclude by integration over the interval $0 \leq t < \infty$ that for $|z| \leq 1/\sqrt{2}$,

$$|\arg f'(z)| \leq \int_0^{|z|} \frac{4\,d|f|}{\sqrt{1 - |f|^2}} = 4\sin^{-1}|z|;$$

while for $1/\sqrt{2} < |z| < 1$,

$$\begin{aligned} |\arg f'(z)| &\leq \int_0^{1/\sqrt{2}} \frac{4\,d|f|}{\sqrt{1 - |f|^2}} + \int_{1/\sqrt{2}}^{|z|} \frac{2\,d|f|}{|f|\sqrt{1 - |f|^2}} \\ &= \pi + \log\frac{|z|^2}{1 - |z|^2}. \end{aligned}$$

This is the desired estimate.

To prove the sharpness of the estimate for each fixed z, it is enough to establish the existence of a pair of functions $\kappa(t)$ and $f(z, t)$ related through the Loewner differential equation, such that $f(z, 0) = z$ and

$$\sin(2\arg\{1 - \kappa f\}) = \begin{cases} 2|f|\sqrt{1 - |f|^2}, & |f| \leq 1/\sqrt{2}, \\ 1, & |f| > 1/\sqrt{2} \end{cases}$$

for all t. An equivalent condition is

$$\sin(\arg\{1-\kappa f\}) = \begin{cases} |f|, & |f| \le 1/\sqrt{2}, \\ 1/\sqrt{2}, & |f| > 1/\sqrt{2}. \end{cases} \tag{27}$$

Geometric considerations (see Figure 3.6) show that equation (27) determines κf as a function of $|f|$, uniquely for $|f| \le 1/\sqrt{2}$, but with two possibilities (one branch of which may be chosen) for $|f| > 1/\sqrt{2}$. Substitute this function for κf in the two Loewner equations (17) and (18). Equation (17) then determines $|f(z,t)|$ as a function of z and t, whereupon equation (18) determines $\arg\{f(z,t)/z\}$. The function $f(z,t)$ thus constructed (with $f(z,0) = z$) may now be combined with the previous expression for κf in terms of $|f|$, as determined by equation (27), to define a continuous function $\kappa(t)$ with $|\kappa(t)| = 1$ for all t. By construction, $\kappa(t)$ and $f(z,t)$ are related through the Loewner equation, and equation (27) is satisfied for all t. This proves the sharpness of the rotation theorem.

For $|z| \le 1/\sqrt{2}$, it is actually possible to exhibit an elementary extremal function. Let $\rho = \cos\theta$ for $0 < \theta < \pi/2$, and consider the function

$$f(z) = \frac{z - \rho z^2}{(1 - e^{-i\theta}z)^2} = z + a_2 z^2 + \cdots,$$

with derivative

$$f'(z) = \frac{1 - e^{i\theta}z}{(1 - e^{-i\theta}z)^3}.$$

To verify that f is univalent, we shall show it is close-to-convex (§2.6). For each real number α, the function

$$\varphi(z) = \frac{e^{i\alpha}z}{(1 - e^{-i\theta}z)}$$

is convex, and

$$\frac{f'(z)}{\varphi'(z)} = e^{-i\alpha}\frac{1 - e^{i\theta}z}{1 - e^{-i\theta}z}.$$

This last expression maps the unit disk onto a half-plane bounded by a line through the origin, so for some choice of α it will have positive real part.

This proves that f is close-to-convex; hence $f \in S$. (Actually, f maps $\mathbb{D}$ onto the complement of a half-line.) On the other hand,

$$\begin{aligned} \arg f'(\rho) &= \arg(1 - \rho e^{i\theta}) - 3 \arg(1 - \rho e^{-i\theta}) \\ &= -4 \sin^{-1} \rho. \end{aligned}$$

This proves the sharpness of the rotation theorem in the range $0 \leq |z| \leq 1/\sqrt{2}$, even within the subclass K of close-to-convex functions.

For $f \in K$, the estimate

$$|\arg f'(z)| \leq 4 \sin^{-1} r$$

persists and is sharp even for $|z| > 1/\sqrt{2}$, as Krzyż [3] has shown. Ogawa [1] showed that within the subclass K_0 of normalized close-to-convex functions, where the associated convex function belongs to C, the sharp bound is

$$2 \sin^{-1} r + 2 \tan^{-1} r.$$

Stroganoff [1] and A. W. Goodman [1] proved that the Koebe function is extremal in the class S^* of starlike functions. (Here the expression for the sharp bound is fairly complicated.) Bieberbach [4] found the sharp bound $2 \sin^{-1} r$ for the class C of convex functions.

Goluzin [1] was unable to establish the sharpness of the rotation theorem in S for the range $|z| > 1/\sqrt{2}$. This was first done by Bazilevich [2], whose proof Goluzin [2] simplified.

The rotation theorem and the distortion theorem (Theorem 2.5) suggest the problem of describing the full region of values of $f'(z)$ for fixed $z \in \mathbb{D}$ and for all $f \in S$. Both this and the region of values of $\log f'(z)$ are rather complicated. A discussion of this problem may be found in Grad [1]. Other solutions were found by Kufarev [6] and by Jenkins [7]. Kufarev applied the variational-parametric method, a combination of Loewner's method and a variational method. Some years later, Aleksandrov and Kopanev [1] used the same approach to find an explicit expression for the boundary curve of the region of values of $\log f'(z)$. Their solution is based on an earlier result of Aleksandrov [1] and is presented in the book of Aleksandrov [3].

§3.8. Coefficients of Odd Functions

The theorem of Littlewood and Paley (Theorem 2.23) asserts that the coefficients of all odd univalent functions

$$h(z) = z + c_3 z^3 + c_5 z^5 + \cdots$$

are bounded by an absolute constant. Because the square-root transform of the Koebe function is

$$\frac{z}{1-z^2} = z + z^3 + z^5 + \cdots,$$

Littlewood and Paley [1] conjectured that $|c_n| \le 1$ for all $h \in S^{(2)}$. This was promptly disproved by Fekete and Szegö [1] through an ingenious application of the Loewner method.

Each $h \in S^{(2)}$ has the form $h(z) = \sqrt{f(z^2)}$ for some function

$$f(z) = z + a_2 z^2 + a_3 z^3 + \cdots$$

of class S. An easy calculation gives

$$c_3 = \tfrac{1}{2}a_2, \qquad c_5 = \tfrac{1}{2}(a_3 - \tfrac{1}{4}a_2^2).$$

Because $|a_2| \le 2$, it is clear that $|c_3| \le 1$ for all $h \in S^{(2)}$. However, Fekete and Szegö found that the maximum of $|c_5|$ is larger than 1. In fact, they obtained the sharp bound for $|a_3 - \alpha a_2^2|$ in S, for each fixed α in the interval $0 \le \alpha \le 1$. Their result may be viewed as an interpolation between Loewner's inequality $|a_3| \le 3$ and the elementary inequality (Chapter 2, Exercise 1) $|a_3 - a_2^2| \le 1$.

Theorem 3.8 (Fekete–Szegö Theorem). *For each $f \in S$,*

$$|a_3 - \alpha a_2^2| \le 1 + 2e^{-2\alpha/(1-\alpha)}, \qquad 0 < \alpha < 1.$$

This bound is sharp for each α.

The choice $\alpha = \frac{1}{4}$ demonstrates the existence of an odd univalent function with $|c_5| > 1$.

Corollary. *For each $h \in S^{(2)}$,*

$$|c_5| \le \tfrac{1}{2} + e^{-2/3} = 1.013\ldots.$$

This bound is sharp.

The proof of the Fekete–Szegö theorem rests on the following lemma, essentially due to Valiron and Landau (see Landau [3]).

Valiron–Landau Lemma. *Let $\varphi(t)$ be real-valued and continuous for $t \ge 0$, with $|\varphi(t)| \le e^{-t}$ and*

$$\int_0^\infty [\varphi(t)]^2\, dt = (\lambda + \tfrac{1}{2})e^{-2\lambda}, \qquad 0 \le \lambda < \infty.$$

Then

$$\left| \int_0^\infty \varphi(t)\, dt \right| \le (\lambda + 1)e^{-\lambda},$$

with equality occurring only for $\varphi(t) = \pm\psi(t)$, *where*

$$\psi(t) = \begin{cases} e^{-\lambda}, & 0 \le t \le \lambda, \\ e^{-t}, & \lambda < t < \infty. \end{cases}$$

Proof. Note first that the constraint $|\varphi(t)| \le e^{-t}$ guarantees that the integral of φ^2 lies between 0 and $\frac{1}{2}$. The value of this integral then uniquely determines λ, since the function $(x + \frac{1}{2})e^{-2x}$ decreases from $\frac{1}{2}$ to 0 as x increases from 0 to ∞. Note also that ψ is continuous and has the properties $|\psi(t)| \le e^{-t}$,

$$\int_0^\infty [\psi(t)]^2\, dt = (\lambda + \tfrac{1}{2})e^{-2\lambda},$$

and

$$\int_0^\infty \psi(t)\, dt = (\lambda + 1)e^{-\lambda}.$$

Now observe that for all $t \ge 0$, the function

$$F(t) = [\psi(t) - |\varphi(t)|][2e^{-\lambda} - \psi(t) - |\varphi(t)|]$$

is nonnegative, so that

$$\begin{aligned} 0 \le \int_0^\infty F(t)\, dt &= 2e^{-\lambda}\left\{\int_0^\infty \psi(t)\, dt - \int_0^\infty |\varphi(t)|\, dt\right\} \\ &\quad - \int_0^\infty [\psi(t)]^2\, dt + \int_0^\infty [\varphi(t)]^2\, dt \\ &= 2e^{-\lambda}\left\{(\lambda + 1)e^{-\lambda} - \int_0^\infty |\varphi(t)|\, dt\right\}. \end{aligned}$$

It follows that

$$\left| \int_0^\infty \varphi(t)\, dt \right| \le \int_0^\infty |\varphi(t)|\, dt \le (\lambda + 1)e^{-\lambda},$$

with equality only for $\varphi(t) = \pm\psi(t)$.

Proof of Theorem. Since a_3 and a_2^2 "turn together" under every rotation, it is enough to estimate $\operatorname{Re}\{a_3 - \alpha a_2^2\}$. The Loewner theory provides the formulas (see §3.5)

$$a_2 = -2\int_0^\infty e^{-t}\kappa(t)\,dt$$

and

$$a_3 = -2\int_0^\infty e^{-2t}[\kappa(t)]^2\,dt + 4\left\{\int_0^\infty e^{-t}\kappa(t)\,dt\right\}^2,$$

where $\kappa(t)$ is a piecewise continuous complex-valued function with $|\kappa(t)| = 1$ for all t. Setting $\kappa(t) = e^{i\theta(t)}$, we obtain

$$\begin{aligned}\operatorname{Re}\{a_3 - \alpha a_2^2\} &= 4(1-\alpha)\left\{\left(\int_0^\infty e^{-t}\cos\theta(t)\,dt\right)^2 - \left(\int_0^\infty e^{-t}\sin\theta(t)\,dt\right)^2\right\}\\ &\quad - 4\int_0^\infty e^{-2t}\cos^2\theta(t)\,dt + 1\\ &\le 4(1-\alpha)\left\{\int_0^\infty \varphi(t)\,dt\right\}^2 - 4\int_0^\infty [\varphi(t)]^2\,dt + 1,\end{aligned}$$

where $\varphi(t) = e^{-t}\cos\theta(t)$. Therefore, if

$$\int_0^\infty [\varphi(t)]^2\,dt = (\lambda + \tfrac{1}{2})e^{-2\lambda},$$

an appeal to the Valiron–Landau lemma gives

$$\operatorname{Re}\{a_3 - \alpha a_2^2\} \le 4e^{-2\lambda}[(1-\alpha)(\lambda+1)^2 - (\lambda + \tfrac{1}{2})] + 1.$$

But this bound attains its maximum value

$$1 + 2e^{-2\alpha/(1-\alpha)}$$

for $\lambda = \alpha/(1-\alpha)$. This gives the desired inequality.

To prove the sharpness, we now take $\lambda = \alpha/(1-\alpha)$ and try to determine a function $\theta(t)$ for which $-\pi/2 < \theta(t) < \pi/2$ and

$$e^{-t}\cos\theta(t) = \psi(t);$$

that is,

$$\cos\theta(t) = \begin{cases} e^{t-\lambda}, & 0 \le t \le \lambda,\\ 1, & \lambda < t < \infty.\end{cases}$$

This function $\theta(t)$ must be piecewise continuous and must also satisfy the condition

$$\int_0^{\infty} e^{-t} \sin \theta(t)\, dt = 0. \tag{28}$$

To achieve this, we choose a parameter τ in the interval $0 < \tau < \lambda$ and require that $\theta(t)$ satisfy

$$0 < \theta(t) < \frac{\pi}{2} \quad \text{for } 0 \leq t < \tau;$$

$$-\frac{\pi}{2} < \theta(t) \leq 0 \quad \text{for } \tau \leq t \leq \lambda.$$

Then

$$\sin \theta(t) = \begin{cases} +\{1 - e^{2(t-\lambda)}\}^{1/2}, & 0 \leq t < \tau, \\ -\{1 - e^{2(t-\lambda)}\}^{1/2}, & \tau \leq t \leq \lambda, \\ 0, & \lambda < t < \infty. \end{cases}$$

It is now clear by continuity that the equation (28) holds for a suitable choice of τ. This completes the proof.

§3.9. An Elementary Counterexample

Although the method of Fekete and Szegö leads to the sharp bound for $|c_5|$ in the class $S^{(2)}$ of odd univalent functions, it fails to describe the extremal functions. In order to gain a better geometric understanding of the problem, Schaeffer and Spencer [1] applied a variational method. They found that the function $h \in S^{(2)}$ with largest $|c_5|$ is unique apart from rotations, that it may be chosen to have all of its coefficients real, and that it then maps the disk onto the complex plane slit along the real axis from $\pm\infty$ to the points $\pm\frac{1}{2}e^{1/6}$, with symmetric forks issuing from these points. (This is indicated in Figure 3.7, the "Fekete–Szegö function.") With this information as a guide, they were then able to give an independent and entirely elementary construction to disprove the Littlewood–Paley conjecture for each coefficient c_n with $n \geq 5$. A version of their construction is presented below.

Theorem 3.9. *To each odd integer $n \geq 5$ there corresponds a function $h \in S^{(2)}$ with all of its coefficients real and with $|c_n| > 1$.*

Proof. Let $f \in S$ be a function which maps the disk onto the complex plane slit along two conjugate rays $w = \beta t$ and $w = \bar{\beta}t$, $1 \leq t < \infty$, where β is a nonreal complex number. Let λ and $\bar{\lambda}$ be the points on the unit circle which correspond to the tips of the slits: $f(\lambda) = \beta$ and $f(\bar{\lambda}) = \bar{\beta}$. Then the function

$$g(z) = \frac{f(z)}{zf'(z)}$$

is analytic in $\bar{\mathbb{D}}$ except for simple poles at λ and $\bar{\lambda}$, with $\mathrm{Re}\{g(z)\} = 0$ on the unit circle except at these two singular points. By the Schwarz reflection principle, the function g is easily seen to have the form

$$g(z) = a\frac{1 + \lambda z}{1 - \lambda z} + b\frac{1 + \bar{\lambda}z}{1 - \bar{\lambda}z},$$

where a and b are real numbers. Thus $g(0) = a + b = 1$, since $f \in S$. Also, g has real coefficients because f does. In particular,

$$g'(0) = 2(a\lambda + b\bar{\lambda})$$

is real. It follows that $a = b = \frac{1}{2}$.

Now let t be a small positive parameter, and consider the function

$$w = \varphi(z) = \varphi(z, t) = f^{-1}(e^{-t}f(z)),$$

which maps $\mathbb{D}$ onto $\mathbb{D}$ minus two short conjugate arcs terminating at λ and $\bar{\lambda}$. Expanding φ in powers of t, we find

$$\begin{aligned}\varphi(z) &= z - \frac{f(z)}{f'(z)}t + O(t^2)\\ &= z - \tfrac{1}{2}z\left(\frac{1 + \lambda z}{1 - \lambda z} + \frac{1 + \bar{\lambda}z}{1 - \bar{\lambda}z}\right)t + O(t^2).\end{aligned}$$

Hence the square-root transform is an odd univalent function of the form

$$\psi(z) = \sqrt{\varphi(z^2)} = z - \tfrac{1}{4}z\left(\frac{1 + \lambda z^2}{1 - \lambda z^2} + \frac{1 + \bar{\lambda}z^2}{1 - \bar{\lambda}z^2}\right)t + O(t^2).$$

This function ψ is now composed with the odd univalent function $\zeta/(1 + \zeta^2)$ and normalized to produce

$$h(z) = \frac{\psi(z)}{\psi'(0)\{1 + [\psi(z)]^2\}} = z + c_3 z^3 + c_5 z^5 + \cdots,$$

which is the desired function of class $S^{(2)}$, depending on the parameters t and λ. It is clear that the functions f, φ, ψ, and h all have real coefficients.

The next step is to calculate the coefficient c_n up to the linear term in t. Setting $\lambda = e^{i\theta}$, we obtain the expansion

$$\psi(z) = (1 - \tfrac{1}{2}t)z - t \sum_{k=1}^{\infty} \cos k\theta\, z^{2k+1} + O(t^2),$$

so that for $v = 1, 2, 3, \ldots,$

$$[\psi(z)]^v = z^v\left\{1 - vt\left(\frac{1}{2} + \sum_{k=1}^{\infty} \cos k\theta\, z^{2k}\right) + O(t^2)\right\}.$$

But h has the expansion

$$h(z) = (1 + \tfrac{1}{2}t) \sum_{j=0}^{\infty} (-1)^j[\psi(z)]^{2j+1} + O(t^2),$$

so a short calculation gives

$$c_{2m+1} = (-1)^m\{1 - s_m t + O(t^2)\},$$

where

$$s_m = m + \sum_{k=1}^{m} (-1)^k[2(m-k) + 1]\cos k\theta.$$

This last sum may be evaluated by computing the real and imaginary parts of

$$\sum_{k=1}^{m} (-1)^k e^{ik\theta} = \frac{(-1)^m e^{i(m+1)\theta} - e^{i\theta}}{1 + e^{i\theta}},$$

then differentiating the imaginary part. After simplifying the result, we find

$$2(1 + \cos\theta)^2 s_m = (-1)^{m+1} \sin\theta\{\sin(m+1)\theta + \sin m\theta\}.$$

This expression shows that $s_m < 0$ if θ is chosen in the interval

$$0 < \theta < \frac{\pi}{m+1}, \qquad m = 2, 4, 6, \ldots;$$

$$\frac{\pi}{m} < \theta < \frac{2\pi}{m+1}, \qquad m = 3, 5, 7, \ldots.$$

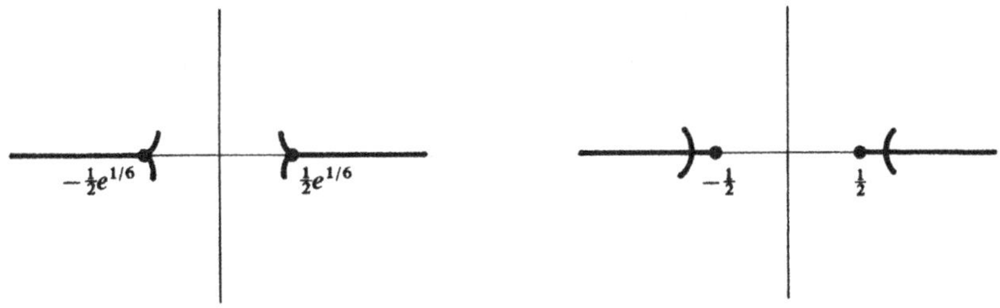

Fekete–Szegö function Schaeffer–Spencer function

Figure 3.7

With such a choice of θ, it is now clear that $|c_{2m+1}| > 1$ for all t in some small interval $0 < t < \delta_m$. This completes the proof.

Figure 3.7 shows schematically, for purpose of comparison, the images of the Fekete–Szegö extremal function for c_5 and of the Schaeffer–Spencer function h just constructed.

For odd *starlike* functions, the Littlewood–Paley conjecture $|c_n| \leq 1$ is true (see Chapter 2, Exercise 9). More generally, Goluzin [3] has shown that if $f \in S^*$ and $a_2 = 0$, then $|a_n| \leq 1$ for all n.

The best general estimate currently known is $|c_n| \leq 1.14$, due to V. I. Milin [1]. This represents a slight improvement over the previous result $|c_n| \leq 1.17$, due to I. M. Milin [3, 6], a proof of which is given in Chapter 5.

The Fekete–Szegö function shows that the bound $|c_5| \leq \frac{1}{2} + e^{-2/3}$ remains sharp for odd functions with real coefficients. More recently, Leeman [5] has shown that for functions in $S^{(2)}$ with real coefficients the sharp upper bound for $|c_7|$ is 1090/1083. The appearance of a *rational* bound was quite unexpected.

§3.10. Robertson's Conjecture

In 1936, Robertson [4] conjectured that the coefficients of odd univalent functions

$$h(z) = \sqrt{f(z^2)} = z + c_3 z^3 + c_5 z^5 + \cdots$$

satisfy the inequality

$$\sum_{k=1}^{n} |c_{2k-1}|^2 \leq n, \qquad n = 1, 2, \ldots,$$

where $c_1 = 1$. This is a weaker version of the Littlewood–Paley conjecture which, as we have already observed (§2.11), also implies the Bieberbach conjecture. Since $|c_3| \leq 1$, the inequality clearly holds for $n = 2$. It has been established by Robertson [4] for $n = 3$ and by Friedland [1] for $n = 4$, but for larger values of n the problem remains open. We shall now give Robertson's proof, which uses the Loewner method and is quite similar to the Fekete–Szegö estimation of $|c_5|$.

Theorem 3.10. *For each function* $h \in S^{(2)}$,

$$|c_1|^2 + |c_3|^2 + |c_5|^2 \leq 3.$$

Proof. Since $c_1 = 1$ and we may assume $c_5 \geq 0$, it is sufficient to prove

$$|c_3|^2 + [\mathrm{Re}\{c_5\}]^2 \leq 2.$$

As in the proof of the Fekete–Szegö theorem (§3.8), the Loewner formulas give

$$c_3 = \tfrac{1}{2}a_2 = -\int_0^\infty e^{-t}\kappa(t)\,dt$$

and

$$c_5 = \tfrac{1}{2}(a_3 - \tfrac{1}{4}a_2^2) = \tfrac{3}{2}\left\{\int_0^\infty e^{-t}\kappa(t)\,dt\right\}^2 - \int_0^\infty e^{-2t}[\kappa(t)]^2\,dt.$$

Setting $\kappa(t) = e^{i\theta(t)}$ and

$$\int_0^\infty e^{-t}\kappa(t)\,dt = u + iv,$$

we find $|c_3|^2 = u^2 + v^2$ and

$$c_5 = \mathrm{Re}\{c_5\} = \tfrac{3}{2}(u^2 - v^2) - 2\int_0^\infty e^{-2t}\cos^2\theta(t)\,dt + \tfrac{1}{2}.$$

We now set

$$\int_0^\infty e^{-2t}\cos^2\theta(t)\,dt = (\lambda + \tfrac{1}{2})e^{-2\lambda}, \qquad 0 \leq \lambda \leq \infty,$$

and apply the Valiron–Landau lemma (§3.8) to obtain

$$|u| \leq (\lambda + 1)e^{-\lambda}.$$

Since $u^2 + v^2 \leq 1$, it follows that

$$|c_3|^2 + |c_5|^2 \leq B(\lambda),$$

where

$$B(\lambda) = \min\{1 - v^2, (\lambda + 1)^2 e^{-2\lambda}\} + v^2 + \tfrac{1}{4}\{\beta(\lambda) + 1 - 3v^2\}^2,$$

$$\beta(\lambda) = (3\lambda^2 + 2\lambda + 1)e^{-2\lambda}.$$

It is easily seen that

$$0 \leq \beta(\lambda) \leq 2e^{-2/3} = 1.026\ldots, \qquad 0 \leq \lambda \leq \infty.$$

We now distinguish two cases.

Case 1: $0 \leq 1 - v^2 \leq (\lambda + 1)^2 e^{-2\lambda}$. Then

$$B(\lambda) = 1 + \tfrac{1}{4}\{\beta(\lambda) + 1 - 3v^2\}^2.$$

For fixed λ, this is a quadratic polynomial in v^2 which attains its maximum at one of the endpoints of the allowed interval; that is, either for

$$v^2 = 1 - (\lambda + 1)^2 e^{-2\lambda}$$

or for $v^2 = 1$. For the former value,

$$B(\lambda) = 1 + \{(3\lambda^2 + 4\lambda + 2)e^{-2\lambda} - 1\}^2 \leq 2;$$

while for $v^2 = 1$,

$$B(\lambda) = 1 + \tfrac{1}{4}\{\beta(\lambda) - 2\}^2 \leq 2.$$

Case 2: $(\lambda + 1)^2 e^{-2\lambda} \leq 1 - v^2 \leq 1$. Then

$$B(\lambda) = (\lambda + 1)^2 e^{-2\lambda} + v^2 + \tfrac{1}{4}\{\beta(\lambda) + 1 - 3v^2\}^2.$$

This again is a quadratic polynomial in v^2 which attains its maximum either for $v^2 = 0$ or for

$$v^2 = 1 - (\lambda + 1)^2 e^{-2\lambda}.$$

For the latter value we have already observed that $B(\lambda) \leq 2$; while if $v^2 = 0$,

$$B(\lambda) = (\lambda + 1)^2 e^{-2\lambda} + \tfrac{1}{4}\{\beta(\lambda) + 1\}^2.$$

A calculation then gives

$$B'(\lambda) = \lambda(1 - 3\lambda)(3\lambda^2 + 2\lambda + 1)e^{-4\lambda} - \lambda(5\lambda + 1)e^{-2\lambda}.$$

Thus $B'(\lambda) < 0$ for $\frac{1}{3} \leq \lambda < \infty$. For $0 < \lambda < \frac{1}{3}$, we use the inequality $e^{-4\lambda} < e^{-2\lambda}$ to obtain

$$B'(\lambda) < -3\lambda^2(3\lambda^2 + \lambda + 2)e^{-2\lambda} < 0.$$

Thus $B(\lambda) \leq B(0) = 2$ for $0 \leq \lambda \leq \infty$, and the proof is complete.

§3.11. Successive Coefficients

Another coefficient problem which has attracted considerable attention is to estimate

$$d_n = ||a_{n+1}| - |a_n||, \qquad n = 2, 3, \ldots,$$

the difference of the moduli of successive coefficients of a function $f \in S$. Then the Koebe function suggests the conjecture that $d_n \leq 1$ for all n, which would imply the Bieberbach conjecture. For each odd index $n \geq 5$, however, we know (Theorem 3.9) that there is an odd function with $|a_n| > 1$. It follows that for each $n \geq 4$, some function $f \in S^{(2)}$ produces a difference $d_n > 1$.

It is not at all obvious, of course, that d_n is bounded in S. Goluzin [12] showed

$$d_n \leq Cn^{1/4} \log n, \qquad n = 2, 3, \ldots,$$

and Biernacki [6] improved this to

$$d_n \leq C(\log n)^{3/2}, \qquad n = 2, 3, \ldots,$$

before Hayman [5] proved that d_n is bounded by an absolute constant. A simplified proof of Hayman's result, based on the general method of I. M. Milin, is given in Chapter 5. The best bound currently known is about 3.61, due to Grinspan [2]. For starlike functions, Leung [1] has proved that $d_n \leq 1$ for all n (see §5.10).

It seems reasonable to suspect that d_n attains its maximum for an odd function, so that the successive coefficient problem is equivalent to the Littlewood–Paley problem of estimating the coefficients of odd functions. For $n = 2$, however, this is not true. We know that $d_2 \leq 1$ for all $f \in S^{(2)}$, but we shall now see that $d_2 > 1$ for certain other functions $f \in S$. The following theorem gives the sharp upper and lower bounds for the difference between $|a_3|$ and $|a_2|$ in the class S.

Theorem 3.11. *For each $f \in S$,*

$$-1 \leq |a_3| - |a_2| \leq \tfrac{3}{4} + e^{-\lambda_0}(2e^{-\lambda_0} - 1) = 1.029\ldots,$$

where λ_0 is the unique value of λ satisfying $0 < \lambda < 1$ and $4\lambda e^{-\lambda} = 1$. Both bounds are sharp.

Proof. The lower bound is elementary. If $|a_2| \leq 1$, then obviously $|a_2| - |a_3| \leq 1$. If $|a_2| \geq 1$, the elementary inequality $|a_3 - a_2^2| \leq 1$ (Chapter 2, Exercise 1) gives

$$|a_2| - |a_3| = |a_2|^2 - |a_3| + |a_2|(1 - |a_2|) \leq 1.$$

Equality is attained, for example, by the starlike function

$$f(z) = \frac{z}{1 + z + z^2} = z - z^2 + z^4 + \cdots.$$

The upper bound lies deeper. We shall again use the Loewner formulas (see §3.5)

$$a_2 = -2\int_0^\infty e^{-t}\kappa(t)\,dt = -2(u + iv)$$

and

$$a_3 = -2\int_0^\infty e^{-2t}[\kappa(t)]^2\,dt + 4\left\{\int_0^\infty e^{-t}\kappa(t)\,dt\right\}^2.$$

After a rotation, we may assume $a_3 \geq 0$. Setting $\kappa(t) = e^{i\theta(t)}$, we then find

$$\begin{aligned}|a_3| - |a_2| &= \mathrm{Re}\{a_3\} - |a_2| \\ &= -4\int_0^\infty e^{-2t}\cos^2\theta(t)\,dt + 1 + 4(u^2 - v^2) - 2\{u^2 + v^2\}^{1/2} \\ &\leq 1 + 4u^2 - 2|u| - 4\int_0^\infty e^{-2t}\cos^2\theta(t)\,dt.\end{aligned}$$

If this last integral has the value $(\lambda + \frac{1}{2})e^{-2\lambda}$, $0 \leq \lambda < \infty$, then the Valiron–Landau lemma (§3.8) gives the estimate $|u| \leq (\lambda + 1)e^{-\lambda}$. We now distinguish two cases.

Case 1: $(\lambda + 1)e^{-\lambda} \leq \frac{1}{2}$. Then $|u| \leq \frac{1}{2}$, and $4u^2 - 2|u| \leq 0$. Thus $|a_3| - |a_2| \leq 1$.

Case 2: $(\lambda + 1)e^{-\lambda} > \frac{1}{2}$. Then because $(4u^2 - 2|u|)$ is positive and increasing in the interval $\frac{1}{2} < |u| < \infty$, we obtain

$$\begin{aligned}|a_3| - |a_2| &\leq 1 + 4(\lambda + 1)^2 e^{-2\lambda} - 2(\lambda + 1)e^{-\lambda} - 4(\lambda + \tfrac{1}{2})e^{-2\lambda}\\ &= 1 + 2(2\lambda^2 + 2\lambda + 1)e^{-2\lambda} - 2(\lambda + 1)e^{-\lambda}.\end{aligned}$$

The problem is now to find the maximum of the function

$$\varphi(\lambda) = (2\lambda^2 + 2\lambda + 1)e^{-2\lambda} - (\lambda + 1)e^{-\lambda}$$

in the given interval. The derivative is

$$\varphi'(\lambda) = \lambda e^{-\lambda}(1 - 4\lambda e^{-\lambda}).$$

Thus φ has critical points at $\lambda = 0$ and at the two points λ_0 and λ_1 where $4\lambda e^{-\lambda} = 1$, defined so that $0 < \lambda_0 < 1$ and $\lambda_1 > 2$. (Observe that $4e^{-1} > 1$ and $8e^{-2} > 1$.) The point λ_1 may be disregarded because $(\lambda_1 + 1)e^{-\lambda_1} < \frac{1}{2}$. The point λ_0 is a local maximum for φ, because

$$\varphi''(\lambda) = (1 - \lambda)e^{-\lambda}(1 - 4\lambda e^{-\lambda}) + 4\lambda(\lambda - 1)e^{-2\lambda},$$

and so

$$\varphi''(\lambda_0) = (\lambda_0 - 1)e^{-\lambda_0} < 0.$$

Thus φ attains its maximum value at λ_0. A numerical calculation gives $\lambda_0 = 0.35740\ldots$. Since $\lambda_0 e^{-\lambda_0} = \frac{1}{4}$, we conclude that

$$|a_3| - |a_2| \leq \tfrac{3}{4} + e^{-\lambda_0}(2e^{-\lambda_0} - 1) = 1.029\ldots.$$

This bound is sharp, as is seen by considerations similar to those in the proof of the Fekete–Szegö theorem (Theorem 3.8).

This method of proof is essentially due to Goluzin [12, 18], who made an algebraic error and was led to a spurious solution. Jenkins [8] used a different method to obtain the correct result.

Exercises

1. Apply the Loewner method to obtain the growth and distortion theorems (Theorems 2.5 and 2.6) for functions in S.

2. Show that if $\kappa(t)$ is identically constant, the Loewner equation produces a rotation of the Koebe function.

3. Show that if a function κ generates a function $f \in S$ through the Loewner differential equation, then for each constant $e^{i\theta}$ the function $\tilde{\kappa} = e^{i\theta}\kappa$ generates the rotation $\tilde{f}(z) = e^{-i\theta}f(e^{i\theta}z)$.

4. Show that each piecewise continuous function κ which takes only the values ± 1 generates a function $f \in S^*$ of the form $f(z) = z(z^2 + cz + 1)^{-1}$, where c is a real constant with $|c| \leq 2$. (Kung Sun [1].)

5. Bazilevich [6] exhibited a class of functions $p(w, t)$ with positive real part for which the Loewner–Kufarev equation

$$\frac{\partial f}{\partial t} = -f(z, t)p(f, t)$$

can be solved by quadrature. The resulting functions are known as *Bazilevich functions*. The basic steps are as follows. Let $\tau = e^{-t}$ and transform the equation to $p(f, t)(d\tau/\tau) = df/f$. Next let a be a real parameter and set $\xi = e^{-iat}f$ to obtain

$$[p(f, t) + ia]\frac{d\tau}{\tau} = \frac{d\xi}{\xi}.$$

Now express

$$[p(f, t) + ia]^{-1} = \frac{q(\xi, \tau)}{1 + a^2} - \frac{ia}{1 + a^2},$$

and observe that q has positive real part if and only if p does. Let q have the form

$$q(\xi, \tau) = p_0(\xi)(1 - \tau^m) + p_1(\xi)\tau^m,$$

where p_0 and p_1 are functions with positive real part and $m > 0$. Transform the differential equation to

$$\frac{d\tau}{d\xi} = F(\xi)\tau + G(\xi)\tau^{m+1},$$

where

$$F(\xi) = \frac{p_0(\xi) - ia}{(1 + a^2)\xi}, \qquad G(\xi) = \frac{p_1(\xi) - p_0(\xi)}{(1 + a^2)\xi}.$$

This is a Bernoulli differential equation. It can be transformed to a linear equation by the substitution $s = \tau^{-m}$. (There is a large literature on Bazilevich functions. See, for instance, Zamorski [1], Bazilevich [8], Thomas [1], Keogh and Miller [1], Sheil-Small [2], Mocanu, Reade, and Złotkiewicz [1], Prohorov [1], Singh [1].)

6. Let a function $f \in S$ map the unit disk onto the complement of an analytic Jordan arc with an asymptotic direction at infinity, and let $g(z, t)$ denote the associated chain of subordinate functions. Prove that as $t \to \infty$, the functions $e^{-t}g(z, t)$ tend uniformly on compact subsets to some rotation of the Koebe function.

7. Show that if $f \in S$ and

$$|a_2| \leq 2(\lambda + 1)e^{-\lambda}, \qquad 0 \leq \lambda \leq \infty,$$

then

$$|a_3| \leq 1 + 4(\lambda^2 + \lambda + \tfrac{1}{2})e^{-2\lambda},$$

and this estimate is sharp. (Jenkins [8].)

8. Let $g(z, t)$ be the mappings defined in §3.3, where Γ has the standard parametrization. Show that $g(z, t)$ satisfies the partial differential equation

$$\frac{\partial g}{\partial t} = \frac{\partial g}{\partial z} z \frac{1 + \kappa z}{1 - \kappa z}.$$

9. Let C_r be the image of the circle $|z| = r$ under the mapping $f \in S$, and let K_z be the curvature of C_r at the point $w = f(z)$. Use Loewner's method to prove

$$K_z \geq \left(\frac{1 + r}{1 - r}\right)^2 \frac{1 - 4r + r^2}{r},$$

and show that this estimate is sharp for each point $z \in \mathbb{D}$ (*cf.* Chapter 2, Exercise 29). Note that this lower bound changes sign, as it must, at the radius of convexity $2 - \sqrt{3}$. (Korickiĭ [1].)

10. Show that each function $f \in S$ satisfies the sharp inequality

$$|f(z)| + |f(-z)| \leq \frac{r}{(1 - r)^2} + \frac{r}{(1 + r)^2} = \frac{2r(1 + r^2)}{(1 - r^2)^2}, \qquad r = |z| < 1.$$

Deduce that $|a_3| \leq 3$ for every $f \in S$. (*Hint*: Apply the inequality to $f(z) - f(-z)$ and let $r \to 0$.) (Goluzin [10], Jenkins [2].)

Chapter 4

Generalizations of the Area Principle

The proof of the area theorem can be generalized to produce a system of inequalities called the *Grunsky inequalities* which are necessary and sufficient for the univalence of the associated function. These inequalities contain a wealth of useful information about the coefficients of univalent functions. For example, they lead to an elementary proof of the Bieberbach conjecture for the fourth coefficient. They also lead to general inequalities on the values of a univalent function at prescribed points. These are basically the *Lebedev inequalities* and a special case, the *Goluzin inequalities*, which have a number of important consequences.

§4.1. Faber Polynomials

Given a function

$$g(z) = z + b_0 + b_1 z^{-1} + b_2 z^{-2} + \cdots$$

of class Σ, consider the expansion

$$\frac{\zeta g'(\zeta)}{g(\zeta) - w} = \sum_{n=0}^{\infty} F_n(w)\zeta^{-n}, \tag{1}$$

valid for all ζ in some neighborhood of ∞. The function

$$F_n(w) = w^n + \sum_{k=1}^{n} a_{nk} w^{n-k}$$

is a monic polynomial of degree n, called the nth *Faber polynomial* of the function g. In particular, straightforward calculations produce the formulas

$$F_0(w) = 1, \qquad F_1(w) = w - b_0,$$

$$F_2(w) = w^2 - 2b_0 w + (b_0^2 - 2b_1),$$

$$F_3(w) = w^3 - 3b_0 w^2 + (3b_0^2 - 3b_1)w + (b_0^3 + 3b_1 b_0 - 3b_2).$$

Now observe that since g is univalent, the function

$$\frac{\zeta g'(\zeta)}{g(\zeta) - g(z)} - \frac{\zeta}{\zeta - z} = \sum_{n=1}^{\infty} \sum_{k=1}^{\infty} \beta_{nk} z^{-k} \zeta^{-n} \tag{2}$$

is analytic for $|z| > 1$ and $|\zeta| > 1$. In view of (1), the relation (2) gives

$$\sum_{n=0}^{\infty} F_n(g(z))\zeta^{-n} = 1 + \sum_{n=1}^{\infty} \left\{ z^n + \sum_{k=1}^{\infty} \beta_{nk} z^{-k} \right\} \zeta^{-n}.$$

Thus the Faber polynomials satisfy

$$F_n(g(z)) = z^n + \sum_{k=1}^{\infty} \beta_{nk} z^{-k}, \qquad n = 1, 2, \ldots. \tag{3}$$

The coefficients β_{nk} are known as the *Grunsky coefficients* of g.

The property (3) actually characterizes the nth Faber polynomial of g among all polynomials of degree n. In fact, F_n can be computed by successively subtracting from $[g(z)]^n$ suitable multiples of $[g(z)]^{n-1}$, $[g(z)]^{n-2}$, etc., chosen to eliminate the terms in z^k for $k = n - 1, n - 2, \ldots, 0$. This process is also a convenient method for calculating the Grunsky coefficients. For example, one finds

$$\beta_{11} = b_1, \quad \beta_{12} = b_2, \quad \beta_{13} = b_3,$$

$$\beta_{21} = 2b_2, \quad \beta_{22} = 2b_3 + b_1^2, \quad \beta_{23} = 2(b_4 + b_1 b_2),$$

$$\beta_{31} = 3b_3, \quad \beta_{32} = 3(b_4 + b_1 b_2),$$

$$\beta_{33} = 3(b_5 + b_1 b_3 + b_0^2 b_3 + b_2^2) + b_1^3.$$

These results suggest the symmetry property

$$k\beta_{nk} = n\beta_{kn}, \qquad k, n = 1, 2, \ldots. \tag{4}$$

This is not difficult to prove. Indeed, let C_r be the image of the circle $|z| = r > 1$ under the mapping $w = g(z)$. Then by Cauchy's theorem and the relation (3),

$$\begin{aligned} 0 &= \frac{1}{2\pi i} \int_{C_r} F_n(w) F_k'(w)\, dw \\ &= \frac{1}{2\pi i} \int_{|z|=r} F_n(g(z)) F_k'(g(z)) g'(z)\, dz \\ &= k\beta_{nk} - n\beta_{kn}. \end{aligned}$$

This proves (4).

§4.2. Polynomial Area Theorem

We now turn to a direct generalization of the area theorem which may be considered a primitive form of the Grunsky inequalities. For want of a better name, we shall call it the *polynomial area theorem*.

Recall that each function $g \in \Sigma$ maps the exterior of the unit disk onto the complement of a compact set E, and g is called a *full mapping* if E has measure zero. The subclass of full mappings is denoted by $\tilde{\Sigma}$.

Theorem 4.1 (Polynomial Area Theorem). *Let $g \in \Sigma$, let p be an arbitrary nonconstant polynomial, and let*

$$p(g(z)) = \sum_{k=-N}^{\infty} c_k z^{-k}, \qquad |z| > 1,$$

where N is the degree of p. Then

$$\sum_{k=-N}^{\infty} k|c_k|^2 \leq 0, \tag{5}$$

with equality if and only if $g \in \tilde{\Sigma}$.

Equivalently, the theorem states that

$$\sum_{k=1}^{\infty} k|c_k|^2 \leq \sum_{k=1}^{N} k|c_{-k}|^2.$$

Of course, the ordinary area theorem corresponds to the polynomial $p(w) = w$.

The proof uses the following lemma, a form of Green's theorem which is easily proved by resort to the Cauchy–Riemann equations.

Lemma. *Let C be a smooth Jordan curve bounding a domain D, and let φ and ψ be analytic in the closure of D. Then*

$$\frac{1}{2i}\int_C \overline{\varphi(w)}\psi(w)\,dw = \iint_D \overline{\varphi'(w)}\psi(w)\,du\,dv, \qquad w = u + iv.$$

Proof of Theorem. Let C_r be the image under g of the circle $|z| = r > 1$, and let E_r be the interior of C_r. Then by the lemma,

$$\begin{aligned} 0 \le \iint_{E_r} |p'(w)|^2 \, du \, dv &= \frac{1}{2i} \int_{C_r} \overline{p(w)} p'(w) \, dw \\ &= \frac{1}{2i} \int_{|z|=r} \overline{p(g(z))} p'(g(z)) g'(z) \, dz \\ &= -\pi \sum_{k=-N}^{\infty} k |c_k|^2 r^{-2k}. \end{aligned}$$

Letting $r \to 1$, we obtain

$$\sum_{k=-N}^{\infty} k |c_k|^2 = -\frac{1}{\pi} \iint_E |p'(w)|^2 \, du \, dv \le 0,$$

with equality if and only if E has measure zero; that is, if and only if $g \in \tilde{\Sigma}$.

Actually, the proof does not make full use of the information that p is a polynomial. It is required only that p be analytic on E. Thus the proof extends to give the following more general result.

Corollary. *Let $g \in \Sigma$ and let F be a nonconstant function analytic in some domain E_ρ, $\rho > 1$. Let*

$$F(g(z)) = \sum_{k=-\infty}^{\infty} c_k z^{-k}, \qquad 1 < |z| < \rho.$$

Then

$$\sum_{k=1}^{\infty} k |c_k|^2 \le \sum_{k=1}^{\infty} k |c_{-k}|^2.$$

If the right-hand sum converges, then equality occurs if and only if $g \in \tilde{\Sigma}$.

We shall see in the next section that the condition of the polynomial area theorem is also sufficient for the univalence of g.

§4.3. The Grunsky Inequalities

The polynomial area theorem will now be used to derive a system of inequalities due to Grunsky [3], known as the *Grunsky inequalities.* It is customary to distinguish two apparently different formulations as the *weak* and *strong Grunsky inequalities*, although they turn out to be equivalent.

Theorem 4.2 (Strong Grunsky Inequalities). *Let β_{nk} be the Grunsky coefficients of a function $g \in \Sigma$. Then*

$$\sum_{k=1}^{\infty} k \left| \sum_{n=1}^{N} \beta_{nk} \lambda_n \right|^2 \le \sum_{n=1}^{N} n |\lambda_n|^2 \tag{6}$$

for each positive integer N and for all complex numbers $\lambda_1, \lambda_2, \ldots, \lambda_N$. Strict inequality holds for all nonzero vectors $(\lambda_1, \ldots, \lambda_N)$ unless $g \in \tilde{\Sigma}$, in which case equality holds for all vectors.

Proof. Let

$$p(w) = \sum_{n=1}^{N} \lambda_n F_n(w)$$

be a polynomial of degree N, expressed in terms of the Faber polynomials of g. In view of (3),

$$p(g(z)) = \sum_{n=1}^{N} \lambda_n z^n + \sum_{k=1}^{\infty} \sum_{n=1}^{N} \lambda_n \beta_{nk} z^{-k}.$$

Thus the theorem follows directly from the polynomial area theorem.

Corollary (Weak Grunsky Inequalities).

$$\left| \sum_{n=1}^{N} \sum_{k=1}^{N} k \beta_{nk} \lambda_n \lambda_k \right| \le \sum_{n=1}^{N} n |\lambda_n|^2. \tag{7}$$

Proof. We shall actually derive an important generalization which will be called the *generalized weak Grunsky inequalities*:

$$\left| \sum_{n=1}^{N} \sum_{k=1}^{N} k \beta_{nk} \lambda_n \mu_k \right|^2 \le \sum_{n=1}^{N} n |\lambda_n|^2 \sum_{k=1}^{N} k |\mu_k|^2, \tag{8}$$

where λ_n and μ_k are arbitrary complex parameters. To prove (8), let

$$\nu_k = \sum_{n=1}^{N} \beta_{nk} \lambda_n, \qquad k = 1, \ldots, N,$$

and observe that the strong Grunsky inequalities imply

$$\sum_{k=1}^{N} k|v_k|^2 \le \sum_{n=1}^{N} n|\lambda_n|^2.$$

Hence the Cauchy–Schwarz inequality gives

$$\begin{aligned}\left|\sum_{n=1}^{N}\sum_{k=1}^{N} k\beta_{nk}\lambda_n\mu_k\right|^2 &= \left|\sum_{k=1}^{N} kv_k\mu_k\right|^2\\ &\le \sum_{k=1}^{N} k|v_k|^2 \sum_{k=1}^{N} k|\mu_k|^2\\ &\le \sum_{n=1}^{N} n|\lambda_n|^2 \sum_{k=1}^{N} k|\mu_k|^2,\end{aligned}$$

as asserted.

There is another approach to the Grunsky inequalities which sheds more light on the symmetry of the coefficients β_{nk}. If $g \in \Sigma$, then

$$\log \frac{g(\zeta) - g(z)}{\zeta - z} = -\sum_{n=1}^{\infty}\sum_{k=1}^{\infty} \gamma_{nk} z^{-k}\zeta^{-n} \tag{9}$$

is an analytic function of the two variables z and ζ in the region

$$\Delta^2 = \{(z, \zeta): |z| > 1, |\zeta| > 1\}.$$

Since this function is symmetric in z and ζ, it is clear that $\gamma_{nk} = \gamma_{kn}$. On the other hand, dividing equation (2) by ζ and integrating, one obtains

$$\log \frac{g(\zeta) - g(z)}{\zeta - z} = -\sum_{n=1}^{\infty}\sum_{k=1}^{\infty} \frac{1}{n}\beta_{nk} z^{-k}\zeta^{-n}.$$

Thus $\beta_{nk} = n\gamma_{nk}$ and $k\beta_{nk} = nk\gamma_{nk}$. This gives a new proof of the symmetry property (4).

In terms of the coefficients γ_{nk} the Grunsky inequalities (6), (7), and (8) take the more symmetric forms

$$\sum_{k=1}^{\infty} k\left|\sum_{n=1}^{N} \gamma_{nk}\lambda_n\right|^2 \le \sum_{n=1}^{N} \frac{1}{n}|\lambda_n|^2, \tag{10}$$

$$\left|\sum_{n=1}^{N}\sum_{k=1}^{N} \gamma_{nk}\lambda_n\lambda_k\right| \le \sum_{n=1}^{N} \frac{1}{n}|\lambda_n|^2, \tag{11}$$

and

$$\left|\sum_{n=1}^{N}\sum_{k=1}^{N}\gamma_{nk}\lambda_n\mu_k\right|^2 \le \sum_{n=1}^{N}\frac{1}{n}|\lambda_n|^2\sum_{k=1}^{N}\frac{1}{k}|\mu_k|^2, \tag{12}$$

respectively.

Now let

$$g(z) = z + b_0 + b_1 z^{-1} + b_2 z^{-2} + \cdots \tag{13}$$

be an arbitrary function analytic in the annulus $1 < |z| < \infty$ and having a simple pole with residue one at ∞. We have shown that if g is univalent, then its coefficients b_n satisfy the Grunsky inequalities. It is a surprising fact that the converse is also true: the Grunsky inequalities, even in their "weak" form (11), are a sufficient condition for the univalence of g. To see this, observe that if g has the form (13), then the logarithmic expression (9) is well defined and analytic for $|z|$ and $|\zeta|$ sufficiently large. In particular, the coefficients γ_{nk} are defined. Suppose they satisfy the weak Grunsky inequalities (11). Choose $\lambda_j = \delta_{nj}$ (the Kronecker delta) to conclude that $|\gamma_{nn}| \le 1/n$. Next choose $\lambda_j = \delta_{nj} + \delta_{kj}$ $(k \ne n)$ to obtain

$$|\gamma_{nk}| \le \frac{1}{n} + \frac{1}{k} < 2.$$

Since the coefficients γ_{nk} are bounded, the double power series (9) converges throughout Δ^2. But since

$$\log\frac{g(\zeta) - g(z)}{\zeta - z}$$

is analytic in Δ^2, g must be univalent in Δ.

Note that we have proved the univalence of g using only a special case of the weak Grunsky inequalities. In particular, the weak Grunsky inequalities imply the strong Grunsky inequalities. This shows that the three formulations (6), (7), and (8) are all equivalent.

Because we derived the Grunsky inequalities from the polynomial area theorem, the result also shows that the condition of the polynomial area theorem is sufficient for univalence. More precisely, if g is analytic in $|z| > 1$ apart from a simple pole at ∞, and if the inequality (5) holds for every polynomial p, then g is univalent. This observation is due to Wolibner [1], who gave a geometric proof independent of the Grunsky inequalities.

§4.4. Inequalities of Goluzin and Lebedev

In 1947, Goluzin [14] obtained a system of inequalities giving precise information on the values of functions $g \in \Sigma$ at prescribed points. His proof was based upon a variational method. Goluzin [17, 26, 28] later gave another proof using the Loewner method. It is not necessary, however, to use such powerful methods. The Goluzin inequalities are equivalent to the Grunsky inequalities and can be derived from them quite easily. With a view to certain applications, we shall first derive them in a more general form due to Lebedev [1].

Theorem 4.3 (Lebedev Inequalities). *Let $g \in \Sigma$, let N be a positive integer, and let $(z_1, z_2, \ldots, z_N)$ and $(\zeta_1, \zeta_2, \ldots, \zeta_N)$ be arbitrary N-tuples of distinct points in Δ. Then*

$$\left| \sum_{i=1}^{N} \sum_{j=1}^{N} \lambda_i \mu_j \log \frac{g(z_i) - g(\zeta_j)}{z_i - \zeta_j} \right|^2$$

$$\leq \left\{ -\sum_{i=1}^{N} \sum_{j=1}^{N} \lambda_i \overline{\lambda_j} \log\left(1 - \frac{1}{z_i \overline{z_j}}\right)\right\}\left\{ -\sum_{i=1}^{N} \sum_{j=1}^{N} \mu_i \overline{\mu_j} \log\left(1 - \frac{1}{\zeta_i \overline{\zeta_j}}\right)\right\} \tag{14}$$

for all complex numbers λ_i and μ_j.

Remarks. In case $z_i = \zeta_j$ for some i and j, the difference quotient is to be interpreted as the derivative $g'(z_i)$. Each of the double sums on the right-hand side of (14) is a Hermitian form and is therefore real. The proof will show that each is nonnegative.

Proof of Theorem. Introduce the expansion (9) into the left-hand side of (14) to obtain, after slight rearrangement,

$$\left| \sum_{n=1}^{\infty} \sum_{k=1}^{\infty} \gamma_{nk} \sum_{i=1}^{N} \lambda_i z_i^{-k} \sum_{j=1}^{N} \mu_j \zeta_j^{-n} \right|^2 .$$

Now apply the generalized weak Grunsky inequalities (12) to estimate this expression by

$$\left\{ \sum_{k=1}^{\infty} \frac{1}{k} \left| \sum_{i=1}^{N} \lambda_i z_i^{-k} \right|^2 \right\}\left\{ \sum_{n=1}^{\infty} \frac{1}{n} \left| \sum_{j=1}^{N} \mu_j \zeta_j^{-n} \right|^2 \right\}.$$

But the first factor is

$$\begin{aligned}\sum_{k=1}^{\infty}\frac{1}{k}\left|\sum_{i=1}^{N}\lambda_i z_i^{-k}\right|^2 &= \sum_{k=1}^{\infty}\frac{1}{k}\sum_{i=1}^{N}\lambda_i z_i^{-k}\sum_{j=1}^{N}\overline{\lambda_j}\overline{z_j}^{-k}\\ &= \sum_{i=1}^{N}\sum_{j=1}^{N}\lambda_i\overline{\lambda_j}\sum_{k=1}^{\infty}\frac{1}{k}(z_i\overline{z_j})^{-k}\\ &= -\sum_{i=1}^{N}\sum_{j=1}^{N}\lambda_i\overline{\lambda_j}\log\left(1-\frac{1}{z_i\overline{z_j}}\right).\end{aligned}$$

A similar reduction applies to the second factor, and the theorem is proved.

There are many interesting corollaries.

Corollary 1 (Goluzin Inequalities). *Let $g \in \Sigma$ and let $z_1, z_2, \ldots, z_N$ be distinct points in Δ. Then*

$$\left|\sum_{i=1}^{N}\sum_{j=1}^{N}\lambda_i\lambda_j\log\frac{g(z_i)-g(z_j)}{z_i-z_j}\right| \le -\sum_{i=1}^{N}\sum_{j=1}^{N}\lambda_i\overline{\lambda_j}\log\left(1-\frac{1}{z_i\overline{z_j}}\right) \tag{15}$$

for all complex numbers $\lambda_1, \lambda_2, \ldots, \lambda_N$.

Corollary 2. *For each odd function $g \in \Sigma$,*

$$\left|\log\frac{zg'(z)}{g(z)}\right| \le \log\frac{r^2+1}{r^2-1}, \qquad |z| = r > 1.$$

Proof. Apply the Goluzin inequality with $N = 2$, $z_1 = z$, $z_2 = \zeta$, $\lambda_1 = 1$, and $\lambda_2 = -1$. This gives

$$\left|\log\frac{g'(z)g'(\zeta)(z-\zeta)^2}{[g(z)-g(\zeta)]^2}\right| \le \log\frac{|1-z\overline{\zeta}|^2}{(|z|^2-1)(|\zeta|^2-1)}$$

for every $g \in \Sigma$. Now specialize this inequality to the case where $\zeta = -z$, and assume g is odd. This produces the desired result.

Corollary 3. *For each $f \in S$,*

$$\left|\log\frac{zf'(z)}{f(z)}\right| \le \log\frac{1+r}{1-r}, \qquad |z| = r < 1.$$

Proof. Apply Corollary 2 to the odd function $g(z) = \{f(z^{-2})\}^{-1/2}$ and replace z^{-2} by z.

Corollary 3 is due to Grunsky [1], whose proof was not so simple. It represents a common generalization of the two sharp inequalities

$$\frac{1-r}{1+r} \leq \left|\frac{zf'(z)}{f(z)}\right| \leq \frac{1+r}{1-r}$$

and

$$\left|\arg\frac{zf'(z)}{f(z)}\right| \leq \log\frac{1+r}{1-r}.$$

We proved the first in §2.3 by elementary means, but we used the Loewner method in §3.6 to prove the second.

Corollary 4. *For each $g \in \Sigma$,*

$$\left|\log\frac{g(z)-g(\zeta)}{z-\zeta}\right|^2 \leq \log\frac{r^2}{r^2-1}\log\frac{\rho^2}{\rho^2-1},$$

where $|z| = r > 1$ and $|\zeta| = \rho > 1$.

Proof. Apply the Lebedev inequality with $N = 1$ and $\lambda_1 = \mu_1 = 1$.

Corollary 5. *For each $g \in \Sigma$,*

$$\frac{r^2-1}{r^2} \leq \left|\frac{g(z)-g(\zeta)}{z-\zeta}\right| \leq \frac{r^2}{r^2-1}, \qquad |z| = |\zeta| = r > 1.$$

Proof. Apply Corollary 4, observing that $|\mathrm{Re}\{w\}| \leq |w|$.

Corollary 6. *For each $g \in \Sigma$,*

$$\frac{r^2-1}{r^2} \leq |g'(z)| \leq \frac{r^2}{r^2-1}, \qquad |z| = r > 1.$$

Corollary 6 was obtained first by Loewner [2] and Corollary 5 by Goluzin [12].

Corollary 7. *For each $f \in S$,*

$$\frac{1-r^2}{r^2}|f(z)f(\zeta)| \leq \left|\frac{f(z)-f(\zeta)}{z-\zeta}\right| \leq \frac{|f(z)f(\zeta)|}{r^2(1-r^2)}, \qquad |z| = |\zeta| = r < 1.$$

Proof. Apply Corollary 5 to $g(z) = 1/f(1/z)$, and replace $1/z$ and $1/\zeta$ by z and ζ.

The following more technical result, due to Bazilevich [4], will be needed in §6.4.

Corollary 8. *For each odd function $g \in \Sigma$,*

$$\left| \log\left(\frac{g(z) + g(\zeta)}{g(z) - g(\zeta)} \cdot \frac{z - \zeta}{z + \zeta}\right) \right|^2 \leq \log \frac{r^2 + 1}{r^2 - 1} \log \frac{\rho^2 + 1}{\rho^2 - 1},$$

where $|z| = r > 1$ and $|\zeta| = \rho > 1$.

Proof. Apply the Lebedev inequality with $N = 2$, $z_1 = -z_2 = z$, $\zeta_1 = -\zeta_2 = \zeta$, $\lambda_1 = \mu_1 = 1$, and $\lambda_2 = \mu_2 = -1$. This gives the result after some manipulation.

As a final corollary, we record the Goluzin inequalities for functions $f \in S$. They are obtained from the inequalities for $g \in \Sigma$ (Corollary 1) by a simple inversion: $g(z) = 1/f(1/z)$.

Corollary 9 (Goluzin Inequalities for S). *Let $f \in S$ and let $z_1, z_2, \ldots, z_N$ be distinct points in $\mathbb{D}$. Then*

$$\left| \sum_{i=1}^{N} \sum_{j=1}^{N} \lambda_i \lambda_j \log \frac{z_i z_j [f(z_i) - f(z_j)]}{f(z_i) f(z_j)(z_i - z_j)} \right| \leq - \sum_{i=1}^{N} \sum_{j=1}^{N} \lambda_i \overline{\lambda_j} \log(1 - z_i \overline{z_j})$$

for all complex numbers $\lambda_1, \lambda_2, \ldots, \lambda_N$.

It should be observed that the Goluzin inequalities, like the Grunsky inequalities, imply univalence. Indeed, let g be analytic in $1 < |z| < \infty$ and have an expansion of the form (13) about infinity. If g satisfies the Goluzin inequalities (15), then certainly

$$\log \frac{g(z_1) - g(z_2)}{z_1 - z_2}$$

is finite for every pair of points z_1 and z_2 in Δ. Hence g is univalent in Δ.

§4.5. Unitary Matrices

We shall now interpret the strong Grunsky inequalities from an operator-theoretic viewpoint. This will allow us to associate a unitary matrix with every full mapping.

Each function $g \in \Sigma$ generates through equation (9) certain coefficients γ_{nk} which satisfy the strong Grunsky inequalities (10):

$$\sum_{k=1}^{\infty} k \left| \sum_{n=1}^{N} \gamma_{nk} \lambda_n \right|^2 \le \sum_{n=1}^{N} \frac{1}{n} |\lambda_n|^2.$$

Equality holds for every complex vector $(\lambda_1, \lambda_2, \ldots, \lambda_N)$ if $g \in \tilde{\Sigma}$; that is, if g is a full mapping. If $g \notin \tilde{\Sigma}$, then strict inequality holds for every nonzero vector. With the substitution

$$c_{nk} = \sqrt{nk}\, \gamma_{nk},$$

these inequalities take the more suggestive form

$$\sum_{k=1}^{\infty} \left| \sum_{n=1}^{N} c_{kn} \lambda_n \right|^2 \le \sum_{n=1}^{N} |\lambda_n|^2, \tag{16}$$

after $\lambda_n/\sqrt{n}$ is replaced by λ_n. (Note that $c_{kn} = c_{nk}$.)

Now let C denote the infinite symmetric matrix with entries c_{kn}. This matrix C determines a linear operator T on the Hilbert space ℓ^2 of all square-summable sequences $\lambda = (\lambda_1, \lambda_2, \ldots)$, with the norm

$$\|\lambda\| = \left\{ \sum_{n=1}^{\infty} |\lambda_n|^2 \right\}^{1/2}.$$

Initially, T is defined only for the sequences

$$\lambda = (\lambda_1, \lambda_2, \ldots, \lambda_N, 0, 0, \ldots)$$

with finitely many nonzero entries. The inequality (16) says that on this dense subspace of ℓ^2, T is a *contraction*: $\|T(\lambda)\| \le \|\lambda\|$. Thus T has a unique extension to the whole space ℓ^2, mapping each $\lambda \in \ell^2$ into a sequence

$$\mu = T(\lambda) = (\mu_1, \mu_2, \ldots)$$

defined by

$$\mu_k = \sum_{n=1}^{\infty} c_{kn} \lambda_n, \qquad k = 1, 2, \ldots.$$

The convergence of this series for each $\lambda \in \ell^2$ is assured by the fact that

$$\sum_{n=1}^{\infty} |c_{kn}|^2 \le 1, \qquad k = 1, 2, \ldots.$$

This may be deduced from (16) by choosing $\lambda_n = \delta_{nj}$. The same choice of λ shows that

$$\sum_{n=1}^{\infty} |c_{kn}|^2 = 1, \qquad k = 1, 2, \ldots \tag{17}$$

if $g \in \tilde{\Sigma}$; while

$$\sum_{n=1}^{\infty} |c_{kn}|^2 < 1, \qquad k = 1, 2, \ldots$$

if $g \notin \tilde{\Sigma}$. It follows that for $g \notin \tilde{\Sigma}$, the operator T is a *proper contraction* in the sense that $\|T(\lambda)\| < \|\lambda\|$ for every $\lambda \in \ell^2$, $\lambda \neq 0$. For $g \in \tilde{\Sigma}$, on the other hand, the condition for equality in (16) shows that T is an *isometry*: $\|T(\lambda)\| = \|\lambda\|$ for every $\lambda \in \ell^2$. We shall now prove that in this case T is actually a *unitary operator*; that is, T is an isometry whose range is the whole space ℓ^2.

Theorem 4.4. *If* $g \in \tilde{\Sigma}$, *then its associated operator* T *is unitary, and*

$$\sum_{n=1}^{\infty} c_{jn} \overline{c_{nk}} = \delta_{jk}, \qquad j, k = 1, 2, \ldots. \tag{18}$$

If $g \notin \tilde{\Sigma}$, *then* T *is a proper contraction.*

Proof. We have already verified the statement regarding $g \notin \tilde{\Sigma}$. For $g \in \tilde{\Sigma}$, we have shown that T is an isometry and that (17) holds for each k. But C is a symmetric matrix, so (17) is equivalent to (18) when $j = k$. For $j \neq k$, consider the sequence $\lambda = \{\lambda_n\}$ defined by

$$\lambda_n = \begin{cases} e^{i\theta}, & n = j, \\ 1, & n = k, \\ 0, & n \neq j, k. \end{cases}$$

In terms of the matrix C, the isometric relation $\|T(\lambda)\|^2 = \|\lambda\|^2$ then reduces to

$$\sum_{i=1}^{\infty} |e^{i\theta} c_{ij} + c_{ik}|^2 = 2.$$

In view of (17), it follows that

$$\operatorname{Re}\left\{ e^{i\theta} \sum_{i=1}^{\infty} c_{ij} \overline{c_{ik}} \right\} = 0.$$

Since θ is arbitrary, this implies that

$$\sum_{i=1}^{\infty} c_{ij}\overline{c_{ik}} = 0,$$

which is equivalent to (18), since $c_{ij} = c_{ji}$.

In order to show that T is unitary, we must show that each $\mu \in \ell^2$ is the image under T of some $\lambda \in \ell^2$. But we need only define

$$\lambda_n = \sum_{k=1}^{\infty} \overline{c_{nk}}\mu_k, \qquad n = 1, 2, \ldots.$$

Then $\overline{\lambda} = T(\overline{\mu}) \in \ell^2$, and $T(\lambda) = \nu$, where

$$\nu_j = \sum_{n=1}^{\infty} c_{jn}\lambda_n = \sum_{n=1}^{\infty}\sum_{k=1}^{\infty} c_{jn}\overline{c_{nk}}\mu_k = \mu_j, \qquad j = 1, 2, \ldots,$$

by the orthogonality relations (18). The interchange of the order of summation is justified by absolute convergence. Hence $T(\lambda) = \mu$, which shows that the range of T is the whole space ℓ^2.

In matrix terminology, the theorem makes the statement (which for general infinite matrices is weaker) that C has an inverse which coincides with the conjugate transpose of C.

§4.6. The Fourth Coefficient

Loewner proved the Bieberbach conjecture for the third coefficient in the year 1923. Over 30 years later, in 1955, the fourth coefficient finally yielded to the efforts of Garabedian and Schiffer [1]. Their proof was a *tour de force* involving a variational method, Loewner's equation, and a long series of laborious numerical estimates. Then in 1960 Charzyński and Schiffer [1] astonished everyone with the discovery of an elementary proof based only upon the Grunsky inequalities. This striking development focused new attention on the Grunsky inequalities and inspired a number of further advances.

We shall present a simplified version of the Charzyński–Schiffer proof, as given by G. V. Kuzmina in the supplement to the second edition of Goluzin [28].

Theorem 4.5. *The fourth coefficient of each function $f \in S$ satisfies $|a_4| \leq 4$, with equality only for a rotation of the Koebe function.*

Proof. Assume without loss of generality that $a_4 > 0$. Transform the function

$$f(z) = z + a_2 z^2 + a_3 z^3 + \cdots$$

to

$$g(z) = \{f(1/z^2)\}^{-1/2} = z + b_1 z^{-1} + b_3 z^{-3} + \cdots.$$

Then $g \in \Sigma$, and a calculation gives

$$b_1 = -\tfrac{1}{2}a_2, \quad b_3 = \tfrac{3}{8}a_2^2 - \tfrac{1}{2}a_3,$$

$$b_5 = \tfrac{3}{4}a_2 a_3 - \tfrac{5}{16}a_2^3 - \tfrac{1}{2}a_4.$$

Let γ_{nk} be the logarithmic Grunsky coefficients of g, as defined by (9). Then

$$\gamma_{11} = b_1 = -\tfrac{1}{2}a_2,$$

$$\gamma_{13} = \gamma_{31} = b_3 = \tfrac{3}{8}a_2^2 - \tfrac{1}{2}a_3,$$

$$\gamma_{33} = \tfrac{1}{3}b_1^3 + b_1 b_3 + b_5 = -\tfrac{13}{24}a_2^3 + a_2 a_3 - \tfrac{1}{2}a_4.$$

With the choices $N = 3$, $\lambda_1 = \lambda$, $\lambda_2 = 0$, and $\lambda_3 = 1$, the weak Grunsky inequality (11) is

$$|\gamma_{11}\lambda^2 + 2\gamma_{13}\lambda + \gamma_{33}| \leq |\lambda|^2 + \tfrac{1}{3}.$$

Introduce the above expressions for γ_{nk} to obtain, after a little manipulation,

$$|a_4 + 4b_3(a_2 - \lambda) - \tfrac{5}{12}a_2^3 + a_2\lambda^2| \leq 2|\lambda|^2 + \tfrac{2}{3}. \tag{19}$$

But the area theorem applied to g gives

$$|b_1|^2 + 3|b_3|^2 \leq 1,$$

or

$$4|b_3| \leq \frac{2}{\sqrt{3}}\{4 - |a_2|^2\}^{1/2}.$$

Combined with (19), this yields

$$a_4 \leq \frac{2}{\sqrt{3}}\{4 - |a_2|^2\}^{1/2}|a_2 - \lambda| + \operatorname{Re}\{\tfrac{5}{12}a_2^3 - a_2\lambda^2\} + 2|\lambda|^2 + \tfrac{2}{3}. \tag{20}$$

Now let $a_2 = 2xe^{i\theta}$, $0 \le x \le 1$. Choose

$$\lambda = 2xe^{-i\theta/2}\cos(3\theta/2),$$

and let $y = |\sin(3\theta/2)|$. Calculations give

$$|a_2 - \lambda| = 2xy, \qquad \mathrm{Re}\{\tfrac{5}{12}a_2^3 - a_2\lambda^2\} = \tfrac{4}{3}x^3y^2 - \tfrac{14}{3}x^3.$$

The inequality (20) therefore becomes

$$a_4 \le \tfrac{2}{3} + 8x^2 - \tfrac{14}{3}x^3 + \frac{8}{\sqrt{3}}(1 - x^2)^{1/2}xy - \tfrac{4}{3}x^2(6 - x)y^2. \tag{21}$$

For each fixed x in the interval $0 < x \le 1$, the right-hand side of (21) is a quadratic function of y which achieves its maximum for

$$y = \frac{\sqrt{3}(1 - x^2)^{1/2}}{x(6 - x)}. \tag{22}$$

Substitute (22) into (21) to conclude that

$$a_4 \le \tfrac{2}{3} + 8x^2 - \tfrac{14}{3}x^3 + \frac{4(1 - x^2)}{6 - x}. \tag{23}$$

It is now to be shown that the right-hand side of (23) is not larger than 4 for $0 \le x \le 1$, which is equivalent to the inequality

$$0 \le 24 - 5x - 66x^2 + 54x^3 - 7x^4,$$

or

$$0 \le 3x^2(1 - x) + (24 + 43x - 4x^2)(1 - x)^2.$$

This last inequality clearly holds for $0 \le x \le 1$, with equality only for $x = 1$. Thus $|a_4| \le 4$ for every $f \in S$, with equality only if $|a_2| = 2$, which implies that f is a rotation of the Koebe function (see Theorem 2.2).

The proof is obviously rather crude when $|a_2| = 2x$ is small. For instance, the maximum value of y as given by (22) does not lie in the relevant interval $0 \le y \le 1$ when x is small. Baranova [1] sharpened the proof to obtain several estimates for $|a_4|$ in terms of $|a_2|$, of which the simplest (and weakest) is

$$|a_4| \le \tfrac{4}{15}(11 + 2|a_2|),$$

an improvement on an earlier result of Ahlfors [2, 3].

It follows from (21) that $|a_4| \le \frac{2}{3}$ if $a_2 = 0$. This estimate is sharp, as is shown by $\{k(z^3)\}^{1/3}$, where k is the Koebe function.

§4.7. Coefficient Problem in the Class Σ

Analogous to the Bieberbach conjecture is the problem of finding sharp bounds for the coefficient b_n of functions

$$g(z) = z + b_0 + b_1 z^{-1} + b_2 z^{-2} + \cdots, \qquad |z| > 1,$$

of class Σ. Of course, nothing can be said about b_0, because the range of g can be translated arbitrarily without disturbing the univalence. It is convenient to consider the subclass Σ_0 of functions $g \in \Sigma$ for which $b_0 = 0$. The area theorem shows that $|b_n| \le n^{-1/2}$, $n = 1, 2, \ldots$. This is sharp only for $n = 1$, in which case the extremal functions are rotations of

$$g(z) = \{k(z^{-2})\}^{-1/2} = z - \frac{1}{z},$$

where k is the Koebe function. These extremal functions map the region $|z| > 1$ onto the complement of a line segment centered at the origin. In 1938, Schiffer [3] extended this result to the second coefficient. He used a variational method to obtain the sharp estimate $|b_2| \le \frac{2}{3}$, with equality only for rotations of

$$g(z) = \{k(z^{-3})\}^{-1/3} = z - \tfrac{2}{3} z^{-2} + \cdots,$$

whose range is the complement of a system of three line segments of equal length, meeting at the origin with equal angles. This evidence naturally suggests the conjecture

$$|b_n| \le \frac{2}{n+1}, \qquad n = 1, 2, \ldots,$$

with equality occurring for given n only for rotations of

$$g(z) = \{k(z^{-n-1})\}^{-1/(n+1)} = z - \frac{2}{n+1} z^{-n} + \cdots, \tag{24}$$

which map $|z| > 1$ onto the complement of an "asterisk" composed of $n + 1$ line segments of equal length, having the origin as a common endpoint and meeting there with equal angles.

Unfortunately, this conjecture is very far from true. It fails for the third coefficient, and even the order of magnitude $O(1/n)$ is incorrect. It is true, however, for certain subclasses of Σ. Before turning to these questions, we shall apply the Grunsky inequalities to derive Schiffer's estimate $|b_2| \leq \frac{2}{3}$.

Theorem 4.6. *The second coefficient of each function $g \in \Sigma_0$ satisfies $|b_2| \leq \frac{2}{3}$, with equality only for $g(z) = \{k(z^{-3})\}^{-1/3}$ and its rotations.*

Proof. Let E be the compact set omitted by g. If $\alpha \in E$, then

$$\begin{aligned} g^*(z) &= \{g(z^2) - \alpha\}^{1/2} \\ &= z\{1 - \alpha z^{-2} + b_1 z^{-4} + b_2 z^{-6} + \cdots\}^{1/2} \\ &= z + b_1^* z^{-1} + b_3^* z^{-3} + b_5^* z^{-5} + \cdots \end{aligned}$$

also belongs to Σ_0. An easy calculation leads to the expressions

$$\begin{aligned} &b_1^* = -\tfrac{1}{2}\alpha, \quad b_3^* = \tfrac{1}{2}b_1 - \tfrac{1}{8}\alpha^2, \\ &b_5^* = \tfrac{1}{2}b_2 + \tfrac{1}{4}b_1\alpha - \tfrac{1}{16}\alpha^3. \end{aligned}$$

Let F_n^* denote the Faber polynomials of g^*, and let β_{nk}^* denote its Grunsky coefficients. Then

$$F_n^*(g^*(z)) = z^n + \sum_{k=1}^{\infty} \beta_{nk}^* z^{-k}$$

and

$$\beta_{33}^* = 3b_5^* + 3b_1^* b_3^* + b_1^{*3} = \tfrac{3}{2}b_2 - \tfrac{1}{8}\alpha^3.$$

The Grunsky inequalities (6) imply

$$\sum_{k=1}^{\infty} k|\beta_{nk}^*|^2 \leq n, \qquad n = 1, 2, \ldots.$$

In particular, $|\beta_{33}^*| \leq 1$, with equality only if $\beta_{3k}^* = 0$ for all $k \neq 3$. Thus

$$|\tfrac{3}{2}b_2 - \tfrac{1}{8}\alpha^3| \leq 1 \tag{25}$$

for each $\alpha \in E$. If $0 \in E$, this implies $|b_2| \leq \frac{2}{3}$, with equality only if

$$F_3^*(g^*(z)) = z^3 + e^{i\theta} z^{-3}, \qquad e^{i\theta} = \beta_{33}^*.$$

But if $\alpha = 0$, then $b_0^* = b_1^* = b_2^* = 0$, and so $F_3^*(w) = w^3$. Therefore,

$$\{g(z^2)\}^{3/2} = \{g^*(z)\}^3 = z^3 + e^{i\theta}z^{-3},$$

which implies

$$g(z) = z\{1 + e^{i\theta}z^{-3}\}^{2/3},$$

a rotation of $\{k(z^{-3})\}^{-1/3}$.

Now suppose $0 \notin E$. We may assume, after a rotation, that b_2 is real and positive. It follows from (25) that

$$b_2 \leq \tfrac{2}{3} + \tfrac{1}{12}\operatorname{Re}\{\alpha^3\}$$

for every $\alpha \in E$. We shall prove $b_2 < \frac{2}{3}$ by showing that $\operatorname{Re}\{\alpha^3\} < 0$ for some $\alpha \in E$. If $\operatorname{Re}\{\alpha^3\} \geq 0$ for all $\alpha \in E$, then E is confined to one of the three sectors where $\operatorname{Re}\{\alpha^3\} \geq 0$, since $0 \notin E$ and E is connected. But this is impossible, because the relation

$$\frac{1}{2\pi}\int_0^{2\pi} g(re^{i\theta})\,d\theta = b_0 = 0, \qquad r > 1,$$

shows that 0 lies in the closed convex hull of the compact set E. Thus $\alpha \in E$ may be chosen so that $\operatorname{Re}\{\alpha^3\} < 0$, as claimed. The proof is complete.

The sharp bound for $|b_3|$ in the class Σ is not $\frac{1}{2}$ as the conjecture asserts, but $\frac{1}{2} + e^{-6}$. This remarkable result is due to Garabedian and Schiffer [2]. Much earlier, Bazilevich [3] had observed that $\frac{1}{2} + e^{-6}$ is the sharp bound within the subclass of odd functions.

Theorem 4.7. *The third coefficient of each odd function in* Σ *satisfies* $|b_3| \leq \frac{1}{2} + e^{-6}$. *This bound is sharp.*

Proof. The most general odd function $g \in \Sigma$ has the form

$$g(z) = \{f(z^{-2})\}^{-1/2} = z + b_1 z^{-1} + b_3 z^{-3} + \cdots,$$

where

$$f(z) = z + a_2 z^2 + a_3 z^3 + \cdots$$

belongs to the class S. A simple calculation gives

$$b_3 = -\tfrac{1}{2}(a_3 - \tfrac{3}{4}a_2^2).$$

But for each α in the interval $0 \leq \alpha < 1$, the Fekete–Szegö theorem (§3.8) gives the sharp inequality

$$|a_3 - \alpha a_2^2| \leq 1 + 2e^{-2\alpha/(1-\alpha)}.$$

Choosing $\alpha = \frac{3}{4}$, we obtain the desired estimate $|b_3| \leq \frac{1}{2} + e^{-6}$.

The sharp estimates for the higher coefficients b_n are not known. As Bazilevich observed, the argument just given for odd functions can be generalized to the class of functions $g \in \Sigma$ with k-fold symmetry; that is, to the class of functions

$$g(z) = \{f(z^{-k})\}^{-1/k} = z + b_{k-1}z^{-k+1} + b_{2k-1}z^{-2k+1} + \cdots$$

obtained from some $f \in S$. An easy calculation gives

$$b_{2k-1} = \frac{1}{k}\left(\frac{k+1}{2k}a_2^2 - a_3\right).$$

Thus for $k \geq 2$, the Fekete–Szegö theorem gives the sharp inequality

$$|b_{2k-1}| \leq \frac{1}{k}(1 + 2e^{-2(k+1)/(k-1)}).$$

In particular, this disproves the conjecture $|b_n| \leq 2/(n+1)$ for every odd index $n \geq 3$.

In fact, the conjecture is false asymptotically. Clunie [1] constructed functions in Σ with $b_n \neq O(n^{-0.98})$, and Pommerenke [8] gave examples where $b_n \neq O(n^{-0.83})$. On the positive side, the area theorem implies $b_n = O(n^{-1/2})$, and Clunie and Pommerenke [2] improved this to $b_n = O(n^{-1/2-1/300})$. The precise order of growth is not known. These matters are discussed further in §8.1.

The conjectured extremal function (24) actually belongs to the class Σ^* of functions in Σ whose omitted set E is starlike with respect to the origin. This suggests that the estimates $|b_n| \leq 2/(n+1)$ may hold within the subclass Σ^*. This was proved by Clunie [2] for $\Sigma_0^* = \Sigma^* \cap \Sigma_0$, and by Pommerenke [3] for the full class Σ^*. (Since $0 \in E$ by the definition of Σ^*, the assumption $b_0 = 0$ entails a slight loss of generality.)

Theorem 4.8. *The* nth *coefficient of every function in* Σ^* *satisfies* $|b_n| \leq 2/(n+1)$, *with equality only for the function* (24) *and its rotations*, $n = 1, 2, \ldots$.

Proof. First note that each $g \in \Sigma^*$ has the property

$$\operatorname{Re}\left\{\frac{zg'(z)}{g(z)}\right\} > 0, \qquad |z| > 1.$$

Indeed, each $g \in \Sigma^*$ has the form $g(z) = 1/f(\zeta)$ for some $f \in S^*$, and

$$\varphi(z) = \frac{zg'(z)}{g(z)} = \frac{\zeta f'(\zeta)}{f(\zeta)}, \qquad \zeta = \frac{1}{z}.$$

Because $\operatorname{Re}\{\varphi(z)\} > 0$, it follows from the maximum modulus principle that

$$\psi(z) = z\,\frac{\varphi(z) - 1}{\varphi(z) + 1} = \frac{z^2 g'(z) - zg(z)}{zg'(z) + g(z)}$$

satisfies $|\psi(z)| < 1$ in $|z| > 1$, since ψ is analytic at ∞. But

$$z^2 g'(z) - zg(z) = -\sum_{k=0}^{\infty} (k+1) b_k z^{-k+1}$$

and

$$zg'(z) + g(z) = 2z - \sum_{k=0}^{\infty} (k-1) b_k z^{-k}.$$

Thus it follows that

$$\left\{2 - \sum_{k=0}^{n-1} (k-1) b_k z^{-k-1}\right\}\psi(z) = -\sum_{k=0}^{n} (k+1) b_k z^{-k} + \sum_{k=n+1}^{\infty} c_k z^{-k} \tag{26}$$

for certain coefficients c_k. Since $|\psi(z)| < 1$, Parseval's formula now gives

$$\sum_{k=0}^{n} (k+1)^2 |b_k|^2 + \sum_{k=n+1}^{\infty} |c_k|^2 \le 4 + \sum_{k=0}^{n-1} (k-1)^2 |b_k|^2,$$

which implies

$$(n+1)^2 |b_n|^2 \le 4 - 4\sum_{k=0}^{n-1} k|b_k|^2 \le 4, \qquad n = 1, 2, \ldots.$$

Hence $|b_n| \le 2/(n+1)$, with equality only if $b_1 = b_2 = \cdots = b_{n-1} = 0$ and $c_k = 0$ for all $k \ge n+1$. An inspection of (26) now shows that equality implies $b_0 = 0$ and $\psi(z) = e^{i\theta} z^{-n}$, which implies that g is a rotation of (24).

The conjectured extremal function (24) also has the property $b_1 = b_2 = \cdots = b_{n-1} = 0$, which suggests that the inequality $|b_n| \le 2/(n+1)$

may hold within this subclass of Σ. In fact, somewhat more is true. The following theorem was stated by Goluzin [3], but Jenkins [9] gave the first correct proof. Later, Duren [1, 7] gave an elementary proof based upon the Grunsky inequalities. The theorem may be viewed as a generalization of Schiffer's result that $|b_2| \leq \frac{2}{3}$ for all $g \in \Sigma$. The proof is omitted.

Theorem 4.9. *If* $g \in \Sigma$ *and if* $b_1 = b_2 = \cdots = b_{m-1} = 0$ *for some* $m \geq 1$, *then* $|b_n| \leq 2/(n+1)$ *for* $n = m, m+1, \ldots, 2m$.

Notes

The equivalence of the two definitions (1) and (3) of Faber polynomials is a result of Schiffer [10]. The symmetry property (4) of the Grunsky coefficients is due to Grunsky [3] and Schur [3]. The logarithmic generating function (9) was introduced by Schiffer [10]. Explicit formulas expressing the Grunsky coefficients in terms of the coefficients of g were given by Schur [3] and Hummel [4]. Further information on Faber polynomials may be found in Pommerenke [4, 14], Kövari and Pommerenke [1], Smirnov and Lebedev [1], and Curtiss [1]. The polynomial area theorem and its generalizations may be traced to Goluzin [6], Biernacki [5], Lebedev and Milin [1], Wolibner [1], Shah [1], and Jenkins [13]. That the Grunsky matrix is unitary if and only if it is generated by a full mapping (Theorem 4.4) was discovered by I. M. Milin [1], Pommerenke [4], Bazilevich [10], and Pederson [1].

Pederson [2] and Ozawa [1, 2] used the Grunsky inequalities to prove the Bieberbach conjecture for the sixth coefficient: $|a_6| \leq 6$. Their proofs were basically similar to that of Charzyński and Schiffer [1] for a_4, but much more complicated. For a technical reason concerning the effect of a square-root transformation, this method seems to apply to the Bieberbach conjecture more readily for even indices than for odd. Ozawa and Kubota [1, 2, 3] applied it to a_8 with partial success. Ilina and Kolomoiceva [1] improved the results of Baranova [1] estimating $|a_4|$ in terms of $|a_2|$. (This is discussed at the end of §4.6.)

Many generalizations of the Grunsky inequalities have been given. Garabedian [2] and Garabedian and Schiffer [3] used a variational method to derive a system of inequalities involving omitted values as parameters. These inequalities are now called the *Garabedian–Schiffer inequalities.* Pederson and Schiffer [1] used them to prove the Bieberbach conjecture for the fifth coefficient. Pommerenke [14] derived the Garabedian–Schiffer inequalities in generalized form by an area method and gave a number of applications. Duren [10] gave further applications. Schiffer and Schmidt [1] obtained a generalization of the Garabedian–Schiffer inequalities.

The *local Bieberbach conjecture* states that $\mathrm{Re}\{a_n\} \leq n$ for all $f \in S$ in a sufficiently small neighborhood of the Koebe function. Duren and Schiffer [2] developed a formula for the second variation and applied it to reduce a certain relatively weak version of the local Bieberbach conjecture for given

n to the positive-definiteness of a certain quadratic form. They verified the positive-definiteness for $n \leq 9$, and Bombieri [3] later verified it for all n. The question remained whether the Koebe function is a local maximum for the nth coefficient in the strong sense that $\operatorname{Re}\{a_n\} \leq n$ whenever $|a_2 - 2| < \varepsilon_n$ for some $\varepsilon_n > 0$. Garabedian, Ross, and Schiffer [1] used a version of the Grunsky inequalities to prove this for all even n. Garabedian [1] proved it for $n = 5$, and Garabedian and Schiffer [3] used the Garabedian–Schiffer inequalities to prove it for all odd n. Independently, Bombieri [1, 4] combined the theory of the second variation with Loewner's method to obtain a proof for all n, although he presented the details only for odd n. Pederson [3] established the equivalence of various topologies near the Koebe function.

The coefficient problem for the class Σ appears more difficult than the Bieberbach conjecture because there are no elementary candidates for extremal functions. The proof of Schiffer's estimate $|b_2| \leq \frac{2}{3}$ via the Grunsky inequalities was first given by Nehari [3]. (This is essentially the proof of Theorem 4.6 as presented in §4.7.) After Garabedian and Schiffer [2] proved the sharp inequality $|b_3| \leq \frac{1}{2} + e^{-6}$ by a variational method, several other approaches were discovered. Jenkins [8] gave a proof by means of the general coefficient theorem. Bombieri [5] and Schmidt [1] found more elementary proofs along similar lines, with simplified analyses of the global structure of the trajectories of the associated quadratic differential. Pommerenke [14] gave a relatively simple proof based only on the Garabedian–Schiffer inequalities. Kubota [2, 3] applied the general coefficient theorem to show that for functions in Σ with real coefficients, the sharp inequalities

$$b_5 \leq \tfrac{1}{3} + \tfrac{4}{507} \quad \text{and} \quad b_4 \leq \tfrac{2}{5} + \tfrac{729}{163840} \quad \text{for } b_1 \geq 0$$

hold. For $b_1 < 0$ the sharp upper bound for b_4 is slightly larger and is presumably no longer a rational number. Kubota [3] gives formulas for its calculation. In particular, this disproves the conjecture $|b_n| \leq 2/(n+1)$ for $n = 4$. (It was disproved in §4.7 for all odd $n \geq 3$.) Tsao [1] has applied the second variation to disprove the conjecture for all $n \geq 5$. This approach reveals that the conjectured extremal function (24) does not provide even a local maximum, but is a saddle point. Chang, Schiffer, and Schober [1] obtained the same result by a similar method. Although the asymptotic formula $b_n = O(1/n)$ fails for the full class Σ (see §8.1), Pommerenke [1] proved it for the subclass of close-to-convex functions.

Exercises

1. Show that for each $g \in \Sigma$,

$$|\arg g'(\zeta)| \leq \log \frac{|\zeta|^2}{|\zeta|^2 - 1}, \qquad |\zeta| > 1.$$

Conclude that for each $f \in S$,

$$\left| \arg f'(z) - 2 \arg \frac{f(z)}{z} \right| \leq -\log(1 - |z|^2), \qquad |z| < 1.$$

2. Show that each $f \in S$ satisfies the inequality

$$\left| \log \frac{f(z)}{z} + \log(1 - |z|^2) \right| \leq \log \frac{1 + |z|}{1 - |z|}, \qquad |z| < 1.$$

(*Hint*: Apply the transformation used in the proof of Theorem 3.6.)

3. Prove that for functions $f \in S$ of the form $f(z) = z + a_{n+1}z^{n+1} + \cdots$, the sharp inequality $|a_{n+1}| \leq 2/n$ holds, and the bound is sharp for each n. (Goluzin [3].)

4. Prove Theorem 4.9: If $g \in \Sigma$ and $b_k = 0$ for all k in the range $1 \leq k < n/2$, then $|b_n| \leq 2/(n + 1)$. (Goluzin [3], Jenkins [9], Duren [1, 7].)

5. Let $g \in \Sigma$, and let $z_1, z_2, \ldots, z_n$ be distinct points in Δ. Show that

$$\prod_{i=1}^{n} \prod_{j=1}^{n} \left| 1 - \frac{1}{z_i \bar{z}_j} \right| \leq \prod_{i=1}^{n} \prod_{j=1}^{n} \left| \frac{g(z_i) - g(z_j)}{z_i - z_j} \right| \leq \prod_{i=1}^{n} \prod_{j=1}^{n} \left| 1 - \frac{1}{z_i \bar{z}_j} \right|^{-1}.$$

6. Let $f(z) = z + a_2 z^2 + \cdots$ be analytic in $\mathbb{D}$. Show that f is univalent in $\mathbb{D}$ if and only if the function

$$F(z, \zeta) = \frac{z - \zeta}{f(z) - f(\zeta)} = \sum_{n=0}^{\infty} \sum_{m=0}^{\infty} d_{nm} z^n \zeta^m$$

is analytic in the bidisk $\mathbb{D}^2 = \{(z, \zeta) : |z| < 1, |\zeta| < 1\}$. Conclude that $f \in S$ if and only if

$$\limsup_{(n+m) \to \infty} |d_{nm}|^{1/(n+m)} \leq 1.$$

Use this result to express the Bieberbach conjecture in purely arithmetic form. (Wolibner [1].)

Chapter 5

Exponentiation of the Grunsky Inequalities

Roughly speaking, the Grunsky inequalities provide information about the coefficients of the logarithm of a univalent function. During the 1960s, I. M. Milin systematically developed the idea of exponentiating these inequalities to obtain information about the coefficients of the univalent function itself. This chapter contains an account of Milin's theory. The method rests on some general inequalities (the *Lebedev–Milin inequalities*) which give sharp bounds on the coefficients of an exponentiated power series in terms of the coefficients of the original series. Some of the final results are improved estimates for the nth coefficient of a function of class S, for the difference of successive coefficients, and for the coefficients of an odd univalent function. A further consequence is Milin's simplified proof of Hayman's regularity theorem, which asserts that $|a_n|/n \to \alpha \leq 1$ for each function in S. The chapter concludes with a presentation of a related method introduced by FitzGerald in 1972, culminating in a derivation of FitzGerald's estimate $|a_n| < \sqrt{\tfrac{7}{6}}n$. This method is similar to Milin's insofar as its point of departure is an exponentiation (in another sense) of the Goluzin inequalities, which are essentially equivalent to the Grunsky inequalities.

§5.1. Exponentiation of Power Series

When a power series is exponentiated, the coefficients of the new series have a complicated dependence upon those of the original series. This section is devoted to three general inequalities, due to Lebedev and Milin, which estimate the new coefficients in terms of the old. Although these inequalities have nothing to do *per se* with univalent functions, they are the basic tool in Milin's exponentiation of the Grunsky inequalities.

Let

$$\varphi(z) = \sum_{k=1}^{\infty} \alpha_k z^k$$

be an arbitrary power series with a positive radius of convergence, normalized so that $\varphi(0) = 0$. Denote the exponentiated power series by

$$\psi(z) = e^{\varphi(z)} = \sum_{k=0}^{\infty} \beta_k z^k.$$

The Lebedev–Milin inequalities may be stated as follows.

First Lebedev–Milin Inequality. *If $\sum_{k=1}^{\infty} k|\alpha_k|^2 < \infty$, then*

$$\sum_{k=0}^{\infty} |\beta_k|^2 \le \exp\left\{\sum_{k=1}^{\infty} k|\alpha_k|^2\right\}, \tag{1}$$

with equality if and only if $\alpha_k = \gamma^k/k$, $k = 1, 2, \ldots$, for some complex constant γ with $|\gamma| < 1$.

Second Lebedev–Milin Inequality. *For $n = 1, 2, \ldots$,*

$$\sum_{k=0}^{n} |\beta_k|^2 \le (n+1)\exp\left\{\frac{1}{n+1}\sum_{m=1}^{n}\sum_{k=1}^{m}\left(k|\alpha_k|^2 - \frac{1}{k}\right)\right\}. \tag{2}$$

Equality occurs for a given integer n if and only if $\alpha_k = \gamma^k/k$, $k = 1, 2, \ldots, n$, for some complex constant γ with $|\gamma| = 1$.

Third Lebedev–Milin Inequality. *For $n = 1, 2, \ldots$,*

$$|\beta_n|^2 \le \exp\left\{\sum_{k=1}^{n}\left(k|\alpha_k|^2 - \frac{1}{k}\right)\right\}, \tag{3}$$

with equality if and only if $\alpha_k = \gamma^k/k$, $k = 1, 2, \ldots, n$, for some complex constant γ with $|\gamma| = 1$.

Proof of First Inequality. Differentiation of the equation $\psi(z) = e^{\varphi(z)}$ gives $\psi'(z) = \psi(z)\varphi'(z)$. Comparing coefficients, we obtain

$$\beta_n = \frac{1}{n}\sum_{k=0}^{n-1}(n-k)\alpha_{n-k}\beta_k, \qquad \beta_0 = 1. \tag{4}$$

Thus by the Cauchy–Schwarz inequality,

$$|\beta_n|^2 \le \frac{1}{n}\sum_{k=0}^{n-1}(n-k)^2|\alpha_{n-k}|^2|\beta_k|^2. \tag{5}$$

Now let $a_k = k|\alpha_k|^2$, and inductively define

$$b_n = \frac{1}{n}\sum_{k=0}^{n-1}(n-k)a_{n-k}b_k, \qquad b_0 = 1. \tag{6}$$

In view of the inequality (5), it is clear by induction that $|\beta_n|^2 \le b_n$, $n = 0, 1, 2, \ldots$. On the other hand, a comparison of (6) with (4) shows that

$$\sum_{k=0}^{\infty} b_k z^k = \exp\left\{\sum_{k=1}^{\infty} a_k z^k\right\}.$$

Since $a_k \ge 0$ and $b_k \ge 0$, it follows that

$$\sum_{k=0}^{\infty}|\beta_k|^2 \le \sum_{k=0}^{\infty} b_k = \exp\left\{\sum_{k=1}^{\infty} a_k\right\} = \exp\left\{\sum_{k=1}^{\infty} k|\alpha_k|^2\right\},$$

which is the inequality (1). Equality can occur only if $|\beta_k|^2 = b_k$ for all k, which implies equality in (5) for each value of n. Recalling the conditions for equality in the Cauchy–Schwarz inequality, we have

$$(n-k)\alpha_{n-k}\beta_k = \lambda_n, \qquad k = 0, 1, \ldots, n-1, \tag{7}$$

where the λ_n are complex constants, $n = 1, 2, \ldots$. In particular, $n\alpha_n = \lambda_n$, since $\beta_0 = 1$. Inserting the relation (7) into (4), we also find $\beta_n = \lambda_n$. Using (7), we conclude by induction that $\lambda_n = \lambda_1^n$, $n = 2, 3, \ldots$. With $\gamma = \lambda_1$, this gives $\alpha_n = \gamma^n/n$ and $\beta_n = \gamma^n$. In other words, equality occurs only if

$$\varphi(z) = -\log(1-\gamma z); \qquad \psi(z) = (1-\gamma z)^{-1}.$$

The condition $|\gamma| < 1$ is needed to ensure that $\sum k|\alpha_k|^2 < \infty$.

Proof of Second Inequality. Applying the Cauchy–Schwarz inequality to (4), we find

$$n^2|\beta_n|^2 \le \sum_{k=1}^{n} k^2|\alpha_k|^2 \sum_{k=0}^{n-1}|\beta_k|^2. \tag{8}$$

For convenience, let

$$A_n = \sum_{k=1}^{n} k^2|\alpha_k|^2; \qquad B_n = \sum_{k=0}^{n}|\beta_k|^2. \tag{9}$$

Then (8) gives

$$
\begin{aligned}
B_n = B_{n-1} + |\beta_n|^2 &\leq \left\{1 + \frac{1}{n^2} A_n\right\} B_{n-1} \\
&= \frac{n+1}{n}\left\{1 + \frac{A_n - n}{n(n+1)}\right\} B_{n-1} \\
&\leq \frac{n+1}{n} \exp\left\{\frac{A_n - n}{n(n+1)}\right\} B_{n-1}.
\end{aligned}
$$

Introducing the corresponding inequalities for $B_{n-1}, B_{n-2}, \ldots$, we obtain

$$
\begin{aligned}
B_n &\leq (n+1) \exp\left\{\sum_{k=1}^{n} \frac{A_k - k}{k(k+1)}\right\} \\
&= (n+1) \exp\left\{\sum_{k=1}^{n} \frac{A_k}{k(k+1)} + 1 - \sum_{k=1}^{n+1} \frac{1}{k}\right\}.
\end{aligned}
$$

But since

$$
s_n = \sum_{k=1}^{n} \frac{1}{k(k+1)} = \sum_{k=1}^{n} \left(\frac{1}{k} - \frac{1}{k+1}\right) = 1 - \frac{1}{n+1},
$$

a summation by parts gives

$$
\begin{aligned}
\sum_{k=1}^{n} A_k \cdot \frac{1}{k(k+1)} &= A_n s_n - \sum_{k=1}^{n} k^2 |\alpha_k|^2 s_{k-1} \\
&= \sum_{k=1}^{n} k|\alpha_k|^2 - \frac{1}{n+1} \sum_{k=1}^{n} k^2 |\alpha_k|^2.
\end{aligned}
$$

Thus

$$
\begin{aligned}
B_n &\leq (n+1) \exp\left\{\frac{1}{n+1} \sum_{k=1}^{n} (n+1-k)\left(k|\alpha_k|^2 - \frac{1}{k}\right)\right\} \\
&= (n+1) \exp\left\{\frac{1}{n+1} \sum_{m=1}^{n} \sum_{k=1}^{m} \left(k|\alpha_k|^2 - \frac{1}{k}\right)\right\},
\end{aligned}
$$

which completes the proof of (2).

If equality occurs in (2) for some n, then equality must occur in the Cauchy–Schwarz inequality and in the inequality $1 + x \leq e^x$ wherever they are used. Thus $A_k = k$ for $k = 1, 2, \ldots, n$; and

$$
\beta_k = \lambda_m (m-k)\overline{\alpha_{m-k}}, \qquad k = 0, 1, \ldots, m-1, \tag{10}
$$

for some complex constants λ_m, $m = 1, \ldots, n$. Substitution of (10) into (4) gives

$$m\beta_m = \lambda_m A_m = m\lambda_m;$$

or $\beta_m = \lambda_m$, $m = 1, 2, \ldots, n$. Since $\beta_0 = 1$, the relation (10) also gives

$$\lambda_m m\overline{\alpha_m} = 1, \qquad m = 1, 2, \ldots, n;$$

and

$$\lambda_1 = \beta_1 = \lambda_m(m-1)\overline{\alpha_{m-1}} = \lambda_m/\lambda_{m-1}.$$

Thus $\beta_m = \lambda_m = \lambda_1^m$, and $m\alpha_m = \overline{\lambda_1}^{-m}$, $m = 1, \ldots, n$. Finally, since $A_k = k$ for each k, we must have $|\lambda_1| = 1$.

Proof of Third Inequality. Introducing the Lebedev–Milin inequality (2) into (8) and using the notation (9), we find

$$|\beta_n|^2 \le e\frac{A_n}{n}\exp\left\{-\frac{A_n}{n} + \sum_{k=1}^{n}\left(k|\alpha_k|^2 - \frac{1}{k}\right)\right\}.$$

Thus we have only to use the inequality $xe^{-x} \le 1/e$, with $x = A_n/n$, to obtain (3). If equality holds for some n, then equality must hold in (8) for the same n, and in (2) for $n-1$. This again implies $\alpha_k = \gamma^k/k$ for $k = 1, 2, \ldots, n$, where $|\gamma| = 1$.

The Lebedev–Milin inequalities have an unusual history. The first inequality appeared in a short paper of Lebedev and Milin [2], with a proof that has little to do with the exponential function. As Milin [6] later observed, the proof depends only on the fact that the exponential function has non-negative coefficients. The proof given above is due to Clunie and first appeared in lecture notes of Pommerenke [9]. Milin [3] presented the second and third inequalities without proof, stating that they were obtained jointly with Lebedev. Their proofs appeared several years later in the book of Milin [6]. Meanwhile, Aharonov [5] had independently discovered the proofs given above.

§5.2. Reformulation of the Grunsky Inequalities

Our next aim is to recast the Grunsky inequalities, or rather some special cases, into a form suitable for certain applications. For each function $g \in \Sigma$ we write

$$\log\frac{g(z) - g(\zeta)}{z - \zeta} = -\sum_{n=1}^{\infty}\sum_{k=1}^{\infty}\gamma_{nk}\zeta^{-k}z^{-n} = -\sum_{n=1}^{\infty}A_n(1/\zeta)z^{-n}, \qquad (11)$$

where

$$A_n(w) = \sum_{k=1}^{\infty} \gamma_{nk} w^k, \qquad |w| < 1.$$

Recall that the strong Grunsky inequalities have the form

$$\sum_{n=1}^{\infty} n \left| \sum_{k=1}^{\infty} \gamma_{nk} \lambda_k \right|^2 \leq \sum_{n=1}^{\infty} \frac{1}{n} |\lambda_n|^2.$$

With the choice $\lambda_k = w^k$, this inequality takes the form

$$\sum_{n=1}^{\infty} n |A_n(w)|^2 \leq \sum_{n=1}^{\infty} \frac{1}{n} |w|^{2n} = -\log(1 - |w|^2). \tag{12}$$

We wish to find a similar inequality for the νth derivative

$$A^{(\nu)}(w) = \sum_{k=\nu}^{\infty} \gamma_{nk} k(k-1)\dots(k-\nu+1)w^{k-\nu}, \qquad \nu = 1, 2, \dots.$$

The obvious choice is

$$\lambda_k = \begin{cases} 0, & k < \nu; \\ k(k-1)\dots(k-\nu+1)w^{k-\nu}, & k \geq \nu. \end{cases}$$

Then the Grunsky inequalities give

$$\begin{aligned} \sum_{n=1}^{\infty} n |A_n^{(\nu)}(w)|^2 &\leq \sum_{k=\nu}^{\infty} \frac{1}{k} k^2(k-1)^2 \dots (k-\nu+1)^2 |w|^{2(k-\nu)} \\ &= -\left[\frac{\partial^{2\nu}}{\partial w^\nu \partial \bar{z}^\nu} \log(1 - w\bar{z})\right]_{z=w}. \end{aligned}$$

In particular, the first derivative satisfies the inequality

$$\sum_{n=1}^{\infty} n |A_n'(w)|^2 \leq (1 - |w|^2)^{-2}. \tag{13}$$

We can now derive an exponentiated form of the Grunsky inequalities. For $|z| < 1$ and $|w| < 1$, let

$$\exp\left\{\sum_{n=1}^{\infty} A_n(w) z^n\right\} = \sum_{n=0}^{\infty} B_n(w) z^n. \tag{14}$$

Then by the first Lebedev–Milin inequality (1),

$$\sum_{n=0}^{\infty} |B_n(w)|^2 \leq \exp\left\{\sum_{n=1}^{\infty} n|A_n(w)|^2\right\}.$$

Combining this with the inequality (12), we obtain

$$\sum_{n=0}^{\infty} |B_n(w)|^2 \leq (1 - |w|^2)^{-1}, \qquad |w| < 1. \tag{15}$$

The functions A_n are closely related to the Faber polynomials. Recall that the nth Faber polynomial F_n of a function $g \in \Sigma$ is defined by the generating relation

$$\frac{g'(z)}{g(z) - w} = \sum_{n=0}^{\infty} F_n(w) z^{-n-1}. \tag{16}$$

Integration with respect to z gives

$$\log \frac{z}{g(z) - w} = \sum_{n=1}^{\infty} \frac{1}{n} F_n(w) z^{-n}. \tag{17}$$

Note that the constant of integration has been correctly chosen, since both sides vanish at ∞. On the other hand, by the definition (11) of the functions A_n,

$$\log \frac{z}{g(z) - g(\zeta)} = \sum_{n=1}^{\infty} A_n(1/\zeta) z^{-n} + \log \frac{z}{z - \zeta}.$$

Comparing this with (17), we obtain

$$F_n(g(\zeta)) = nA_n(1/\zeta) + \zeta^n, \qquad n = 1, 2, \ldots. \tag{18}$$

There is also a close connection between the Faber polynomials and the functions B_n. In view of (11) and (14), we have

$$\frac{z - \zeta}{g(z) - g(\zeta)} = \sum_{n=0}^{\infty} B_n(1/\zeta) z^{-n}, \qquad |z| > 1, \quad |\zeta| > 1. \tag{19}$$

On the other hand, differentiation of (17) with respect to w leads to

$$\frac{1}{g(z) - w} = \sum_{n=1}^{\infty} \frac{1}{n} F_n'(w) z^{-n}. \tag{20}$$

Combining this with (19), we find

$$\sum_{n=1}^{\infty} \frac{1}{n} F_n'(g(\zeta))z^{-n} = \sum_{k=0}^{\infty} \zeta^k z^{-k-1} \sum_{n=0}^{\infty} B_n(1/\zeta)z^{-n}.$$

A comparison of coefficients now gives the interesting relation

$$F_n'(g(\zeta)) = n \sum_{k=0}^{n-1} \zeta^{n-k-1} B_k(1/\zeta). \tag{21}$$

§5.3. Estimation of the nth Coefficient

We now return to the problem of finding the sharp bound for the nth coefficient a_n among all functions

$$f(z) = z + a_2 z^2 + a_3 z^3 + \cdots$$

of class S. Littlewood's proof (§2.4) that $|a_n| < en$ relies on the Cauchy integral representation of a_n and the crude estimate

$$M_1(r, f) \leq r(1 - r)^{-1} \tag{22}$$

of the integral mean. We have already noted that even if (22) is replaced by the sharp estimate

$$M_1(r, f) \leq M_1(r, k) = r(1 - r^2)^{-1},$$

where k is the Koebe function, the method can give nothing better than $|a_n| < (e/2)n$.

In 1965, I. M. Milin [2, 6] became the first to penetrate the $e/2$-barrier. His point of departure is not the Cauchy formula, but a new representation for the nth coefficient of a function $f \in S$ in terms of Faber polynomials. An inspection of the expansion (20) gives

$$a_n = \frac{1}{n} F_n'(0), \qquad n = 2, 3, \ldots, \tag{23}$$

where F_n is the nth Faber polynomial of the function $g \in \Sigma$ associated with f by inversion: $g(z) = 1/f(1/z)$. Milin's result is as follows.

Theorem 5.1 (Milin's Theorem). *For each $f \in S$,*

$$|a_n| < 1.243\, n, \qquad n = 2, 3, \ldots.$$

Proof. The argument is based on the formula (23) for a_n. In order to estimate $F_n'(0)$, we apply the Cauchy–Schwarz inequality to (21) and invoke the exponentiated Grunsky inequality (15) to obtain

$$\begin{aligned} |F_n'(g(\zeta))|^2 &\le n^2 \sum_{k=0}^{n-1} |\zeta|^{2k} \sum_{k=0}^{n-1} |B_k(1/\zeta)|^2 \\ &\le n^2 \frac{|\zeta|^{2n}-1}{|\zeta|^2-1} \cdot \frac{|\zeta|^2}{|\zeta|^2-1}, \qquad |\zeta|>1. \end{aligned} \tag{24}$$

But since g omits the value 0, the image under g of the circle $|\zeta| = \rho > 1$ is a Jordan curve C_ρ which surrounds the origin. Hence it follows from the maximum modulus principle that

$$|F_n'(0)| \le \max_{w\in C_\rho} |F_n'(w)| = \max_{|\zeta|=\rho} |F_n'(g(\zeta))|.$$

Combining this with (24) and using the representation (23), we conclude that

$$|a_n| \le \frac{\rho}{\rho^2-1}(\rho^{2n}-1)^{1/2}, \qquad \rho > 1.$$

For the optimum result, we must choose ρ to minimize this last expression. The substitution $\rho = e^{x/2n}$ transforms the inequality to

$$|a_n| \le \frac{e^{x/2n}}{e^{x/n}-1}(e^x-1)^{1/2} < \frac{(e^x-1)^{1/2}}{x} n$$

for all $x > 0$, since

$$\frac{e^t - e^{-t}}{2t} = 1 + \frac{1}{3!}t^2 + \cdots > 1, \qquad t > 0.$$

Differentiation shows that the function $x^{-1}(e^x-1)^{1/2}$ has a minimum at the point where

$$x = \log 2 - \log(2-x).$$

The solution to this equation is approximately $x = 1.594$, where the function has a value slightly less than 1.243. This completes the proof of Milin's theorem.

One possible source of improvement would be to estimate the partial sum

$$\sum_{k=0}^{n-1} |B_k(w)|^2$$

by the second Lebedev–Milin inequality instead of the first, but this would require an improved estimate for the partial sums of the series $\sum k|A_k(w)|^2$, which the Grunsky inequalities do not seem to provide. In any event, FitzGerald [1] in 1972 introduced a new method which reduces the constant to $\sqrt{\frac{7}{6}} < 1.081$. A version of his proof appears later in this chapter.

§5.4. Logarithmic Coefficients

Associated with each function f in S are its *logarithmic coefficients* γ_n defined by

$$\log \frac{f(z)}{z} = 2 \sum_{n=1}^{\infty} \gamma_n z^n, \qquad |z| < 1.$$

The Grunsky inequalities provide a natural method for estimating the logarithmic coefficients. By means of the Lebedev–Milin inequalities, these bounds can then be transferred to bounds on the coefficients of f and related functions.

The logarithmic coefficients of the Koebe function are $\gamma_n = 1/n$. The inequality $|\gamma_n| \leq 1/n$ persists for starlike functions (see Exercise 2) but fails in general, even in order of magnitude (see §8.1). Nevertheless, I. M. Milin [3, 6] has shown that in a certain average sense, γ_n cannot be much larger than $1/n$. This result, known as *Milin's lemma*, occupies a central position in his theory.

Milin's Lemma. *For some constant* $\delta < 0.312$,

$$\sum_{k=1}^{n} \frac{1}{k} \leq \sup_{f \in S} \sum_{k=1}^{n} k|\gamma_k|^2 \leq \sum_{k=1}^{n} \frac{1}{k} + \delta, \qquad n = 1, 2, \ldots .$$

Proof. The lower estimate is obtained simply by choosing f to be the Koebe function, for which $\gamma_k = 1/k$.

The upper estimate is not so simple. First observe that in view of (17),

$$2\gamma_n = \frac{1}{n} F_n(0), \tag{25}$$

where F_n is the nth Faber polynomial of $g(z) = 1/f(1/z)$. Next recall the relation (18) between F_n and A_n, and apply the trivial inequality $(a + b)^2 \leq 2(a^2 + b^2)$ to obtain

$$\frac{1}{n}|F_n(g(\zeta))|^2 \leq 2n|A_n(1/\zeta)|^2 + \frac{2}{n}|\zeta|^{2n}. \tag{26}$$

The identity (25) gives

$$4\sum_{k=1}^{n} k|\gamma_k|^2 = \sum_{k=1}^{n} \frac{1}{k}|F_k(0)|^2.$$

On the other hand, g maps each circle $|z| = \rho > 1$ onto a Jordan curve C_ρ which surrounds the origin, and the function

$$\sum_{k=1}^{n} \frac{1}{k}|F_k(w)|^2$$

is subharmonic inside C_ρ. We may therefore apply the maximum principle for subharmonic functions to conclude that

$$4\sum_{k=1}^{n} k|\gamma_k|^2 \le \max_{w\in C_\rho} \sum_{k=1}^{n} \frac{1}{k}|F_k(w)|^2. \tag{27}$$

Combining (27) with (26) and invoking the Grunsky inequalities in the form (12), we obtain

$$2\sum_{k=1}^{n} k|\gamma_k|^2 \le \sum_{k=1}^{n} \frac{1}{k}\rho^{2k} - \log(1 - \rho^{-2}), \qquad \rho > 1. \tag{28}$$

The final step should now be to choose ρ to minimize the right-hand side of (28), but this is not so easy. We shall resort instead to estimation, applying the elementary inequalities

$$\log(n + \tfrac{1}{2}) < \sum_{n=1}^{n} \frac{1}{k} - \gamma, \tag{29}$$

where $\gamma = 0.577\ldots$ is Euler's constant; and

$$m\sum_{k=1}^{n} k^{m-1} < (n + \tfrac{1}{2})^m, \qquad m = 1, 2, \ldots. \tag{30}$$

Postponing the proofs of (29) and (30), we now use them to complete the proof of Milin's lemma. Returning to the inequality (28), we make the substitutions

$$\rho^2 = e^t \quad \text{and} \quad t = \frac{2x}{2n+1}, \qquad t > 0.$$

We then find, using (29),

$$\begin{aligned} -\log(1-\rho^{-2}) &= \frac{t}{2} - \log(e^{t/2} - e^{-t/2}) < \frac{t}{2} - \log t \\ &= \frac{x}{2n+1} - \log x + \log(n + \tfrac{1}{2}) \\ &< \frac{x}{2n+1} - \log x + \sum_{k=1}^{n} \frac{1}{k} - \gamma. \end{aligned} \tag{31}$$

In view of (30), the other term in (28) is

$$\begin{aligned} \sum_{k=1}^{n} \frac{1}{k} \rho^{2k} &= \sum_{k=1}^{n} \frac{1}{k} \sum_{m=0}^{\infty} \frac{1}{m!} (kt)^m \\ &= \sum_{m=0}^{\infty} \frac{1}{m!} t^m \sum_{k=1}^{n} k^{m-1} \\ &< \sum_{k=1}^{n} \frac{1}{k} + nt + \sum_{m=2}^{\infty} \frac{1}{m!\, m} t^m (n + \tfrac{1}{2})^m \\ &= \sum_{k=1}^{n} \frac{1}{k} + \frac{2nx}{2n+1} + \sum_{m=2}^{\infty} \frac{x^m}{m!\, m}. \end{aligned} \tag{32}$$

Adding (31) and (32), we see that the right-hand side of (28) is smaller than

$$G_n(x) = \int_0^x \frac{e^\xi - 1}{\xi}\, d\xi - \log x + 2 \sum_{k=1}^{n} \frac{1}{k} - \gamma.$$

But this function G_n attains its minimum where

$$G_n'(x) = \frac{e^x - 1}{x} - \frac{1}{x} = 0;$$

that is, at $x = \log 2$. Thus the inequality (28) gives

$$\sum_{k=1}^{n} k|\gamma_k|^2 \le \tfrac{1}{2} G_n(\log 2) = \sum_{k=1}^{n} \frac{1}{k} + \delta,$$

where

$$\delta = \frac{1}{2} \int_0^{\log 2} \frac{e^\xi - 1}{\xi}\, d\xi - \tfrac{1}{2} \log \log 2 - \frac{\gamma}{2} < 0.312.$$

This completes the proof of Milin's lemma, except for the verification of the inequalities (29) and (30).

One can prove (29) by observing that

$$y_n = \sum_{k=1}^{n} \frac{1}{k} - \log(n + \tfrac{1}{2})$$

decreases to γ. Indeed, $y_n - y_{n-1} = \psi(n)$, where

$$\psi(x) = \frac{1}{x} - \log \frac{x + \frac{1}{2}}{x - \frac{1}{2}}.$$

Easy calculations show that $\psi(1) < 0$, $\psi'(x) > 0$, and $\psi(x) \to 0$ as $x \to +\infty$. Thus $y_n < y_{n-1}$, $n = 2, 3, \ldots$ which proves (29). The inequality (30) may be proved by induction on n. It is true for $n = 1$, as one sees by induction on m. If (30) is true for any integer $(n - 1)$, then

$$\sum_{k=1}^{n} k^{m-1} < \frac{1}{m}(n - \tfrac{1}{2})^m + n^{m-1} < \frac{1}{m}(n + \tfrac{1}{2})^m,$$

since binomial expansion shows

$$(n + \tfrac{1}{2})^m - (n - \tfrac{1}{2})^m > mn^{m-1}.$$

This proves (30).

The number δ in Milin's lemma is known as *Milin's constant*.

Milin's lemma can be applied to obtain an improved universal bound for the coefficients of odd univalent functions

$$h(z) = z + c_3 z^3 + c_5 z^5 + \cdots, \qquad |z| < 1.$$

The argument of Littlewood and Paley (§2.11) gives the estimate $|c_n| \leq 14$ for all n. The bound cannot be reduced to 1 (see §3.8 and §3.9), but it was improved by Levin [3] to 3.39 and by Kung Sun [2] to 2.54. The work of I. M. Milin [3, 6] improves it to 1.17, and V. I. Milin [1] has further improved it to 1.14, the best value currently known.

Theorem 5.2. *For each odd function $h \in S$,*

$$|c_n| < e^{\delta/2} < 1.17, \qquad n = 3, 5, 7, \ldots,$$

where δ is Milin's constant.

Proof. Each odd univalent function h has the form

$$h(z) = \sqrt{f(z^2)}, \qquad f \in S.$$

Thus

$$\log \frac{h(\sqrt{z})}{\sqrt{z}} = \tfrac{1}{2} \log \frac{f(z)}{z} = \sum_{n=1}^{\infty} \gamma_n z^n.$$

In other words,

$$\sum_{n=0}^{\infty} c_{2n+1} z^n = \exp\left\{ \sum_{n=1}^{\infty} \gamma_n z^n \right\}, \qquad c_1 = 1. \tag{33}$$

Hence by the third Lebedev–Milin inequality,

$$|c_{2n+1}|^2 \leq \exp\left\{ \sum_{k=1}^{n} k|\gamma_k|^2 - \sum_{k=1}^{n} \frac{1}{k} \right\}. \tag{34}$$

Now Milin's lemma gives

$$|c_{2n+1}| \leq e^{\delta/2} < e^{0.156} < 1.17, \qquad n = 1, 2, \ldots,$$

as claimed.

It is an open problem of some interest to find the best possible value of the Milin constant δ. Milin's estimate of δ is the best obtainable from the inequality (28). Observe that δ cannot be reduced to zero, since by Theorem 5.2 this would imply the Littlewood–Paley conjecture $|c_n| \leq 1$, which is known to be false (see §3.8 and §3.9). Nevertheless, I. M. Milin [6] has proposed the following conjecture, which asserts that $\delta = 0$ in an average sense.

Milin Conjecture. *For each* $f \in S$,

$$\sum_{m=1}^{n} \sum_{k=1}^{m} \left(k|\gamma_k|^2 - \frac{1}{k} \right) \leq 0, \qquad n = 1, 2, \ldots.$$

In view of the representation (33) and the second Lebedev–Milin inequality, Milin's conjecture implies Robertson's conjecture (§2.11)

$$\sum_{k=0}^{n} |c_{2k+1}|^2 \leq n + 1,$$

which implies the Bieberbach conjecture $|a_{n+1}| \leq n + 1$. Milin's conjecture is trivially true for $n = 1$, since $\gamma_1 = \frac{1}{2}a_2$. Grinspan [1] has proved it for $n = 2$ and for $n = 3$.

By a simple modification of Milin's method, it is possible to establish the Bieberbach conjecture for functions with sufficiently small second coefficient. The following result is due to Aharonov [4].

Theorem 5.3. *If $f \in S$ and $|a_2| < 0.867$, then $|a_n| < n$, $n = 3, 4, 5, \ldots$.*

Proof. The proof of Milin's lemma is easily adjusted to yield the inequality

$$\begin{aligned} 2\sum_{k=2}^{n} k|\gamma_k|^2 &\leq \sum_{k=2}^{n} \frac{1}{k}\rho^{2k} - \log(1 - \rho^{-2}) \\ &< \sum_{k=1}^{n} \frac{1}{k}\rho^{2k} - \log(1 - \rho^{-2}) - 1 \end{aligned}$$

instead of (28). Choosing ρ to minimize this last expression, we find as in the proof of Milin's lemma

$$\sum_{k=1}^{n} k|\gamma_k|^2 < \sum_{k=1}^{n} \frac{1}{k} + \delta - \tfrac{1}{2} + \tfrac{1}{4}|a_2|^2,$$

since $2\gamma_1 = a_2$. In view of the inequality (34) for the coefficients of

$$h(z) = \sqrt{f(z^2)} = \sum_{n=0}^{\infty} c_{2n+1}z^{2n+1},$$

we obtain

$$|c_{2n+1}|^2 < \exp\{\delta - \tfrac{1}{2} + \tfrac{1}{4}|a_2|^2\}.$$

Because $\delta < 0.312$, a numerical calculation now shows that $|c_{2n+1}| < 1$ for all $n > 0$ if $|a_2| < 0.867$. This gives the desired result, by the familiar argument leading from the Littlewood–Paley conjecture to the Bieberbach conjecture (*cf.* §2.11).

Theorem 5.3 can be improved by appeal to the third Lebedev–Milin inequality instead of the second, together with the argument leading from Robertson's conjecture to Bieberbach's (*cf.* §2.11). Using a strengthened form of the third Lebedev–Milin inequality, Aharonov [6] and Ilina [2] independently improved the constant from 0.867 to 1.05. Later, Bshouty [1, 3] used FitzGerald's method to make further improvements to 1.55 and 1.59. Ehrig [1, 2] had obtained related results by similar techniques.

§5.5. Radial Growth

In 1955, Hayman [1, 2] discovered that for each fixed function in S, thc sequence $\{|a_n|/n\}$ converges to a limit $\alpha \le 1$. This striking result is known as Hayman's regularity theorem. We shall now begin preparations for a simplified proof due to I. M. Milin [5, 6]. The preliminary results of this section are due to Hayman.

The limit α in Hayman's theorem arises first in a more elementary theorem on the radial growth of a univalent function. This will be deduced from the following lemma on the growth of the maximum modulus

$$M_\infty(r,f) = \max_{|z|=r} |f(z)|.$$

Lemma. *For each function $f \in S$,*

$$\lim_{r\to 1}(1 - r)^2 M_\infty(r,f) = \alpha \le 1.$$

If f is not a rotation of the Koebe function, the expression $r^{-1}(1 - r)^2 M_\infty(r,f)$ is strictly decreasing in the interval $(0, 1)$, *and* $\alpha < 1$.

Proof. First recall the inequality (Theorem 2.7)

$$\frac{1-r}{1+r} \le \left|\frac{zf'(z)}{f(z)}\right| \le \frac{1+r}{1-r}, \qquad |z| = r < 1. \tag{35}$$

Strict inequality holds for each z unless f is a suitable rotation of the Koebe function. If f is not a rotation of the Koebe function, it follows that

$$\frac{\partial}{\partial r} \log|f(re^{i\theta})| \le \left|\frac{f'(re^{i\theta})}{f(re^{i\theta})}\right| < \frac{1+r}{r(1-r)}.$$

Integrating from r_1 to r_2 $(0 < r_1 < r_2 < 1)$, we obtain

$$\log\left|\frac{f(r_2e^{i\theta})}{f(r_1e^{i\theta})}\right| < \int_{r_1}^{r_2} \frac{1+r}{r(1-r)}\,dr = \log\frac{r_2(1-r_1)^2}{r_1(1-r_2)^2}.$$

In other words,

$$\frac{(1-r_2)^2}{r_2}|f(r_2e^{i\theta})| < \frac{(1-r_1)^2}{r_1}|f(r_1e^{i\theta})|, \qquad 0 < r_1 < r_2 < 1, \tag{36}$$

for each θ. Now choose θ so that

$$|f(r_2e^{i\theta})| = M_\infty(r_2, f).$$

Then

$$\frac{(1-r_2)^2}{r_2} M_\infty(r_2, f) < \frac{(1-r_1)^2}{r_1} |f(r_1 e^{i\theta})| \leq \frac{(1-r_1)^2}{r_1} M_\infty(r_1, f),$$

which shows that $r^{-1}(1-r)^2 M_\infty(r,f)$ decreases to a limit $\alpha \geq 0$. Since the growth theorem (§2.3) gives the bound

$$\frac{(1-r)^2}{r} M_\infty(r,f) \leq 1, \tag{37}$$

it follows that $\alpha \leq 1$. If f is a rotation of the Koebe function, equality holds in (37) for every r, and $\alpha = 1$. Otherwise the expression in (37) is strictly decreasing, so $\alpha < 1$. This completes the proof of the lemma.

In fact, the argument proves more. The inequality (36) shows that in each direction $e^{i\theta}$, the expression $r^{-1}(1-r)^2|f(re^{i\theta})|$ decreases to a limit no larger than α. We shall now show that each f in S has a *direction of maximal growth* $e^{i\theta_0}$, for which this limit is actually equal to α. The number α will be called the *Hayman index* of f.

Theorem 5.4. *If a function $f \in S$ has Hayman index $\alpha > 0$, there is a unique direction $e^{i\theta}$ for which*

$$\lim_{r\to 1}(1-r)^2|f(re^{i\theta_0})| = \alpha. \tag{38}$$

Proof. Let $\{r_n\}$ be a sequence of radii increasing to 1, and choose angles θ_n, $0 \leq \theta_n < 2\pi$, such that

$$|f(r_n e^{i\theta_n})| = M_\infty(r_n, f), \qquad n = 1, 2, \ldots.$$

Then for all $r < r_n$, we have by (36)

$$\alpha \leq \frac{(1-r_n)^2}{r_n} |f(r_n e^{i\theta_n})| \leq \frac{(1-r)^2}{r} |f(re^{i\theta_n})|.$$

If θ_0 is a cluster point of the sequence $\{\theta_n\}$, it follows that

$$\alpha \leq \frac{(1-r)^2}{r} |f(re^{i\theta_0})| \leq \frac{(1-r)^2}{r} M_\infty(r,f).$$

Since $r^{-1}(1-r)^2 M_\infty(r,f) \to \alpha$ as $r \to 1$, this gives the desired result (38).

To prove the uniqueness of θ_0, we invoke the inequality (Theorem 4.3, Corollary 5)

$$1 - \rho^{-2} \le \left| \frac{g(\zeta_1) - g(\zeta_2)}{\zeta_1 - \zeta_2} \right| \le \frac{1}{1 - \rho^{-2}}, \qquad |\zeta_1| = |\zeta_2| = \rho > 1,$$

valid for all functions $g \in \Sigma$. For $f \in S$, we conclude by inversion that

$$\frac{1 - r^2}{r} |e^{i\theta} - e^{i\theta_0}| \le \left| \frac{1}{f(re^{i\theta})} - \frac{1}{f(re^{i\theta_0})} \right| \le \frac{1}{|f(re^{i\theta})|} + \frac{1}{|f(re^{i\theta_0})|}.$$

If $e^{i\theta} \neq e^{i\theta_0}$ *and if the limit* α *in* (38) *is positive, it follows that*

$$\lim_{r \to 1} (1 - r)^2 |f(re^{i\theta})| = 0.$$

In particular, the direction of maximal growth is unique.

If $\alpha = 0$, the limit (38) obviously has the value 0 for every choice of θ_0. In other words, every direction is a direction of maximal growth in this case.

A function $f \in S$ is said to be of *slow growth* if $\alpha = 0$ and of *maximal growth* if $\alpha > 0$. In general, functions of slow growth have less regular behavior.

The Hayman index α of a function $f \in S$ can be estimated in terms of the value of its second coefficient a_2. Jenkins [3] found the sharp upper bound for $|f(z)|$ among all functions $f \in S$ with fixed a_2, thus solving what was called the *Gronwall problem*. (Gronwall [3, 4] had given an erroneous solution.) This enabled Hayman [1] to deduce the sharp bound

$$\alpha \le 4\lambda^2 e^{2-4\lambda}, \qquad \lambda = \{2 - (2 - C)^{1/2}\}^{-1},$$

for functions $f \in S$ with $|a_2| \le C$.

§5.6. Bazilevich's Theorem

Consider again the logarithmic coefficients γ_n defined by

$$\log \frac{f(z)}{z} = 2 \sum_{n=1}^{\infty} \gamma_n z^n, \qquad f \in S.$$

Observe that

$$k_{-\theta_0}(z) = z(1 - e^{-i\theta_0}z)^{-2}$$

is the rotation of the Koebe function whose direction of maximal growth is $e^{i\theta_0}$ and whose logarithmic coefficients are

$$\gamma_n = \frac{1}{n} e^{-in\theta_0}, \qquad n = 1, 2, \ldots .$$

Loosely speaking, it is to be expected that if $f \in S$ has the same direction of maximal growth $e^{i\theta_0}$ and is sufficiently near $k_{-\theta_0}$, in the sense that its Hayman index α is near 1, then its logarithmic coefficients should be close to those of $k_{-\theta_0}$. The theorem of Bazilevich [9, 10] expresses this principle in a precise quantitative form.

Theorem 5.5 (Bazilevich's Theorem). *Let $f \in S$ have Hayman index $\alpha > 0$ and direction of maximal growth $e^{i\theta_0}$. Then*

$$\sum_{n=1}^{\infty} n\left|\gamma_n - \frac{1}{n} e^{-in\theta_0}\right|^2 \leq \tfrac{1}{2} \log \frac{1}{\alpha}. \tag{39}$$

Proof. The proof is based on Milin's reformulation of the Grunsky inequalities (§5.2). Let $g(\zeta) = 1/f(1/\zeta)$, and consider the expansion

$$\log \frac{g(\zeta) - g(w)}{\zeta - w} = -\sum_{n=1}^{\infty} A_n(z) w^{-n}, \qquad z = \frac{1}{\zeta}.$$

According to the inequality (12),

$$\sum_{n=1}^{\infty} n |A_n(z)|^2 \leq -\log(1 - |z|^2), \qquad |z| < 1.$$

Hence

$$\begin{aligned}\sum_{n=1}^{\infty} n\left|A_n(z) - \frac{1}{n}\bar{z}^n\right|^2 &= \sum_{n=1}^{\infty} n|A_n(z)|^2 - 2\,\mathrm{Re}\left\{\sum_{n=1}^{\infty} A_n(z) z^n\right\} + \sum_{n=1}^{\infty} \frac{1}{n}|z|^{2n} \\ &\leq -2\log(1 - |z|^2) - 2\,\mathrm{Re}\left\{\sum_{n=1}^{\infty} A_n(z) z^n\right\}. \end{aligned} \tag{40}$$

But

$$-\sum_{n=1}^{\infty} A_n(z) z^n = \log g'(\zeta) = \log \frac{z^2 f'(z)}{[f(z)]^2}.$$

Thus the inequalities (40) and (35) give

$$\sum_{n=1}^{\infty} n\left|A_n(z) - \frac{1}{n}\bar{z}^n\right|^2 \le 2\log\frac{r^2|f'(z)|}{(1-r^2)|f(z)|^2}$$
$$\le -2\log\left\{\frac{(1-r)^2}{r}|f(z)|\right\}, \qquad r = |z| < 1.$$

Now take $z = re^{i\theta_0}$ and use the relations (36) and (38) to obtain

$$\sum_{n=1}^{\infty} n\left|A_n(re^{i\theta_0}) - \frac{1}{n}r^n e^{-in\theta_0}\right|^2 \le -2\log\alpha, \qquad r < 1. \tag{41}$$

Now recall that the functions A_n are closely related to the Faber polynomials. Equation (18) can be put in the form

$$nA_n(re^{i\theta_0}) + r^{-n}e^{-in\theta_0} = F_n(1/f(re^{i\theta_0})),$$

where F_n is the nth Faber polynomial of g. Since $f(re^{i\theta_0}) \to \infty$ as $r \to 1$, this shows that $A_n(re^{i\theta_0})$ has a limit, and

$$A_n(e^{i\theta_0}) = \lim_{r\to 1} A_n(re^{i\theta_0}) = \frac{1}{n}F_n(0) - \frac{1}{n}e^{-in\theta_0}$$
$$= 2\gamma_n - \frac{1}{n}e^{-in\theta_0},$$

where the relation (25) has been used. The inequality (41) now leads to the conclusion

$$4\sum_{n=1}^{\infty} n\left|\gamma_n - \frac{1}{n}e^{-in\theta_0}\right|^2 \le -2\log\alpha,$$

which is equivalent to (39).

This proof was suggested by I. M. Milin [6]. Bazilevich [9] showed that equality occurs for functions $f \in S$ which map the disk onto the complement of an analytic arc.

The theorem has some interesting consequences.

Corollary 1. *For each $f \in S$ with $\alpha > 0$,*

$$\lim_{n\to\infty}\sum_{k=1}^{n}\left(k|\gamma_k|^2 - \frac{1}{k}\right) \le \tfrac{1}{2}\log\alpha.$$

(In particular, the limit exists.)

Proof. Let f have direction of maximal growth $e^{i\theta_0}$, and observe that

$$2\,\mathrm{Re}\left\{\sum_{k=1}^{\infty}\left(\gamma_k e^{ik\theta_0}-\frac{1}{k}\right)r^k\right\}=\log\left\{\frac{(1-r)^2}{r}\,|f(re^{i\theta_0})|\right\}\to\log\alpha$$

as $r\to 1$. By virtue of Bazilevich's theorem and Fejér's Tauberian theorem (§5.8, Lemma 2), this implies

$$2\,\mathrm{Re}\left\{\sum_{k=1}^{n}\left(\gamma_k e^{ik\theta_0}-\frac{1}{k}\right)\right\}\to\log\alpha$$

as $n\to\infty$. In view of the representation

$$\sum_{k=1}^{n}\left(k|\gamma_k|^2-\frac{1}{k}\right)=2\,\mathrm{Re}\left\{\sum_{k=1}^{n}\left(\gamma_k e^{ik\theta_0}-\frac{1}{k}\right)\right\}+\sum_{k=1}^{n}k\left|\gamma_k e^{ik\theta_0}-\frac{1}{k}\right|^2,$$

another appeal to Bazilevich's theorem now gives the desired result.

Corollary 2. *For each $f\in S$ with $\alpha>0$,*

$$\lim_{n\to\infty}\frac{1}{n}\sum_{m=1}^{n}\sum_{k=1}^{m}\left(k|\gamma_k|^2-\frac{1}{k}\right)\le\tfrac{1}{2}\log\alpha<0.$$

Proof. The Cesàro means of the series in Corollary 1 converge to the same limit.

Corollary 2 shows that for each fixed function $f\in S$ with $\alpha>0$, the Milin conjecture (§5.4) is "eventually true." Grinspan [1] has extended this result to $\alpha=0$ by showing that the expressions in Corollary 2 tend to $-\infty$ in this case.

§5.7. Hayman's Regularity Theorem

The ground is now prepared for a proof of Hayman's regularity theorem, which gives remarkably precise information on the asymptotic behavior of the coefficients of an arbitrary univalent function. The theorem asserts that the modulus of the nth coefficient is asymptotic to a constant multiple of n. Hayman [1, 2] actually proved somewhat more, but we shall be content with this restricted form of the theorem.

Theorem 5.6 (Hayman's Regularity Theorem). *For each $f \in S$,*

$$\lim_{n \to \infty} \frac{|a_n|}{n} = \alpha \leq 1,$$

and $\alpha < 1$ unless f is a rotation of the Koebe function.

At first glance, Hayman's theorem would appear to prove the Bieberbach conjecture for large n. Unfortunately, however, the proof does not establish any uniformity of convergence as f ranges through the class S. In fact, Shirokov [1] has shown that for any prescribed $\alpha < 1$, the sequence $\{|a_n|/n\}$ may approach α arbitrarily slowly.

The theorem should be compared with another result of Hayman [3] that the sequence $\{A_n/n\}$ also converges, where A_n is the maximum of $|a_n|$ for all $f \in S$. The *asymptotic Bieberbach conjecture* (§2.12) asserts that $A_n/n \to 1$.

The limit in the regularity theorem is actually

$$\alpha = \lim_{r \to 1}(1 - r)^2 M_\infty(r, f), \tag{42}$$

the Hayman index of f. We saw in §5.5 that the limit (42) exists for each $f \in S$, and that $\alpha < 1$ unless f is a rotation of the Koebe function. Thus with α defined as the Hayman index of f, the proof reduces to showing that $|a_n|/n \to \alpha$. This is comparatively easy when $\alpha = 0$, but quite difficult when $\alpha > 0$.

Fortunately, I. M. Milin [5, 6] has discovered an alternate approach, substantially simpler than Hayman's, to the case where $\alpha > 0$. We shall present Milin's proof in a somewhat modified form. The crux of his argument is a general theorem on summability of series, which may be viewed as a new Tauberian theorem. In its application to Hayman's theorem, the assumption that α is positive allows Bazilevich's theorem (§5.6) to supply the Tauberian condition. (For general background in summability theory and Tauberian theorems, the reader is referred to the books of Knopp [1], Hardy [1], and Landau [1].)

We shall first dispense with the case $\alpha = 0$. Here the proof rests on a simple lemma relating the growth of $M_\infty(r, f)$ to that of

$$M_1(r, f) = \frac{1}{2\pi} \int_0^{2\pi} |f(re^{i\theta})| \, d\theta.$$

Lemma. *If $f \in S$ and*

$$\lim_{r \to 1}(1 - r)^2 M_\infty(r, f) = 0,$$

then

$$\lim_{r\to 1}(1 - r)M_1(r,f) = 0. \tag{43}$$

Proof of Lemma. By Prawitz' theorem (§2.10),

$$r\frac{d}{dr}M_1(r,f) \le M_\infty(r,f).$$

By the lemma of §5.5 and the present hypothesis,

$$\varepsilon(r) = r^{-1}(1 - r)^2 M_\infty(r,f)$$

decreases to 0 as $r \to 1$. In this notation, Prawitz' inequality may be written

$$\frac{d}{dr}M_1(r,f) \le \varepsilon(r)(1 - r)^{-2}.$$

Integration from r_1 to r_2 ($0 < r_1 < r_2 < 1$) produces

$$\begin{aligned} M_1(r_2,f) - M_1(r_1,f) &\le \varepsilon(r_1)\int_{r_1}^{r_2}(1 - r)^{-2}\,dr \\ &= \varepsilon(r_1)\{(1 - r_2)^{-1} - (1 - r_1)^{-1}\}, \end{aligned}$$

or

$$(1 - r_2)M_1(r_2,f) \le (1 - r_2)M_1(r_1,f) + \varepsilon(r_1).$$

Thus

$$\limsup_{r_2\to 1}(1 - r_2)M_1(r_2,f) \le \varepsilon(r_1)$$

for each $r_1 > 0$. Letting r_1 tend to 1, one obtains the desired conclusion (43). This proves the lemma.

It is now a short step to Hayman's theorem in the case $\alpha = 0$. Represent a_n by the Cauchy formula and make the usual crude estimation to obtain

$$|a_n| \le r^{-n}M_1(r,f), \qquad 0 < r < 1.$$

Choose $r = r_n = 1 - 1/n$ to conclude that

$$\frac{|a_n|}{n} \le \left(1 - \frac{1}{n}\right)^{-n}(1 - r_n)M_1(r_n,f) \le 4(1 - r_n)M_1(r_n,f).$$

Thus the lemma shows that $|a_n|/n \to 0$.

It remains to consider the case in which the Hayman index α, as defined by (42), is positive. Here the proof is based on Milin's Tauberian theorem, which we now proceed to develop. Let

$$\psi(r) = \sum_{n=0}^{\infty} \beta_n r^n, \qquad \beta_0 = 1,$$

be an arbitrary power series with complex coefficients, convergent for $0 < r < 1$. Let

$$s_n = \sum_{k=0}^{n} \beta_k \quad \text{and} \quad \sigma_n = \frac{1}{n+1} \sum_{k=0}^{n} s_k$$

denote the partial sums of the coefficients and the *Cesàro means*, respectively. Finally, let

$$\log \psi(r) = \sum_{n=1}^{\infty} \lambda_n r^n.$$

It is well known (see, for example, Hardy [1] or Knopp [1]) that if a series is convergent or *Cesàro summable*, then it is *Abel summable* to the same limit. In other words, if $s_n \to s$ or if $\sigma_n \to s$ as $n \to \infty$, then $\psi(r) \to s$ as $r \to 1$. Milin's theorem is a partial converse. It asserts that if the *moduli* of the Abel means converge, then the moduli of the partial sums and the moduli of the Cesàro means converge to the same limit, provided the terms satisfy a certain growth condition known as a *Tauberian condition*. This is a rather unusual kind of Tauberian theorem, since it concerns not the means but their moduli.

Theorem 5.7 (Milin's Tauberian Theorem). *Suppose* $|\psi(r)| \to \alpha$ *as* $r \to 1$, *and* $\sum_{n=1}^{\infty} n|\lambda_n|^2 < \infty$. *Then* $|s_n| \to \alpha$ *and* $|\sigma_n| \to \alpha$ *as* $n \to \infty$.

The proof is postponed to §5.8. The result will now be applied to prove Hayman's theorem for $\alpha > 0$. It should be emphasized that Milin's theorem is true even if $\alpha = 0$.

Deduction of Hayman's Regularity Theorem. Let $f \in S$ have Hayman index $\alpha > 0$ and direction of maximal growth $e^{i\theta_0}$. Suppose, without loss of generality, that $\theta_0 = 0$. This can be achieved by a rotation. Then the function

$$\psi(r) = r^{-1}(1-r)^2 f(r) = \sum_{n=1}^{\infty} \beta_n r^n$$

has the property that $|\psi(r)| \to \alpha > 0$ as $r \to 1$. A simple calculation relates the partial sums s_n and the Cesàro means σ_n of the series $\sum \beta_n$ to the coefficients a_n of f:

$$s_n = a_{n+1} - a_n \quad \text{and} \quad \sigma_{n-1} = \frac{a_n}{n}.$$

The logarithmic coefficients of ψ, defined by $\log \psi(r) = \sum \lambda_n r^n$, are

$$\lambda_n = 2\left(\gamma_n - \frac{1}{n}\right),$$

where γ_n are the logarithmic coefficients of f:

$$\log \frac{f(z)}{z} = 2 \sum_{n=1}^{\infty} \gamma_n z^n.$$

Since $\alpha > 0$, the theorem of Bazilevich (§5.6) provides the Tauberian condition $\sum n|\lambda_n|^2 < \infty$, and Milin's Tauberian theorem allows us to conclude that

$$|a_{n+1} - a_n| \to \alpha \quad \text{and} \quad \frac{|a_n|}{n} \to \alpha.$$

This completes the deduction of Hayman's theorem from Milin's in the case $\alpha > 0$. Actually, the proof gives the additional information that $|a_{n+1} - a_n| \to \alpha$, provided $\theta_0 = 0$.

If $\alpha = 0$, Bazilevich's theorem does not imply the convergence of the series $\sum n|\lambda_n|^2$. Indeed, simple examples show that the series may actually diverge in this case, and that $|a_{n+1} - a_n|$ need not tend to zero (see Exercise 5). Hence Milin's Tauberian theorem may not be applicable if $\alpha = 0$.

With additional hypotheses on the behavior of f in its direction of maximal growth, Duren [8, 9] applied Tauberian remainder theorems to deduce correspondingly stronger conclusions on the rate of convergence of $\{|a_n|/n\}$ to α. An elementary argument (see Exercise 6) shows that if f maps the disk onto the complement of an analytic arc, then

$$a_n = \lambda n + \mu + o(n^{-1/2}), \qquad n \to \infty,$$

where λ and μ are complex constants and $|\lambda| = \alpha > 0$.

Hayman's regularity theorem appears to provide overwhelming evidence in favor of the Bieberbach conjecture until it is compared with Hayman's parallel result for odd univalent functions, where the corresponding conjecture $|c_{2n+1}| \leq 1$ is known to be false.

Theorem 5.8. *For each odd univalent function* $h(z) = z + c_3 z^3 + c_5 z^5 + \cdots$,

$$\lim_{n\to\infty} |c_{2n+1}| = \beta \leq 1,$$

and $\beta < 1$ *unless* h *is a rotation of* q, *where* $q(z) = z(1 - z^2)^{-1}$.

Proof. Every $h \in S^{(2)}$ is the square-root transform of some $f \in S$: $h(z) = \sqrt{f(z^2)}$. If f has Hayman index α, we may suppose after a rotation that

$$\lim_{r \to 1}(1 - r)^2 |f(r)| = \alpha,$$

by Theorem 5.4. Thus

$$\lim_{r \to 1}(1 - r^2)|h(r)| = \sqrt{\alpha}.$$

Let

$$\psi(r) = \frac{1}{\sqrt{r}}(1 - r)h(\sqrt{r}) = \sum_{n=0}^{\infty} \beta_n r^n, \qquad \beta_0 = 1.$$

Then $|\psi(r)| \to \sqrt{\alpha}$, while

$$\beta_n = c_{2n+1} - c_{2n-1}, \qquad n = 1, 2, \ldots,$$

and

$$s_n = \sum_{k=0}^{n} \beta_k = c_{2n+1}, \qquad n = 1, 2, \ldots .$$

In order to apply Milin's Tauberian theorem, and thus to conclude that $|c_{2n+1}| \to \sqrt{\alpha}$, we have only to verify the Tauberian condition $\sum n|\lambda_n|^2 < \infty$, where

$$\log \psi(r) = \sum_{n=1}^{\infty} \lambda_n r^n.$$

But a simple calculation shows that

$$\lambda_n = \gamma_n - \frac{1}{n},$$

where γ_n are the logarithmic coefficients of f. Thus if $\alpha > 0$, the Tauberian condition is again a consequence of Bazilevich's theorem, and the proof that $|c_{2n+1}| \to \sqrt{\alpha}$ is complete.

The lemma in §5.5 provides the remaining information that $\sqrt{\alpha} \leq 1$, with equality only if f is a rotation of the Koebe function; that is, only if h is a rotation of q.

The proof in the case $\alpha = 0$ is similar to that for the class S and is left as an exercise.

§5.8. Proof of Milin's Tauberian Theorem

We turn now to the proof of Milin's Tauberian theorem, used in the preceding section to prove Hayman's regularity theorem and its analogue for odd functions. The first two lemmas are familiar results concerning arbitrary series of complex numbers.

Lemma 1. *If the series $\sum_{k=1}^{\infty} c_k$ converges, then $\sum_{k=1}^{n} kc_k = o(n)$.*

Proof. Let $s_n = \sum_{k=1}^{n} c_k$, and suppose $s_n \to s$. A summation by parts (Abel summation) gives

$$\frac{1}{n}\sum_{k=1}^{n} kc_k = \frac{n+1}{n}s_n - \frac{1}{n}\sum_{k=1}^{n} s_k \to s - s = 0.$$

Lemma 2 (Fejér's Tauberian Theorem). *If $\sum_{k=1}^{\infty} k|\lambda_k|^2 < \infty$, then*

$$\left|\sum_{k=1}^{n} \lambda_k - \sum_{k=1}^{\infty} \lambda_k r^k\right| \to 0, \qquad r = 1 - \frac{1}{n}.$$

Proof. Write

$$\Delta_n(r) = \sum_{k=1}^{n} \lambda_k - \sum_{k=1}^{\infty} \lambda_k r^k = \sum_{k=1}^{n} \lambda_k(1 - r^k) - \sum_{k=n+1}^{\infty} \lambda_k r^k$$

and apply the Cauchy–Schwarz inequality to both series to obtain

$$|\Delta_n(r)| \le \left\{\sum_{k=1}^{n} k|\lambda_k|^2(1 - r^k)\right\}^{1/2}\left\{\sum_{k=1}^{n} \frac{1}{k}(1 - r^k)\right\}^{1/2} + \frac{1}{n}\left\{\sum_{k=n+1}^{\infty} k|\lambda_k|^2\right\}^{1/2}\left\{\sum_{k=n+1}^{\infty} kr^{2k}\right\}^{1/2}.$$

Now use the inequality $1 - r^k \le k(1 - r)$ and apply Lemma 1 to obtain

$$\left|\Delta_n\left(1 - \frac{1}{n}\right)\right| \le \left\{\frac{1}{n}\sum_{k=1}^{n} k^2|\lambda_k|^2\right\}^{1/2} + \left\{\sum_{k=n}^{\infty} k|\lambda_k|^2\right\}^{1/2} \to 0.$$

The next lemma is much more subtle. It is the main step in the proof of Milin's theorem. Again let $\{\beta_n\}$ be a complex sequence with $\beta_0 = 1$, for which

$$\psi(r) = \sum_{n=0}^{\infty} \beta_n r^n$$

converges in $0 < r < 1$, and let

$$\varphi(r) = \log \psi(r) = \sum_{n=1}^{\infty} \lambda_n r^n.$$

Let s_n and σ_n denote the partial sums and the Cesàro means of the series $\sum \beta_n$. Fejér's Tauberian theorem asserts that if $\sum n|\lambda_n|^2 < \infty$, then the partial sums and the Abel means of the series $\sum \lambda_n$ are equiconvergent. The next lemma asserts that if also the real parts of the partial sums of $\sum \lambda_n$ are bounded above, then the partial sums, the Cesàro means, and the Abel means of the series $\sum \beta_n$ are all equiconvergent.

Lemma 3. *If* $\sum_{n=1}^{\infty} n|\lambda_n|^2 < \infty$ *and* $\sup_n \operatorname{Re}\{\sum_{k=1}^{n} \lambda_k\} < \infty$, *then* $|s_n - \sigma_n| \to 0$ *and* $|s_n - \psi(r)| \to 0$, $r = 1 - 1/n$.

Proof. Differentiation of the equation $\psi(r) = e^{\varphi(r)}$ *gives* $\psi'(r) = \psi(r)\varphi'(r)$. By comparison of coefficients,

$$k\beta_k = \sum_{j=1}^{k} j\lambda_j \beta_{k-j}, \qquad k = 1, 2, \ldots .$$

Hence

$$\begin{aligned} s_n - \sigma_n &= \sum_{k=0}^{n} \left(1 - \frac{n-k+1}{n+1}\right)\beta_k = \frac{1}{n+1} \sum_{k=1}^{n} k\beta_k \\ &= \frac{1}{n+1} \sum_{k=1}^{n} \sum_{j=1}^{k} j\lambda_j \beta_{k-j} = \frac{1}{n+1} \sum_{j=1}^{n} j\lambda_j \sum_{k=j}^{n} \beta_{k-j} \\ &= \frac{1}{n+1} \sum_{j=1}^{n} j\lambda_j s_{n-j}. \end{aligned}$$

It follows from this representation that

$$|s_n - \sigma_n| \le \frac{1}{n} \left\{\sum_{j=1}^{n} j^2 |\lambda_j|^2\right\}^{1/2} \left\{\sum_{k=0}^{n} |s_k|^2\right\}^{1/2} = o\left(\frac{1}{\sqrt{n}}\right) \left\{\sum_{k=0}^{n} |s_k|^2\right\}^{1/2}, \tag{44}$$

by Lemma 1. On the other hand,

$$\sum_{n=0}^{\infty} s_n r^n = (1 - r)^{-1}\psi(r) = \exp\{\varphi(r) - \log(1 - r)\} = \exp\left\{\sum_{n=1}^{\infty} \left(\lambda_n + \frac{1}{n}\right) r^n\right\}.$$

Thus the third Lebedev–Milin inequality (§5.1) gives

$$|s_n|^2 \le \exp\left\{\sum_{n=1}^{n} k\left|\lambda_k + \frac{1}{k}\right|^2 - \sum_{k=1}^{n}\frac{1}{k}\right\}$$
$$= \exp\left\{\sum_{k=1}^{n} k|\lambda_k|^2 + 2\,\mathrm{Re}\left\{\sum_{k=1}^{n}\lambda_k\right\}\right\}.$$

But since by hypothesis the exponent is bounded above, this implies that the partial sums s_n are bounded. Hence the estimate (44) shows that $|s_n - \sigma_n| \to 0$.

Now consider the Abel means $\psi(r)$. Write

$$s_n - \psi(r) = \sum_{k=1}^{n}\beta_k(1 - r^k) - \sum_{k=n+1}^{\infty}\beta_k r^k = P_n(r) + Q_n(r). \tag{45}$$

With the notation

$$B_n = \sum_{k=1}^{n} k\beta_k \quad \text{and} \quad \rho_k = \frac{1}{k}(1 - r^k),$$

a summation by parts gives

$$P_n(r) = B_n\rho_{n+1} - \sum_{k=1}^{n} B_k(\rho_k - \rho_{k+1}).$$

But for each r in the interval $0 < r < 1$,

$$\rho_{k+1} < \rho_k < 1 - r, \qquad k = 1, 2, \ldots.$$

Thus

$$\left|P_n\left(1 - \frac{1}{n}\right)\right| \le \frac{1}{n}|B_n| + \frac{1}{n}\max_{1\le k\le n}|B_k| \le \frac{2}{n}\max_{1\le k\le n}|B_k|. \tag{46}$$

But it has already been proved that

$$\frac{1}{n+1}|B_n| = |s_n - \sigma_n| \to 0. \tag{47}$$

Let $\nu(n)$ be the smallest index for which

$$|B_{\nu(n)}| = \max_{1\le k\le n}|B_k|.$$

Then the sequence $\{v(n)\}$ is nondecreasing, so either it tends to infinity, or it is eventually constant. In either case, (46) and (47) show $P_n(1 - 1/n) \to 0$.

In order to estimate $Q_n(r)$, consider the doubly indexed expressions

$$C_n^k = \frac{1}{k}\sum_{j=n}^{k} j\beta_j, \qquad n \le k,$$

and observe that

$$k\beta_k = kC_n^k - (k-1)C_n^{k-1}.$$

Hence

$$\sum_{k=n+1}^{m} \beta_k r^k = \sum_{k=n+1}^{m} (C_n^k - C_n^{k-1})r^k + \sum_{k=n+1}^{m} \frac{1}{k} C_n^{k-1} r^k, \qquad n < m. \tag{48}$$

On the other hand,

$$kC_n^k = B_k - B_{n-1}, \qquad n \le k.$$

Since $B_n/n \to 0$, it follows that $C_n^k \to 0$ as $n \to \infty$ and $k \to \infty$ with $n \le k$. Thus for each $\varepsilon > 0$, there is an integer N such that

$$\left|\sum_{k=n+1}^{m} \frac{1}{k} C_n^{k-1} r^k\right| \le \varepsilon \sum_{k=n}^{m} \frac{r^k}{k} < \varepsilon, \qquad r = 1 - \frac{1}{n}, \tag{49}$$

whenever $m > n \ge N$. Finally, another summation by parts gives

$$\sum_{k=n+1}^{m} (C_n^k - C_n^{k-1})r^k = \sum_{k=n}^{m} C_n^k(r^k - r^{k+1}) + C_n^m r^{m+1} - C_n^n r^n,$$

whence it follows that

$$\left|\sum_{k=n+1}^{m} (C_n^k - C_n^{k-1})r^k\right| \le 3 \max_{n \le k \le m} |C_n^k| < 3\varepsilon \tag{50}$$

for all r, $0 < r < 1$, provided $m > n \ge N$. The relations (48), (49), and (50) show that $Q_n(1 - 1/n) \to 0$. In view of (45), this completes the proof of Lemma 3.

Proof of Milin's Tauberian Theorem. With the aid of Lemmas 2 and 3, it is now an easy matter to prove Milin's Tauberian theorem. If $|\psi(r)| \to \alpha$, then

$$\operatorname{Re}\left\{\sum_{n=1}^{\infty} \lambda_n r^n\right\} = \log|\psi(r)| \to \log \alpha \ge -\infty.$$

But $\sum_{n=1}^{\infty} n|\lambda_n|^2 < \infty$, by hypothesis. Thus by Lemma 2,

$$\operatorname{Re}\left\{\sum_{k=1}^{n} \lambda_k\right\} \to \log \alpha.$$

In particular, these expressions have a finite upper bound. Thus the hypotheses of Lemma 3 are satisfied, and it follows that $|s_n| \to \alpha$ and $|\sigma_n| \to \alpha$, as Milin's theorem asserts.

§5.9. Successive Coefficients

As a final application of Milin's method, we shall now prove the absolute boundedness of $||a_{n+1}| - |a_n||$, the difference of the moduli of successive coefficients of a function $f \in S$. This result was first obtained by Hayman [5] in 1963. (The previous history of the problem is discussed in §3.11.) A few years later, I. M. Milin [4, 6] found a simpler approach which also provided a better numerical bound. Ilina [1] then refined the argument and improved the bound to 4.26. More recently, Grinspan [2] modified Milin's proof and obtained the best bounds currently known:

$$-2.97 < |a_{n+1}| - |a_n| < 3.61, \qquad n = 1, 2, \ldots.$$

We shall follow the essential lines of Grinspan's argument, omitting the technicalities which provide the optimal estimates.

Theorem 5.9. *For each function $f(z) = \sum a_n z^n$ of class S,*

$$||a_{n+1}| - |a_n|| \leq A, \qquad n = 1, 2, \ldots,$$

where A is an absolute constant.

By way of orientation, recall that if f has positive Hayman index α and direction of maximal growth $e^{i\theta_0}$, then

$$|e^{i\theta_0}a_{n+1} - a_n| \to \alpha.$$

(See the proof of Hayman's regularity theorem in §5.7.) Thus for each $f \in S$ with $\alpha > 0$,

$$\limsup_{n\to\infty} ||a_{n+1}| - |a_n|| \leq \alpha \leq 1.$$

In fact, Hayman [1] showed that $d_n = ||a_{n+1}| - |a_n|| \to \alpha$ if $\alpha > 0$. For functions of slow growth ($\alpha = 0$), Eke [1, 2] proved that $d_n \to 0$ except for functions having "maximal growth on two rays." Hamilton [1] recently completed this program by proving that

$$\limsup_{n\to\infty} d_n \le 1$$

for every function (with $\alpha = 0$) having maximal growth on two rays. Equality actually occurs for functions of the form

$$f(z) = z(1 - e^{i\theta}z)^{-1}(1 - e^{i\varphi}z)^{-1}.$$

Nevertheless, the absolute bound on d_n cannot be reduced to 1, for reasons discussed in §3.11.

Theorem 5.9 is a consequence of the following rather technical lemma.

Lemma. *Let $h(z) = \sum_{n=0}^{\infty} c_n z^n$ be analytic and univalent in $\mathbb{D}$ and satisfy $|h(z)| \le M(r)$ in $|z| \le r < 1$. Let $\varphi(z) = \sum_{n=1}^{\infty} \alpha_n z^n$ be analytic in $\mathbb{D}$ and let*

$$h(z)e^{\varphi(z)} = \sum_{n=0}^{\infty} \lambda_n z^n, \qquad |z| < 1. \tag{51}$$

Finally, let

$$\sigma_n(r) = \frac{1}{n} \sum_{k=1}^{n} k^2 |\alpha_k|^2 r^{2k}.$$

Then for every $r < 1$ and for $n = 1, 2, \ldots,$

$$|\lambda_n| \le r^{-n} M(r)[1 + \sqrt{\sigma_n(r)}] \exp \tfrac{1}{2}\left\{ \sum_{k=1}^{n} \left(k|\alpha_k|^2 - \frac{1}{k} \right) - \sigma_n(r) + 1 \right\}.$$

Proof of Lemma. Let $e^{\varphi(z)} = \sum_{n=0}^{\infty} \beta_n z^n$. Then differentiation of (51) and comparison of coefficients gives

$$n\lambda_n = \sum_{k=0}^{n-1} (n-k) c_{n-k} \beta_k + \sum_{k=1}^{n} \lambda_{n-k} k \alpha_k, \qquad n = 1, 2, \ldots.$$

Therefore, by the Cauchy–Schwarz inequality,

$$\begin{aligned} n|\lambda_n| r^n \le{}& \left\{ \sum_{k=1}^{n} k^2 |c_k|^2 r^{2k} \right\}^{1/2} \left\{ \sum_{k=0}^{n-1} |\beta_k|^2 r^{2k} \right\}^{1/2} \\ &+ \left\{ \sum_{k=0}^{n-1} |\lambda_k|^2 r^{2k} \right\}^{1/2} \left\{ \sum_{k=1}^{n} k^2 |\alpha_k|^2 r^{2k} \right\}^{1/2}. \end{aligned} \tag{52}$$

But since

$$h(z) \sum_{k=0}^{n-1} \beta_k z^k = \sum_{k=0}^{n-1} \lambda_k z^k + \cdots,$$

it follows that

$$\sum_{k=0}^{n-1} |\lambda_k|^2 r^{2k} \le \frac{1}{2\pi r} \int_{|z|=r} |h(z)|^2 \left| \sum_{k=0}^{n-1} \beta_k z^k \right|^2 |dz|$$
$$\le [M(r)]^2 \sum_{k=0}^{n-1} |\beta_k|^2 r^{2k}.$$

On the other hand, because h is univalent, a comparison of areas gives

$$\pi \sum_{k=1}^{n} k^2 |c_k|^2 r^{2k} \le \pi n \sum_{k=1}^{\infty} k |c_k|^2 r^{2k} \le \pi n [M(r)]^2.$$

Introducing these inequalities into (52), we find

$$|\lambda_n| \le n^{-1/2} r^{-n} M(r)[1 + \sqrt{\sigma_n(r)}] \left\{ \sum_{k=0}^{n-1} |\beta_k|^2 r^{2k} \right\}^{1/2}.$$

The desired result now follows from the second Lebedev–Milin inequality, which can be expressed in the form

$$\sum_{k=0}^{n-1} |\beta_k|^2 \le n \exp\left\{ \sum_{k=1}^{n} \left(1 - \frac{k}{n}\right) k |\alpha_k|^2 - \sum_{k=1}^{n} \frac{1}{k} + 1 \right\}.$$

Proof of Theorem. Given $f \in S$, consider the identity

$$\left(\frac{1}{z} - \frac{1}{\zeta}\right) f(z) = \left\{1 - \frac{f(z)}{f(\zeta)}\right\} \frac{1/z - 1/\zeta}{g(1/z) - g(1/\zeta)}, \qquad |z| < 1, \quad |\zeta| < 1, \quad (53)$$

where $g(1/z) = 1/f(z)$. Fixing $r < 1$, let ζ be chosen as a point on the circle $|z| = r$ where $|f(z)|$ attains its maximum. Then $h(z) = 1 - f(z)/f(\zeta)$ is analytic and univalent, and $|h(z)| \le 2$ in $|z| \le r$. In view of the Grunsky expansion (11), the identity (53) can be recast in the form

$$\sum_{n=0}^{\infty} \lambda_n z^n = h(z) \exp\left\{ \sum_{n=1}^{\infty} A_n(\zeta) z^n \right\},$$

where

$$\lambda_n = a_{n+1} - \zeta^{-1} a_n, \qquad n = 1, 2, \ldots.$$

Consequently, the lemma gives

$$|a_{n+1} - \zeta^{-1}a_n| \le 2r^{-n}\mu \exp \tfrac{1}{2}\left\{\sum_{k=1}^{n} k|A_k(\zeta)|^2 - \sum_{k=1}^{n} \frac{1}{k} + 1\right\},$$

where

$$\mu = \max_{x \ge 0}(1 + x)e^{-x^2/2} < 1.337.$$

Now introduce the Grunsky inequality (12) and the elementary inequality (29) to conclude that

$$\sum_{k=1}^{n} k|A_k(\zeta)|^2 - \sum_{k=1}^{n} \frac{1}{k} \le -\log(1 - r^2) - \log(n + \tfrac{1}{2}) - \gamma,$$

where $\gamma = 0.577\ldots$ is Euler's constant. With the choice $r = 1 - 1/n$ ($n = 2, 3, \ldots$), it follows that

$$|a_{n+1} - \zeta^{-1}a_n| \le 8\mu \exp \tfrac{1}{2}\{1 - \gamma + \log \tfrac{8}{15}\} < 9.66.$$

Thus, in view of Littlewood's estimate $|a_n| < en$,

$$\begin{aligned} ||a_{n+1}| - |a_n|| &\le |a_{n+1} - \zeta^{-1}a_n| + |\zeta^{-1}a_n - |\zeta|\zeta^{-1}a_n| \\ &< 9.66 + 2e < 16, \qquad n = 2, 3, \ldots, \end{aligned}$$

since $|\zeta| = r = 1 - 1/n$. This proves the theorem.

For functions with large Hayman index α, Grinspan's estimate $A < 3.61$ can be improved. The following theorem of Duren [14] is a simple consequence of Milin's theory.

Theorem 5.10. *For each $f \in S$ with Hayman index $\alpha > 0$,*

$$||a_{n+1}| - |a_n|| \le e^{\delta}\alpha^{-1/2} < 1.37\alpha^{-1/2}, \qquad n = 1, 2, \ldots,$$

where δ is Milin's constant.

Proof. Let γ_k be the logarithmic coefficients of f and write

$$\begin{aligned} \log\left\{(1 - \zeta z)\frac{f(z)}{z}\right\} &= \log \sum_{k=0}^{\infty} (a_{k+1} - \zeta a_k)z^k \\ &= \sum_{k=1}^{\infty} \left(2\gamma_k - \frac{\zeta^k}{k}\right)z^k. \end{aligned}$$

Then by the third Lebedev–Milin inequality,

$$\begin{aligned}|a_{n+1} - \zeta a_n|^2 &\le \exp\left\{\sum_{k=1}^{n} k\left|2\gamma_k - \frac{\zeta^k}{k}\right|^2 - \sum_{k=1}^{n}\frac{1}{k}\right\}\\ &= \exp 2\left\{\sum_{k=1}^{n}\left(k|\gamma_k|^2 - \frac{1}{k}\right) + \sum_{k=1}^{n} k\left|\gamma_k - \frac{\zeta^k}{k}\right|^2\right\}.\end{aligned}$$

Now let $\bar{\zeta} = e^{i\theta_0}$ be the direction of maximal growth of f, and apply Milin's lemma (§5.4) and Bazilevich's theorem (§5.6) to obtain

$$|a_{n+1} - \zeta a_n| \le \exp\left\{\delta + \frac{1}{2}\log\frac{1}{\alpha}\right\} = e^{\delta}\alpha^{-1/2}.$$

A similar argument leads to a corresponding theorem for odd univalent functions (*cf.* I. M. Milin [4, 6]).

Theorem 5.11. *Let $f \in S$ have Hayman index $\alpha > 0$, and let*

$$h(z) = \sqrt{f(z^2)} = \sum_{n=0}^{\infty} c_{2n+1} z^{2n+1}$$

be its square-root transform. Then

$$\big||c_{2n+1}| - |c_{2n-1}|\big| < e^{-\gamma/2}\alpha^{-1/4}n^{-1/2}, \qquad n = 1, 2, \ldots,$$

where γ is Euler's constant.

Proof. Again let γ_k be the logarithmic coefficients and $\bar{\zeta} = e^{i\theta_0}$ the direction of maximal growth of f. In view of the identity

$$\begin{aligned}(1 - \zeta z)\frac{h(\sqrt{z})}{\sqrt{z}} &= 1 + \sum_{n=1}^{\infty}(c_{2n+1} - \zeta c_{2n-1})z^n\\ &= \exp\sum_{k=1}^{\infty}\left(\gamma_k - \frac{\zeta^k}{k}\right)z^k,\end{aligned}$$

the third Lebedev–Milin inequality gives

$$|c_{2n+1} - \zeta c_{2n-1}|^2 \le \exp\left\{\sum_{k=1}^{n} k\left|\gamma_k - \frac{\zeta^k}{k}\right|^2 - \sum_{k=1}^{n}\frac{1}{k}\right\}.$$

Thus it follows from Bazilevich's theorem and the elementary inequality (29) that

$$\begin{aligned} |c_{2n+1} - \zeta c_{2n-1}|^2 &\leq \exp\left\{\tfrac{1}{2}\log\frac{1}{\alpha} - \log(n+\tfrac{1}{2}) - \gamma\right\} \\ &= e^{-\gamma}\alpha^{-1/2}(n+\tfrac{1}{2})^{-1}, \end{aligned}$$

which gives the desired result.

It is not known whether the asymptotic estimate

$$\delta_n = \big||c_{2n+1}| - |c_{2n-1}|\big| = O(n^{-1/2})$$

holds generally for each $f \in S$. Certainly nothing better is true, as the example $[k(z^4)]^{1/4}$ shows, where k is the Koebe function. It is known that $\delta_n \to 0$, and Goluzin [19] showed $\delta_n = O(n^{-1/4}\log n)$. The best estimate currently known is $\delta_n = O(n^{1-\sqrt{2}})$, due to Lucas [1]. For functions $h(z) = [f(z^4)]^{1/4}$ with four-fold symmetry, Levin [2] proved long ago that $c_{4n+1} = O(n^{-1/2}\log n)$. It seems likely that the logarithmic factor is superfluous, but even this is an open question.

V. I. Milin [2] has recently shown that

$$\sum_{n=1}^{\infty} n^{-\varepsilon}\delta_n^2 < 50\sum_{n=1}^{\infty} n^{-1-\varepsilon} < \infty$$

for each $f \in S$ and for each $\varepsilon > 0$. This may be regarded as evidence in favor of the conjecture that $\delta_n = O(n^{-1/2+\varepsilon})$. For odd *starlike* functions, Milin proves $\delta_n < 40n^{-1/2}$.

§5.10. Successive Coefficients of Starlike Functions

Although it is not generally true that $\big||a_{n+1}| - |a_n|\big| \leq 1$ for all $f \in S$, the inequality does hold for starlike functions. This was conjectured by Pommerenke [12] and proved by Leung [1].

Theorem 5.12. *For every* $f \in S^*$,

$$\big||a_{n+1}| - |a_n|\big| \leq 1, \qquad n = 1, 2, \ldots .$$

Corollary 1. *For every* $f \in S^*$, $|a_n| \leq n$, $n = 2, 3, \ldots$.

Corollary 2. *For every odd function* $f \in S^*$, $|a_n| \leq 1$, $n = 3, 5, 7, \ldots$.

The second corollary goes back to Privalov [1]. Goluzin [3] showed more generally that $|a_n| \leq 1$, $n = 3, 4, \ldots$, for each $f \in S^*$ with $a_2 = 0$.

The proof of the theorem makes use of a lemma, due to MacGregor [3], concerning the class P of functions

$$\varphi(z) = 1 + \sum_{n=1}^{\infty} c_n z^n$$

of positive real part in $|z| < 1$.

Lemma. *Let $\varphi \in P$, and let $\lambda_n \geq 0$. Suppose $\psi(z) = \sum_{n=1}^{\infty} \lambda_n c_n z^n$ is analytic in $|z| < 1$ and $\mathrm{Re}\{\psi(z)\} \leq M$ for some $M > 0$. Then $\sum_{n=1}^{\infty} \lambda_n |c_n|^2 \leq 2M$.*

Proof of Lemma. Let $c_n = x_n + iy_n$ and write

$$u(r, \theta) = \mathrm{Re}\{\varphi(re^{i\theta})\} = 1 + \sum_{n=1}^{\infty} (x_n \cos n\theta - y_n \sin n\theta) r^n;$$

$$v(r, \theta) = \mathrm{Re}\{\psi(re^{i\theta})\} = \sum_{n=1}^{\infty} \lambda_n (x_n \cos n\theta - y_n \sin n\theta) r^n.$$

Then since $u(r, \theta) \geq 0$ and $v(r, \theta) \leq M$,

$$\begin{aligned} 2\pi M = M \int_0^{2\pi} u(r, \theta)\, d\theta &\geq \int_0^{2\pi} u(r, \theta) v(r, \theta)\, d\theta \\ &= \pi \sum_{n=1}^{\infty} \lambda_n (x_n^2 + y_n^2) r^{2n} = \pi \sum_{n=1}^{\infty} \lambda_n |c_n|^2 r^{2n}. \end{aligned}$$

Let $r \to 1$ to complete the proof.

Corollary. *For each $\varphi \in P$ and for each integer $n > 0$, there is a point ζ on the unit circle such that*

$$\sum_{k=1}^{n} \frac{1}{k} |c_k - \zeta^k|^2 \leq \sum_{k=1}^{n} \frac{1}{k}.$$

Proof of Corollary. Apply the lemma with $\lambda_k = 1/k$ for $1 \leq k \leq n$ and $\lambda_k = 0$ for $k > n$. This gives

$$\begin{aligned} \sum_{k=1}^{n} \frac{1}{k} |c_k - \zeta^k|^2 &= \sum_{k=1}^{n} \frac{1}{k} |c_k|^2 - 2\,\mathrm{Re}\{\psi(\bar{\zeta})\} + \sum_{k=1}^{n} \frac{1}{k} \\ &\leq 2M - 2\,\mathrm{Re}\{\psi(\bar{\zeta})\} + \sum_{k=1}^{n} \frac{1}{k}, \end{aligned}$$

where

$$\psi(z) = \sum_{k=1}^{n} \frac{1}{k} c_k z^k$$

and M is the maximum of $\mathrm{Re}\{\psi(z)\}$ on $|z| = 1$. Choosing ζ so that $\mathrm{Re}\{\psi(\bar{\zeta})\} = M$, we obtain the desired result.

Proof of Theorem. If $f \in S^*$, then $zf'(z)/f(z) = \varphi(z)$ for some $\varphi \in P$. Integration gives

$$\log \frac{f(z)}{z} = \int_0^z \frac{\varphi(t) - 1}{t} \, dt = \sum_{k=1}^{\infty} \frac{1}{k} c_k z^k.$$

Thus for $|\zeta| = 1$,

$$\log\left\{(1 - \zeta z) \frac{f(z)}{z}\right\} = \sum_{k=1}^{\infty} \alpha_k z^k, \quad \text{where } \alpha_k = \frac{1}{k}(c_k - \zeta^k).$$

On the other hand,

$$(1 - \zeta z) \frac{f(z)}{z} = \sum_{k=0}^{\infty} \beta_k z^k, \quad \text{where } \beta_k = a_{k+1} - \zeta a_k.$$

Since

$$\exp\left\{\sum_{k=1}^{\infty} \alpha_k z^k\right\} = \sum_{k=0}^{\infty} \beta_k z^k, \qquad \beta_0 = 1,$$

we may apply the third Lebedev–Milin inequality to conclude that

$$|a_{n+1} - \zeta a_n|^2 \le \exp\left\{\sum_{k=1}^{n} \frac{1}{k}(|c_k - \zeta^k|^2 - 1)\right\}$$

for every ζ with $|\zeta| = 1$. But by the corollary to the lemma, the exponent is nonpositive for some ζ. Hence $|a_{n+1} - \zeta a_n| \le 1$ for some ζ with $|\zeta| = 1$. This completes the proof, since

$$\big||a_{n+1}| - |a_n|\big| \le |a_{n+1} - \zeta a_n|$$

for all ζ with $|\zeta| = 1$.

§5.11. Exponentiation of the Goluzin Inequalities

I. M. Milin's estimate $|a_n| < 1.243\,n$ was the best known until 1972, when FitzGerald [1] introduced a new method to prove

$$|a_n| < \sqrt{\tfrac{7}{6}}n < 1.081\,n, \qquad n = 2, 3, \ldots,$$

for all functions $f(z) = \sum a_n z^n$ of class S. The general approach is somewhat similar to Milin's, insofar as Milin's method is based on an exponentiation of the Grunsky inequalities. FitzGerald's point of departure is an exponentiation of the closely related Goluzin inequalities (§4.4)

$$\left|\sum_{j=1}^{N}\sum_{k=1}^{N} \lambda_j \lambda_k \log \frac{z_j z_k [f(z_j) - f(z_k)]}{f(z_j) f(z_k)(z_j - z_k)}\right| \le \sum_{j=1}^{N}\sum_{k=1}^{N} \lambda_j \overline{\lambda_k} \log \frac{1}{1 - z_j \overline{z_k}}. \tag{54}$$

But now the exponentiation proceeds along different lines. The idea is to replace the inequalities (54) by the same inequalities with logarithms removed. This can be done by purely algebraic methods.

Our derivation of FitzGerald's coefficient estimate will rely only upon the Goluzin inequalities and will be otherwise self-contained. In particular, it will make no reference to Milin's theory.

We begin with some algebraic preliminaries. A real symmetric matrix $A = (a_{jk})$ is said to be *positive semidefinite* (written $A \ge 0$) if its associated quadratic form

$$\sum_{j=1}^{n}\sum_{k=1}^{n} a_{jk} x_j x_k$$

is nonnegative for all real numbers $x_1, \ldots, x_n$. This will be the case if and only if all of the (real) eigenvalues of A are nonnegative.

The *Hadamard product* (or *pointwise product*) of two $n \times n$ matrices $A = (a_{jk})$ and $B = (b_{jk})$ is defined as the $n \times n$ matrix $(a_{jk} b_{jk})$. In order to avoid confusion with the ordinary matrix product, we shall denote the Hadamard product of A and B by $A * B$. This operation is obviously commutative: $A * B = B * A$. It is also clear that $A * B$ is symmetric if A and B are. For any function f, we shall let $*f(A)$ denote the matrix $(f(a_{jk}))$. In particular, $*A^m = (a_{jk}^m)$, $m = 1, 2, \ldots$.

The following theorem, a classical result of Schur [1], will be our main tool in the exponentiation of the Goluzin inequalities.

Theorem 5.13 (Schur's Theorem). *If $A \ge 0$ and $B \ge 0$, then $A * B \ge 0$.*

Before passing to the proof, let us note the following corollary.

Corollary. *If $A \geq 0$ and $\varphi(z) = \sum_{\nu=0}^{\infty} c_\nu z^\nu$ is an entire function with nonnegative coefficients c_ν, then $*\varphi(A) \geq 0$.*

Proof of Corollary. By Schur's theorem, each of the matrices $*A^\nu$ ($\nu = 1, 2, \ldots$) is positive semidefinite. Clearly, the matrix I with 1 in every position is also positive semidefinite. Thus each of the partial sums

$$c_0 I + \sum_{\nu=1}^{m} c_\nu * A^\nu = \left(\sum_{\nu=0}^{m} c_\nu a_{jk}^\nu\right)$$

has the same property, since $c_\nu \geq 0$. Now let $m \to \infty$ to conclude that $*\varphi(A)$ is positive semidefinite.

Proof of Theorem. Every symmetric matrix can be diagonalized. Thus there exist an orthogonal matrix $U = (u_{jk})$ and a diagonal matrix $D = (\lambda_j \delta_{jk})$ such that

$$A = UDU',$$

where $U' = U^{-1}$ is the transpose of U and $\lambda_1, \lambda_2, \ldots, \lambda_n$ are the eigenvalues of A, repeated according to multiplicity. In terms of components,

$$a_{jk} = \sum_{i=1}^{n} \lambda_i u_{ji} u_{ki}, \qquad j, k = 1, 2, \ldots, n.$$

Hence the quadratic form associated with $A * B$ is

$$\sum_{j=1}^{n} \sum_{k=1}^{n} a_{jk} b_{jk} x_j x_k = \sum_{i=1}^{n} \lambda_i \sum_{j=1}^{n} \sum_{k=1}^{n} b_{jk}(u_{ji} x_j)(u_{ki} x_k) \geq 0,$$

since $B \geq 0$ and the hypothesis $A \geq 0$ implies that each of the eigenvalues λ_i is nonnegative.

We need to make one more general observation, which we state as a lemma.

Lemma. *If the real symmetric matrix $A = a_{jk}$ is positive semidefinite, then the Hermitian form*

$$\sum_{j=1}^{n} \sum_{k=1}^{n} a_{jk} \zeta_j \bar{\zeta}_k \geq 0$$

for all complex members $\zeta_1, \zeta_2, \ldots, \zeta_n$.

Proof. If $\zeta_j = x_j + iy_j$, the Hermitian form is simply

$$\sum_{j=1}^{n} \sum_{k=1}^{n} a_{jk}(x_j x_k + y_j y_k) \geq 0,$$

because

$$ia_{jk}(y_j x_k - x_j y_k) + ia_{kj}(y_k x_j - x_k y_j) = 0$$

by the symmetry of A.

We now return to the Goluzin inequalities (54). Choosing the parameters λ_j to be real numbers and using the fact that $-\mathrm{Re}\{\alpha\} \leq |\alpha|$ for any complex number α, we conclude that the symmetric matrix A with elements

$$a_{jk} = \log\left|\frac{z_j z_k[f(z_j) - f(z_k)]}{f(z_j)f(z_k)(z_j - z_k)(1 - z_j\overline{z_k})}\right|$$

is positive semidefinite. Applying the corollary to Schur's theorem with the entire function $\varphi(z) = e^{2z} - 1$, we deduce that the matrix C with elements

$$c_{jk} = e^{2a_{jk}} - 1$$

is positive semidefinite. Thus by the lemma,

$$\sum_{j=1}^{N} \sum_{k=1}^{N} c_{jk}\zeta_j\overline{\zeta_k} \geq 0$$

for all complex $\zeta_1, \zeta_2, \ldots, \zeta_N$. In other words,

$$\left|\sum_{j=1}^{N} \zeta_j\right|^2 \leq \sum_{j=1}^{N} \sum_{k=1}^{N} \left|\frac{z_j z_k[f(z_j) - f(z_k)]}{f(z_j)f(z_k)(z_j - z_k)(1 - z_j\overline{z_k})}\right|^2 \zeta_j\overline{\zeta_k}.$$

If the numbers ζ_j are expressed as

$$\zeta_j = \lambda_j\left|\frac{f(z_j)}{z_j}\right|^2, \qquad j = 1, 2, \ldots, N,$$

these inequalities assume the form

$$\left|\sum_{j=1}^{N} \lambda_j\left|\frac{f(z_j)}{z_j}\right|^2\right|^2 \leq \sum_{j=1}^{N} \sum_{k=1}^{N} \lambda_j\overline{\lambda_k}\left|\frac{f(z_j) - f(z_k)}{(z_j - z_k)(1 - z_j\overline{z_k})}\right|^2 \tag{55}$$

for all complex $\lambda_1, \lambda_2, \ldots, \lambda_N$.

These last inequalities (55) are the *exponentiated Goluzin inequalities*, from which FitzGerald's coefficient inequality will be deduced. As a matter of fact, this method of exponentiation is quite flexible and can be used to obtain other systems of inequalities similar to (55).

§5.12. FitzGerald's Theorem

We are now prepared to derive an important inequality which leads to the best currently available general estimate for the nth coefficient of a univalent function.

Theorem 5.14 (FitzGerald's Theorem). *For each function* $f(z) = \sum a_n z^n$ *of class* S,

$$|a_n|^4 \leq \sum_{k=1}^{n} k|a_k|^2 + \sum_{k=n+1}^{2n-1} (2n-k)|a_k|^2, \qquad n = 2, 3, \ldots. \tag{56}$$

Corollary. *For each* $f \in S$,

$$|a_n| \leq \sqrt{\tfrac{7}{6}}\,n < 1.081\,n, \qquad n = 2, 3, \ldots.$$

Deduction of Corollary. Let us begin by computing the right-hand side of the inequality (56) in the case where $a_k = k$. The familiar formulas

$$\sum_{k=1}^{n} k^2 = \tfrac{1}{6}n(n+1)(2n+1) \quad \text{and} \quad \sum_{k=1}^{n} k^3 = \tfrac{1}{4}n^2(n+1)^2$$

then give

$$\sum_{k=1}^{n} k^3 + \sum_{k=n+1}^{2n-1} (2n-k)k^2 = \tfrac{7}{6}n^4 - \tfrac{1}{6}n^2. \tag{57}$$

Now let

$$C = \sup_n \sup_{f \in S} \frac{|a_n|}{n}.$$

In other words, C is the "smallest" constant such that $|a_n| \leq Cn$ for all n and for all $f \in S$. Thus $1 \leq C \leq e$. Given $\varepsilon > 0$, choose an integer n and a function $f \in S$ such that $|a_n| > (C - \varepsilon)n$. Then by (56) and (57),

$$(C-\varepsilon)^4 n^4 < |a_n|^4 \leq C^2(\tfrac{7}{6}n^4 - \tfrac{1}{6}n^2),$$

which implies $(C - \varepsilon)^4 < \frac{7}{6}C^2$. Now let $\varepsilon \to 0$ to conclude that $C \leq \sqrt{\frac{7}{6}}$.

Proof of Theorem. The first step is to average the exponentiated Goluzin inequalities (55) to obtain

$$\left|\sum_{\nu=1}^{m}\lambda_\nu\int_0^{2\pi}\left|\frac{f(r_\nu e^{i\theta})}{r_\nu}\right|^2 d\theta\right|^2$$
$$\leq \sum_{\nu=1}^{m}\sum_{\mu=1}^{m}\lambda_\nu\overline{\lambda_\mu}\int_0^{2\pi}\int_0^{2\pi}\left|\frac{f(r_\nu e^{i\theta})-f(r_\mu e^{i\varphi})}{r_\nu e^{i\theta}-r_\mu e^{i\varphi}}\cdot\frac{1}{1-r_\nu r_\mu e^{i(\theta-\varphi)}}\right|^2 d\theta\, d\varphi \tag{58}$$

for arbitrary complex parameters λ_ν and distinct numbers r_ν, $0 < r_\nu < 1$. More specifically, apply (55) to the doubly-indexed sequence of points $r_\nu e^{i\theta_j}$, where $\theta_j = 2\pi j/N$, $j = 1, 2, \ldots, N$. This gives

$$\left|\sum_{\nu=1}^{m}\lambda_\nu\sum_{j=1}^{N}g_\nu(\theta_j)\right|^2 \leq \sum_{\nu,\mu=1}^{m}\lambda_\nu\overline{\lambda_\mu}\sum_{j,k=1}^{N}G_{\nu\mu}(\theta_j,\theta_k),$$

where

$$g_\nu(\theta) = \left|\frac{f(r_\nu e^{i\theta})}{r_\nu}\right|^2; \qquad G_{\nu\mu}(\theta,\varphi) = \left|\frac{f(r_\nu e^{i\theta})-f(r_\mu e^{i\varphi})}{r_\nu e^{i\theta}-r_\mu e^{i\varphi}}\cdot\frac{1}{1-r_\nu r_\mu e^{i(\theta-\varphi)}}\right|^2.$$

Now multiply by $(2\pi/N)^2$ and let $N \to \infty$ to obtain

$$\left|\sum_{\nu=1}^{m}\lambda_\nu\int_0^{2\pi}g_\nu(\theta)\,d\theta\right|^2 \leq \sum_{\nu,\mu=1}^{m}\lambda_\nu\overline{\lambda_\mu}\int_0^{2\pi}\int_0^{2\pi}G_{\nu\mu}(\theta,\varphi)\,d\theta\,d\varphi,$$

which is the same as (58).

The next step is to deduce the coefficient inequality (56) from (58). Since the subclass of univalent polynomials is dense in S, we may assume that f has the form

$$f(z) = \sum_{k=1}^{m} a_k z^k, \qquad a_1 = 1,$$

for some $m > 2n$. Given distinct numbers $r_1, r_2, \ldots, r_m$ with $0 < r_\nu < 1$, let the real numbers $\lambda_1, \lambda_2, \ldots, \lambda_m$ be defined by the conditions

$$\sum_{\nu=1}^{m}\lambda_\nu r_\nu^{2(k-1)} = \delta_{kn} = \begin{cases}1, & k = n\\ 0, & k \neq n\end{cases}, \tag{59}$$

for $k = 1, 2, \ldots, m$. Here the matrix of coefficients is a familiar (nonsingular) Vandermonde matrix, so these equations uniquely determine the parameters λ_ν.

With this choice of the λ_ν, the left-hand side of the inequality (58) is the square of

$$\sum_{\nu=1}^{m} \lambda_\nu \int_0^{2\pi} \left(\sum_{k=1}^{m} a_k r_\nu^{k-1} e^{ik\theta} \right) \left(\sum_{j=1}^{m} \overline{a}_j r_\nu^{j-1} e^{-ij\theta} \right) d\theta$$

$$= 2\pi \sum_{k=1}^{m} |a_k|^2 \sum_{\nu=1}^{m} \lambda_\nu r_\nu^{2(k-1)} = 2\pi |a_n|^2,$$

in view of the equations (59).

The right-hand side of (58) is more difficult to compute. First note that

$$\frac{f(z) - f(\zeta)}{z - \zeta} = \sum_{k=1}^{m} a_k \sum_{s=1}^{k} z^{s-1} \zeta^{k-s},$$

with the difference quotient understood to be a derivative when $z = \zeta$. Also,

$$\frac{1}{1 - z\overline{\zeta}} = \sum_{p=0}^{\infty} (z\overline{\zeta})^p.$$

Thus the double integral in the right-hand side of (58) reduces to

$$(2\pi)^2 \sum_{p=0}^{\infty} \sum_{q=0}^{\infty} \sum_{k=1}^{m} \sum_{j=1}^{m} \sum_{s=1}^{k} \sum_{t=1}^{j} a_k \overline{a}_j r_\nu^{p+q+s+t-2} r_\mu^{p+q+k+j-s-t},$$

the sum being restricted to indices for which

$$p - q + s - t = q - p + k - s - j + t = 0;$$

that is, for which $p + s = q + t$ and $k = j$. This six-fold sum reduces to

$$(2\pi)^2 \sum_{k=1}^{m} |a_k|^2 \sum_{s=1}^{k} \sum_{t=1}^{k} \sum_{p} r_\nu^{2(p+s-1)} r_\mu^{2(p+k-t)},$$

the inner sum being restricted to $p \geq 0$ and $p \geq t - s$, since $q \geq 0$. The right-hand side of (58) therefore becomes

$$(2\pi)^2 \sum_{k=1}^{m} |a_k|^2 \sum_{s=1}^{k} \sum_{t=1}^{k} \sum_{p} \sum_{\nu=1}^{m} \lambda_\nu r_\nu^{2(p+s-1)} \sum_{\mu=1}^{m} \lambda_\mu r_\mu^{2(p+k-t)},$$

where the sum on p is taken over the range $\max\{t - s, 0\} \le p < \infty$. But by the equations (59),

$$\sum_{\nu=1}^{m} \lambda_\nu r_\nu^{2(p+s-1)} = \delta_{p+s,n} \quad \text{for } 1 \le p + s \le m;$$

while

$$\sum_{\mu=1}^{m} \lambda_\mu r_\mu^{2(p+k-t)} = \delta_{p+k-t+1,n} \quad \text{for } 1 \le p + k - t + 1 \le m.$$

If r_ν is now replaced by εr_ν, $\nu = 1, 2, \ldots, m$, the preceding considerations show that the left-hand side of (58) becomes

$$(2\pi)^2 \varepsilon^{4(n-1)} |a_n|^4;$$

while the right-hand side becomes

$$(2\pi)^2 \varepsilon^{4(n-1)} \sum_{k=1}^{m} |a_k|^2 S_k + O(\varepsilon^{2m}),$$

where

$$S_k = \sum_{s=1}^{k} \sum_{t=1}^{k} \sum_{p} \delta_{p+s,n} \delta_{p+k-t+1,n},$$

the inner sum extending over those values of p for which

$$0 \le p \le m - s \quad \text{and} \quad t - s \le p \le m + t - k - 1. \tag{60}$$

Dividing by $(2\pi)^2 \varepsilon^{4(n-1)}$, letting $\varepsilon \to 0$, and recalling that $m > 2n$, we find that

$$|a_n|^4 \le \sum_{k=1}^{m} |a_k|^2 S_k.$$

In order to obtain the inequality (56), it now remains only to show that

$$S_k = \begin{cases} k, & 1 \le k \le n \\ 2n - k, & n + 1 \le k \le 2n - 1 \\ 0, & 2n \le k \le m. \end{cases}$$

Suppose first that $1 \le k \le n$. Then whenever $s + t = k + 1$, the equations

$$n = p + s = p + k - t + 1 \tag{61}$$

hold for some p in the range (60), since $m > 2n$. Thus

$$S_k = \sum_{s=1}^{k} 1 = k, \qquad 1 \le k \le n.$$

Next suppose $n + 1 \le k \le 2n - 1$, and let $s + t = k + 1$. Then the equations (61) will hold for some p in the range (60) if and only if $s \le n$ and $t \le n$. Thus

$$S_k = \sum_{s=k+1-n}^{n} 1 = 2n - k.$$

If $2n \le k \le m$, the same considerations show that $S_k = 0$. This completes the proof.

Refinements of the method lead to slight improvements of the constant $\sqrt{\frac{7}{6}}$. FitzGerald [1] established a more general version of the basic inequalities (58) which Horowitz [1] used to prove

$$|a_n| \le \left(\frac{209}{140}\right)^{1/6} n < 1.0691\, n, \qquad n = 2, 3, \ldots .$$

Horowitz [3] later made a further small improvement to

$$|a_n| < 1.0657\, n, \qquad n = 2, 3, \ldots .$$

Exercises

1. Generalize the first Lebedev–Milin inequality by showing that

$$\sum_{k=0}^{\infty} |\beta_k|^p \le \exp\left\{\sum_{k=1}^{\infty} k^{p-1} |\alpha_k|^p\right\}, \qquad 1 \le p < \infty.$$

2. Show that the logarithmic coefficients γ_n of each function $f \in S^*$ satisfy the inequality $|\gamma_n| \le 1/n$, $n = 1, 2, \ldots$. (*Suggestion*: Use the Herglotz representation of starlike functions.)

3. Complete the proof of Hayman's regularity theorem for odd functions (Theorem 5.8) by showing that if $h \in S^{(2)}$ and $(1 - r)M_\infty(r, h) \to 0$, then $|c_{2n+1}| \to 0$.

4. (a) Show that if the image of a function $f \in S$ has finite area, then f has Hayman index $\alpha = 0$.
   (b) More generally, show that $\alpha = 0$ if $A_r = o((1-r)^{-3})$, where A_r is the area of the image of the disk $|z| \le r$ under the mapping f.

5. Show that for $0 < \theta < \pi$, the function

$$f(z) = \frac{z}{1 - 2z\cos\theta + z^2} = \sum_{n=1}^{\infty} \frac{\sin n\theta}{\sin\theta} z^n$$

is of class S and has Hayman index $\alpha = 0$, yet for certain values of θ, $\limsup_{n\to\infty} \| a_{n+1}| - |a_n\| > 0$.

6. Suppose that a function $f \in S$ is meromorphic at $z = 1$, and so has a pole of order at most two there. Suppose also that for each $\varepsilon > 0$, the set of points $z \in \mathbb{D}$ with $|z - 1| > \varepsilon$ is mapped by f onto a region of finite area. (This will be true, for instance, if f maps $\mathbb{D}$ onto the complement of an arc analytic at infinity.) Show that the coefficients of f satisfy

$$a_n = \lambda n + \mu + o(n^{-1/2}), \qquad n \to \infty,$$

for some complex constants λ and μ with $|\lambda| = \alpha \ge 0$. (Duren [9].)

7. Let a function $f \in S$ be analytic in the closed disk $\overline{\mathbb{D}}$ except for a pole of order at most two at $z = 1$. (This will be true, for instance, if f maps $\mathbb{D}$ onto the complement of an analytic arc.) Show that

$$a_n = \lambda n + \mu + O(\rho^{-n}), \qquad n \to \infty,$$

for some complex constants λ and μ with $|\lambda| = \alpha \ge 0$, and for some $\rho > 1$.

8. Let S_α denote the class of functions in S with Hayman index α. For $0 < \alpha < 1$, show that the function

$$f(z) = [z + (\alpha - 1)z^2](1 - z)^{-2}$$

belongs to S_α.

9. Let $f_n \in S_\alpha$ for some fixed α in the interval $0 \le \alpha < 1$ and for $n = 1, 2, \ldots$. Suppose that $f_n \to f$ uniformly on compact subsets of $\mathbb{D}$ as $n \to \infty$. Show that $f \in S_\beta$ for some $\beta \ge \alpha$. (*Note*: The converse is deeper. Lebedev [6] proved that for $0 \le \alpha < \beta \le 1$, each $f \in S_\beta$ can be approximated uniformly on compact subsets of $\mathbb{D}$ by functions $g \in S_\alpha$.)

10. Prove that if $f \in S_\alpha$, then

$$\lim_{r\to 1}(1 - r)M_1(r, f) = \frac{\alpha}{2}, \qquad 0 \le \alpha \le 1.$$

(*Suggestion*: Express the integral mean in terms of the coefficients of the square-root transform of f.) (Hayman [1], I. M. Milin [6].)

11. Prove that if $f \in S_\alpha$, then

$$\lim_{r \to 1}(1 - r)M_\infty(r, f') = 2\alpha, \qquad 0 \le \alpha \le 1.$$

(Krzyż [1], Bazilevich [7].)

Chapter 6

Subordination

The concept of subordination can be traced to Lindelöf [1], but Littlewood [1, 3] and Rogosinski [6, 7] introduced the term and discovered the basic relations. Over the years a substantial theory has been developed, and subordination now plays an important role in complex analysis.

The general question to be considered in this chapter is the extent to which the superordinate function "dominates" the subordinate function. We begin with Littlewood's theorem on the domination of integral means and derive Rogosinski's closely related result on the average domination of coefficients. This is followed by a discussion of the Rogosinski conjecture, also known as the generalized Bieberbach conjecture, for functions subordinate to *univalent* functions. In preparation for the deeper results on pointwise domination, or majorization, we then digress to establish some sharpened versions of the Schwarz lemma. The chapter concludes with Goluzin's applications of the Loewner method to the case in which the subordinate function is also univalent.

§6.1. Basic Principles

Let $f(z)$ and $g(z)$ be analytic in the unit disk $\mathbb{D}$, with $f(0) = g(0)$. Suppose for the moment that f is univalent, and that the range of g is contained in the range of f. Then

$$\omega(z) = f^{-1}(g(z))$$

is analytic in $\mathbb{D}$, $\omega(0) = 0$, and $|\omega(z)| < 1$. Thus by the Schwarz lemma, $|\omega(z)| < |z|$ for $0 < |z| < 1$ unless ω is simply a rotation of the disk. In general, an analytic function g is said to be *subordinate* to an analytic function f (written $g \prec f$) if

$$g(z) = f(\omega(z)), \qquad |z| < 1,$$

for some analytic function ω with $|\omega(z)| \leq |z|$. The *superordinate* function f need not be univalent.

If $g \prec f$, it is clear that $|g'(0)| \leq |f'(0)|$, and that the image under f of each disk $|z| \leq r < 1$ contains the image under g of the same disk. This is sometimes called the *subordination principle*, or *Lindelöf's principle*. In particular, the maximum modulus of f dominates that of g:

$$M_\infty(r, g) \leq M_\infty(r, f), \qquad 0 \leq r < 1.$$

Littlewood [1] generalized this to integral means of order p. Recall the notation

$$M_p(r, f) = \left\{ \frac{1}{2\pi} \int_0^{2\pi} |f(re^{i\theta})|^p \, d\theta \right\}^{1/p}, \qquad 0 < p < \infty.$$

Theorem 6.1 (Littlewood's Subordination Theorem). *Let f and g be analytic in the unit disk, and suppose $g \prec f$. Then for $0 < p < \infty$,*

$$M_p(r, g) \leq M_p(r, f), \qquad 0 \leq r < 1.$$

Strict inequality holds for $0 < r < 1$ unless f is constant or $\omega(z) = \alpha z$, $|\alpha| = 1$.

Proof. The theorem is actually a special case of a more general result on subharmonic functions. Let u be subharmonic in $\mathbb{D}$, and let $v(z) = u(\omega(z))$, where ω is analytic and $|\omega(z)| \leq |z|$. Then we claim

$$\int_0^{2\pi} v(re^{i\theta}) \, d\theta \leq \int_0^{2\pi} u(re^{i\theta}) \, d\theta.$$

Since $|f(z)|^p$ is subharmonic ($0 < p < \infty$) if $f(z)$ is analytic, the theorem will follow. To prove the inequality, fix r ($0 < r < 1$) and let $U(z)$ be the function harmonic in $|z| < r$ and equal to $u(z)$ on $|z| = r$. Then $u(z) \leq U(z)$ in $|z| \leq r$, and $v(z) \leq V(z) = U(\omega(z))$ on $|z| = r$. Thus

$$\begin{aligned} \frac{1}{2\pi} \int_0^{2\pi} v(re^{i\theta}) \, d\theta &\leq \frac{1}{2\pi} \int_0^{2\pi} V(re^{i\theta}) \, d\theta = V(0) \\ &= U(0) = \frac{1}{2\pi} \int_0^{2\pi} U(re^{i\theta}) \, d\theta = \frac{1}{2\pi} \int_0^{2\pi} u(re^{i\theta}) \, d\theta. \end{aligned}$$

If equality holds for some r, $0 < r < 1$, then $v(z) = V(z)$, or $u(\omega(z)) = U(\omega(z))$, on the entire circle $|z| = r$. Unless $\omega(z) = \alpha z$ with $|\alpha| = 1$, this implies that the subharmonic function $u(z)$ agrees with its harmonic majorant $U(z)$ at some interior points of the disk $|z| \leq r$; hence the two functions agree throughout this disk, by the maximum principle. This shows that if ω is not a rotation, equality can occur in Littlewood's theorem only if $u(z) = |f(z)|^p$ is harmonic in $|z| < r$. But this implies that f is constant. Indeed, if

$|f(z)|^p$ is harmonic and $f \neq 0$, it follows from the maximum principle that $f(z) \neq 0$ in $|z| < r$, so that a branch of $[f(z)]^p$ is analytic there. Hence

$$[f(0)]^p = \frac{1}{2\pi} \int_0^{2\pi} [f(\rho e^{i\theta})]^p \, d\theta, \qquad 0 < \rho < r;$$

and

$$|f(0)|^p = \frac{1}{2\pi} \int_0^{2\pi} |f(\rho e^{i\theta})|^p \, d\theta,$$

since $|f(z)|^p$ is harmonic in $|z| \le \rho$. This shows that $f(z)$ has constant argument on $|z| = \rho$, and is therefore constant.

This proof is essentially due to F. Riesz [1].

§6.2. Coefficient Inequalities

If $g(z) = \sum b_n z^n$ is subordinate to $f(z) = \sum a_n z^n$, it does not follow that $|b_n| \le |a_n|$ for all n. For example, z^2 is subordinate to z. Nevertheless, the coefficients of f always dominate those of g in a certain average sense. The basic result of this type is due to Rogosinski [6, 7].

Theorem 6.2 (Rogosinski's Theorem). *Let $f(z) = \sum_{n=1}^{\infty} a_n z^n$ and $g(z) = \sum_{n=1}^{\infty} b_n z^n$ be analytic in $\mathbb{D}$, and suppose $g \prec f$. Then*

$$\sum_{k=1}^{n} |b_k|^2 \le \sum_{k=1}^{n} |a_k|^2, \qquad n = 1, 2, \ldots.$$

Proof. Let $s_n(z) = \sum_{k=1}^{n} a_k z^k$, and write $f(z) = s_n(z) + r_n(z)$. Also let $t_n(z) = \sum_{k=1}^{n} b_k z^k$. Then

$$g(z) = f(\omega(z)) = s_n(\omega(z)) + r_n(\omega(z)).$$

Since $\omega(0) = 0$, it follows that

$$s_n(\omega(z)) = t_n(z) + \sum_{k=n+1}^{\infty} c_k z^k$$

for some c_k, so that by Parseval's relation,

$$\frac{1}{2\pi} \int_0^{2\pi} |s_n(\omega(re^{i\theta}))|^2 \, d\theta = \sum_{k=1}^{n} |b_k|^2 r^{2k} + \sum_{k=n+1}^{\infty} |c_k|^2 r^{2k}.$$

But $s_n \circ \omega \prec s_n$, so by Littlewood's theorem,

$$\frac{1}{2\pi}\int_0^{2\pi} |s_n(\omega(re^{i\theta}))|^2\, d\theta \le \frac{1}{2\pi}\int_0^{2\pi} |s_n(re^{i\theta})|^2\, d\theta = \sum_{k=1}^{n} |a_k|^2 r^{2k}.$$

Thus $\sum_{k=1}^{n} |b_k|^2 r^{2k} \le \sum_{k=1}^{n} |a_k|^2 r^{2k}$, and the proof is completed by letting r tend to 1.

Corollary. *If $a_n = O(1)$, then $b_n = O(\sqrt{n})$ as $n \to \infty$.*

This apparently crude estimate is actually best possible: b_n need not be $o(\sqrt{n})$. Rogosinski [7] constructed an example to demonstrate this remarkable fact. It is perhaps equally remarkable that Rogosinski's theorem holds only for squares and is false for all exponents $p \ne 2$. (See Exercises 5 and 6.)

Goluzin [22] observed that Rogosinski's theorem can be applied to obtain the following more general result, which has some interesting geometric consequences.

Theorem 6.3. *If $g \prec f$ and $\lambda_1 \ge \lambda_2 \ge \cdots \ge 0$, then*

$$\sum_{k=1}^{\infty} \lambda_k |b_k|^2 \le \sum_{k=1}^{\infty} \lambda_k |a_k|^2.$$

If $\lambda_1 > \lambda_2$ and $a_1 \ne 0$, finite equality occurs only for $g(z) = f(\alpha z)$, $|\alpha| = 1$.

Proof. With the notation $A_n = \sum_{k=1}^{n} |a_k|^2$ and $B_n = \sum_{k=1}^{n} |b_k|^2$, a summation by parts and Rogosinski's theorem give

$$\begin{aligned}\sum_{k=1}^{n} \lambda_k |b_k|^2 &= \sum_{k=1}^{n-1} (\lambda_k - \lambda_{k+1}) B_k + \lambda_n B_n \\ &\le \sum_{k=1}^{n-1} (\lambda_k - \lambda_{k+1}) A_k + \lambda_n A_n = \sum_{k=1}^{n} \lambda_k |a_k|^2.\end{aligned}$$

If $\lambda_1 > \lambda_2$, then equality implies $|a_1| = |b_1|$, or $0 \ne |f'(0)| = |g'(0)|$. Thus $|\omega'(0)| = 1$, and $\omega(z) = \alpha z$ with $|\alpha| = 1$, by the Schwarz lemma.

Corollary 1. *If $g \prec f$, then*

$$\iint_{|z| \le r} |g'(\rho e^{i\theta})|^2 \rho\, d\rho\, d\theta \le \iint_{|z| \le r} |f'(\rho e^{i\theta})|^2 \rho\, d\rho\, d\theta, \qquad 0 < r \le 1/\sqrt{2}.$$

If equality occurs for some $r < 1/\sqrt{2}$, *then* $g(z) = f(\alpha z)$ *for* $|\alpha| = 1$. *The constant* $1/\sqrt{2}$ *is best possible.*

Proof. The integrals represent the areas of the (multisheeted) images of $|z| < r$ under f and g. In terms of coefficients, the inequality takes the form

$$\sum_{k=1}^{\infty} k|b_k|^2 r^{2k} \leq \sum_{k=1}^{\infty} k|a_k|^2 r^{2k},$$

which follows from the theorem if the sequence $\{kr^{2k}\}$ is nonincreasing:

$$(k+1)r^{2(k+1)} \leq kr^{2k}, \qquad k = 1, 2, \ldots;$$

that is, if

$$r \leq \left(\frac{k}{k+1}\right)^{1/2}, \qquad k = 1, 2, \ldots.$$

Since the last bound is smallest for $k = 1$, the condition is equivalent to $r \leq 1/\sqrt{2}$. The example $f(z) = z$, $g(z) = z^2$ shows that the inequality may fail for $r > 1/\sqrt{2}$.

Corollary 2. *If* $g \prec f$, *then*

$$M_2(r, g') \leq M_2(r, f'), \qquad 0 < r \leq \tfrac{1}{2}.$$

If equality occurs for some $r < \frac{1}{2}$, *then* $g(z) = f(\alpha z)$ *for* $|\alpha| = 1$. *The constant* $\frac{1}{2}$ *is best possible.*

Proof. In terms of coefficients, it is to be shown that

$$\sum_{k=1}^{\infty} k^2|b_k|^2 r^{2(k-1)} \leq \sum_{k=1}^{\infty} k^2|a_k|^2 r^{2(k-1)}.$$

The proof is similar to that of Corollary 1.

With additional information on the superordinate function f, one can obtain general estimates on the individual coefficients of the subordinate function g. For example, if f is convex or starlike, we have the following result.

Theorem 6.4. *If $g \prec f$, and*

(i) *if $f \in C$, then $|b_n| \leq 1$, $n = 1, 2, \ldots$;*
(ii) *if $f \in S^*$, then $|b_n| \leq n$, $n = 1, 2, \ldots$.*

The proof will make use of a general inequality for complex polynomials, due to S. Bernstein.

Lemma (Bernstein's Theorem). *Let $P(z)$ be a polynomial of degree at most n, with $|P(z)| \leq 1$ in $|z| \leq 1$. Then $|P'(z)| \leq n$ in $|z| \leq 1$.*

Proof of Lemma. A well-known theorem of Lucas asserts that the convex hull of the zeros of a polynomial contains the zeros of the derivative. (This is easily proved by factoring the polynomial Q in terms of its zeros and considering Q'/Q; see the book of M. Marden [1].) For a fixed complex number α with $|\alpha| > 1$, consider

$$Q(z) = P(z) - \alpha z^n.$$

Since the function $z^{-n}P(z)$ is analytic in $1 \leq |z| \leq \infty$ and $|z^{-n}P(z)| \leq 1$ on $|z| = 1$, it follows from the maximum modulus theorem that the inequality holds throughout $|z| \geq 1$. In other words, all of the zeros of Q lie in $|z| < 1$. By Lucas' theorem, then, the same is true of

$$Q'(z) = P'(z) - \alpha n z^{n-1}.$$

The fact that $Q'(z)$ never vanishes in $|z| \geq 1$ for any α with $|\alpha| > 1$ implies $|P'(z)z^{-n+1}| \leq n$ in $|z| \geq 1$. In particular, $|P'(z)| \leq n$ on $|z| = 1$, which gives the desired result.

Proof of Theorem. (i) Suppose $f \in C$, and let D be the image of the unit disk under f. Let $\varepsilon_k = e^{2\pi ik/n}$ denote the nth roots of unity, $k = 1, 2, \ldots, n$. Then $g(\varepsilon_k z) \in D$ for $|z| < 1$, $k = 1, 2, \ldots, n$. Since D is convex, it follows that

$$\varphi(z^n) = \frac{1}{n}\sum_{k=1}^{n} g(\varepsilon_k z) = b_n z^n + \cdots \in D, \qquad |z| < 1.$$

This last expression is an analytic function of z^n, since $\sum_{k=1}^{n} \varepsilon_k^m = 0$ unless m is a multiple of n. Thus $\varphi \prec f$, and $|b_n| = |\varphi'(0)| \leq |f'(0)| = 1$, by the Schwarz lemma.

(ii) If $f \in S^*$, there exists $h \in C$ such that $f(z) = zh'(z)$. Let $g(z) = f(\omega(z))$. Then for arbitrary fixed ζ, $|\zeta| \leq 1$,

$$h(\zeta\omega(z)) = \sum_{n=1}^{\infty} P_n(\zeta)z^n \prec h(\zeta z),$$

so $|P_n(\zeta)| \leq 1$, by part (i). Since $P_n(\zeta)$ is a polynomial of degree n in ζ, it follows from Bernstein's theorem that $|P_n'(\zeta)| \leq n$ for $|\zeta| \leq 1$. But $b_n = P_n'(1)$, because a differentiation with respect to ζ gives

$$f(\zeta\omega(z)) = \zeta\omega(z)h'(\zeta\omega(z)) = \zeta \sum_{n=1}^{\infty} P_n'(\zeta)z^n.$$

Thus $|b_n| \leq n$, $n = 1, 2, \ldots$.

Theorem 6.4 is due to Rogosinski [7], who also showed that $|b_n| \leq n$ if f is typically real. More recently, Robertson [8] proved it under the assumption that f is close-to-convex. These results suggest the following conjecture, known as the *generalized Bieberbach conjecture* or the *Rogosinski conjecture.*

Rogosinski Conjecture. *If $g(z) = \sum_{n=1}^{\infty} b_n z^n$ is analytic in $\mathbb{D}$ and $g \prec f$ for some $f \in S$, then $|b_n| \leq n$ for $n = 1, 2, \ldots$.*

The Rogosinski conjecture is trivially true for $n = 1$, and Littlewood [1] proved it for $n = 2$, but the general conjecture remains an open problem. It follows from a result of Baernstein [1] on integral means that if $g \prec f \in S$, then $|b_n| < (e/2)n$ for all n (see Chapter 7, Exercise 1). V. I. Milin [1] has improved this estimate to $|b_n| < 1.21\, n$.

It is clear that the Rogosinski conjecture implies the Bieberbach conjecture. Robertson [11] observed that the Rogosinski conjecture is implied by Robertson's conjecture (§3.10) on the coefficients of odd univalent functions. More precisely, the following theorem is true.

Theorem 6.5. *Suppose $f \in S$ and $g \prec f$, and let $h(z) = \sqrt{f(z^2)} = z + c_3 z^3 + c_5 z^5 + \cdots$. Then*

$$|b_n| \leq \sum_{k=1}^{n} |c_{2k-1}|^2, \qquad n = 1, 2, \ldots.$$

Remark. Robertson's conjecture asserts that

$$\sum_{k=1}^{n} |c_{2k-1}|^2 \leq n, \qquad n = 1, 2, \ldots,$$

for each $f \in S$. Thus the theorem allows the deduction of Rogosinski's conjecture from Robertson's for each n. Since Robertson's conjecture has been proved up to $n = 4$ (see §3.10), it follows that Rogosinski's conjecture is true for $n \leq 4$. For $n > 4$ both conjectures remain unsettled.

Proof of Theorem. Let $g(z) = f(\omega(z))$ and

$$\varphi(z) = \frac{h(\sqrt{z})}{\sqrt{z}} = 1 + c_3 z + c_5 z^2 + \cdots.$$

Then $[\varphi(z)]^2 = f(z)/z$ and

$$g(z) = \omega(z)\{1 + c_3\omega(z) + c_5[\omega(z)]^2 + \cdots\}^2.$$

Let

$$s_n(z) = \sum_{k=1}^{n} c_{2k-1} z^{k-1}$$

denote the nth partial sum of φ. Then because $\omega(0) = 0$, the Cauchy representation for the coefficients of g can be modified to

$$b_n = \frac{1}{2\pi i}\int_{|z|=r} \frac{\omega(z)[s_n(\omega(z))]^2}{z^{n+1}}\,dz.$$

Since $s_n \circ \omega \prec s_n$, it follows from Littlewood's subordination theorem that

$$|b_n| \le r^{-n} M_2^2(r, s_n \circ \omega) \le r^{-n} M_2^2(r, s_n)$$
$$= r^{-n} \sum_{k=1}^{n} |c_{2k-1}|^2 r^{2k-2}, \qquad r < 1.$$

The proof is completed by letting r tend to 1.

By way of summary, it is interesting to note that six of the coefficient conjectures we have discussed are related by a chain of implications:

MILIN CONJECTURE $\Rightarrow$ ROBERTSON CONJECTURE $\Rightarrow$ ROGOSINSKI CONJECTURE

$\Rightarrow$ BIEBERBACH CONJECTURE $\Rightarrow$ ASYMPTOTIC BIEBERBACH CONJECTURE

$\Leftrightarrow$ LITTLEWOOD CONJECTURE.

§6.3. Sharpened Forms of the Schwarz Lemma

The deeper results in subordination theory require a detailed study of bounded analytic functions. Let f be analytic and satisfy $|f(z)| < 1$ for $|z| < 1$. (At the risk of confusion, we will switch notation in this section only and call such a function f instead of ω.) If $f(0) = 0$, the Schwarz lemma says that $|f(z)| \leq |z|$ and $|f'(0)| \leq 1$. Essentially the same argument, based on the maximum modulus theorem, shows in general that

$$\left|\frac{f(z) - f(\zeta)}{1 - \overline{f(\zeta)}f(z)}\right| \leq \left|\frac{z - \zeta}{1 - \overline{\zeta}z}\right| \tag{1}$$

for all z and ζ in the disk. In particular,

$$|f'(z)| \leq \frac{1 - |f(z)|^2}{1 - |z|^2}. \tag{2}$$

A more careful analysis leads to sharper results on the region of values of both $f(z_0)$ and $f'(z_0)$ at a fixed point z_0 in the disk. We begin with the derivative and state a result essentially due to Dieudonné [3].

Dieudonné's Lemma. *Let z_0 and w_0 be given points in $\mathbb{D}$, with $z_0 \neq 0$. Then for all functions f analytic and satisfying $|f(z)| < 1$ in $\mathbb{D}$, with $f(0) = 0$ and $f(z_0) = w_0$, the region of values of $f'(z_0)$ is the closed disk*

$$\left|w - \frac{w_0}{z_0}\right| \leq \frac{|z_0|^2 - |w_0|^2}{|z_0|(1 - |z_0|^2)}.$$

Proof. According to the inequality (1), the function g defined by

$$\frac{f(z) - w_0}{1 - \overline{w_0}f(z)} = \frac{z - z_0}{1 - \overline{z_0}z}g(z)$$

satisfies $|g(z)| \leq 1$ in $|z| < 1$. Thus an application of (1) to g gives

$$\left|\frac{g(z_0) - g(0)}{1 - \overline{g(0)}g(z_0)}\right| \leq |z_0|.$$

But since $f(0) = 0$ implies $g(0) = w_0/z_0$, this inequality shows (see Titchmarsh [1], p. 192) that $g(z_0)$ lies in the closed circular disk with center γ and radius ρ given by

$$\gamma = \frac{w_0}{z_0}\cdot\frac{1 - |z_0|^2}{1 - |w_0|^2}, \qquad \rho = \frac{|z_0|^2 - |w_0|^2}{|z_0|(1 - |w_0|^2)}.$$

Since

$$f'(z_0) = \frac{1 - |w_0|^2}{1 - |z_0|^2}g(z_0),$$

this is equivalent to saying that $f'(z_0)$ lies in the disk given in the statement of the lemma.

To show that the whole disk is covered, let

$$\beta = \frac{\alpha - w_0}{1 - \overline{w_0}\alpha}, \qquad |\alpha| < 1,$$

and let $f(z)$ be defined implicitly by the equation

$$\frac{f(z) - w_0}{1 - \overline{w_0} f(z)} = \frac{z - z_0}{1 - \overline{z_0}z} \cdot \frac{w_0 + \overline{z_0}\beta z}{z_0 + \overline{w_0}\beta z}.$$

Each of the factors on the right-hand side of this equation has modulus less than one for $|z| < 1$, so it follows that $|f(z)| < 1$. The conditions $f(0) = 0$ and $f(z_0) = w_0$ are obvious, and a calculation gives

$$f'(z_0) = \frac{w_0}{z_0} + \frac{|z_0|^2 - |w_0|^2}{z_0(1 - |z_0|^2)}\alpha.$$

Since α is an arbitrary point with $|\alpha| < 1$, this completes the proof.

Corollary. *If f is analytic in $\mathbb{D}$, $|f(z)| < 1$, and $f(0) = 0$, then*

$$|f'(z)| \le \begin{cases} 1, & r = |z| \le \sqrt{2} - 1 \\ \dfrac{(1 + r^2)^2}{4r(1 - r^2)}, & r \ge \sqrt{2} - 1. \end{cases}$$

This bound is sharp for each $r < 1$.

Proof. For fixed z, let $r = |z|$ and $R = |f(z)|$. Then $R \le r$, and the theorem gives

$$|f'(z)| \le \frac{R}{r} + \frac{r^2 - R^2}{r(1 - r^2)} = \frac{\psi(R)}{r(1 - r^2)},$$

where

$$\psi(R) = -R^2 + (1 - r^2)R + r^2.$$

But $\psi(R)$ reaches its maximum at $R = \frac{1}{2}(1 - r^2)$, which is less than r if and only if $r > \sqrt{2} - 1$. In this case, the sharp upper bound for $|f'(z)|$ is

$$\frac{\psi(\frac{1}{2}(1 - r^2))}{r(1 - r^2)} = \frac{(1 + r^2)^2}{4r(1 - r^2)}.$$

For $r \le \sqrt{2} - 1$, $\psi(R) \le \psi(r) = r(1 - r^2)$ in the interval $0 \le R \le r$, so that $|f'(z)| \le 1$.

The next result, due to Rogosinski [4], may also be viewed as a sharpened form of the Schwarz lemma. For a fixed point z_0 with $0 < |z_0| < 1$, let Δ_{z_0} denote the closed region containing the disk $|w| \le |z_0|^2$ and bounded by an arc of the circle $|w| = |z_0|^2$ and the two circular arcs γ_{z_0} and $\tilde{\gamma}_{z_0}$ joining z_0 to the respective points $i|z_0|z_0$ and $-i|z_0|z_0$, and tangent to the circle $|w| = |z_0|^2$ at these two points. (See Figure 6.1.)

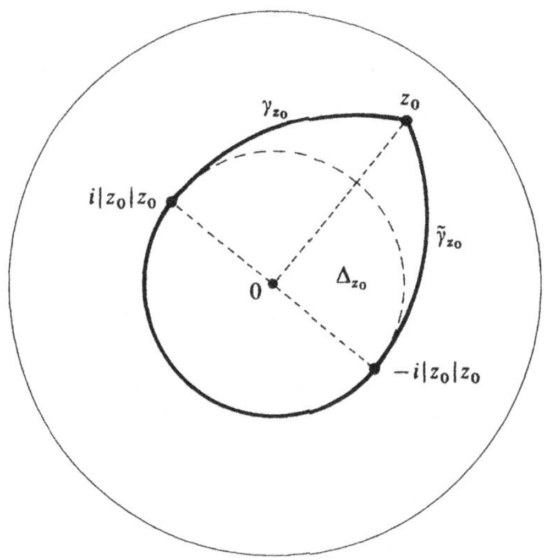

Figure 6.1. The region Δ_{z_0}.

Rogosinski's Lemma. *Let z_0 be a fixed point, $0 < |z_0| < 1$. Then for all functions f analytic and satisfying $|f(z)| < 1$ in $\mathbb{D}$, with $f(0) = 0$ and $f'(0) \ge 0$, the region of values of $f(z_0)$ is precisely Δ_{z_0}.*

Proof. Apply the inequality (1) with $\zeta = 0$ and with $f(z)$ replaced by $f(z)/z$ to obtain

$$\left| \frac{f(z) - f'(0)z}{z - f'(0)f(z)} \right| \le |z|.$$

This places $w_0 = f(z_0)$ in the disk D_t defined by

$$\left| \frac{w - tz_0}{z_0 - tw} \right| \le |z_0|, \qquad t = f'(0),$$

where $0 \le t \le 1$. The disk D_t has radius

$$\frac{(1-t^2)r_0^2}{1-t^2r_0^2}, \qquad r_0 = |z_0|,$$

and center

$$\frac{t(1-r_0^2)}{1-t^2r_0^2}\, z_0,$$

a point which traverses the segment from 0 to z_0 as t increases from 0 to 1. For $t = 0$, the disk D_t reduces to $|w| \le |z_0|^2$.

The union of the disks D_t, $0 \le t \le 1$, coincides with Δ_{z_0}. To see this, write the equation for the boundary of D_t in the form

$$F(w, t) = \mathrm{Re}\left\{\log \frac{w - tz_0}{z_0 - tw}\right\} - \log|z_0| = 0.$$

The envelope of this family of circles (see *e.g.* Struik [1]) is determined by the pair of equations $F(w, t) = 0$ and

$$\frac{\partial F}{\partial t} = \mathrm{Re}\left\{\frac{w}{z_0 - tw} - \frac{z_0}{w - tz_0}\right\} = 0.$$

If $w/(z_0 - tw)$ is replaced by its complex conjugate, this second equation reduces to

$$\mathrm{Re}\left\{\frac{z_0 - t\bar{w}}{w - tz_0}\right\} = 0.$$

Thus the envelope is given by

$$\frac{w - tz_0}{z_0 - tw} = \pm i r_0, \qquad r_0 = |z_0|,$$

or

$$w = \frac{t \pm i r_0}{1 \pm i t r_0}\, z_0, \qquad 0 \le t \le 1.$$

This equation defines the circular arcs γ_{z_0} and $\tilde{\gamma}_{z_0}$ which comprise part of the boundary of Δ_{z_0}.

To see that the entire region Δ_{z_0} is covered, consider the function

$$f(z) = z\frac{\alpha z + t}{1 + t\alpha z}, \qquad |\alpha| \le 1, \quad 0 \le t \le 1.$$

Then $|f(z)| < 1$ for $|z| < 1$, and it is easy to verify that $f(0) = 0$, $f'(0) = t$, and

$$\frac{w_0 - tz_0}{z_0 - tw_0} = \alpha z_0, \qquad w_0 = f(z_0).$$

Since α is an arbitrary point with $|\alpha| \le 1$, this shows that every point in the disk D_t is the image of z_0 under some function f with the required properties.

§6.4. Majorization

If $g \prec f$, then by Littlewood's theorem, f dominates g "in the mean." We now turn to one of the deepest results in subordination theory, which asserts that under an appropriate normalization, f actually dominates g pointwise in a certain region: $|g(z)| \le |f(z)|$. A similar statement can be made for derivatives. For these results it is essential that f be univalent.

We shall say that f *majorizes* g in a certain region if $|g(z)| \le |f(z)|$ there. By way of introduction to the theorems on majorization, we begin with a closely related elementary theorem, first noticed by Schiffer [1].

Theorem 6.6. *If $f \in S$ and $g \prec f$, then*

$$|g(z)| \le \frac{r}{(1-r)^2} \quad \textit{and} \quad |g'(z)| \le \frac{1+r}{(1-r)^3}, \qquad r = |z| < 1.$$

Proof. The first inequality is trivial, since $M_\infty(r, g) \le M_\infty(r, f)$. For the second inequality, write $g(z) = f(\omega(z))$, where $|\omega(z)| \le |z|$. Now apply the general estimate (2) and the distortion theorem for the class S to obtain

$$|g'(z)| = |f'(\omega(z))\omega'(z)| \le \frac{1 + |\omega(z)|}{(1 - |\omega(z)|)^3} \cdot \frac{1 - |\omega(z)|^2}{1 - |z|^2}$$

$$= \left(\frac{1 + |\omega(z)|}{1 - |\omega(z)|}\right)^2 \frac{1}{1 - |z|^2} \le \frac{1 + |z|}{(1 - |z|)^3}.$$

The next theorem is the result already mentioned which identifies the radius of majorization in subordination.

Theorem 6.7. *If $f \in S$, $g \prec f$, and $g'(0) \geq 0$, then $|g(z)| \leq |f(z)|$ for all z in the disk $|z| \leq \frac{1}{2}(3 - \sqrt{5}) = 0.381\ldots$. This radius is best possible.*

This theorem has a long history. Biernacki [2] first proved it for $|z| < \frac{1}{4}$. Goluzin [22] improved the radius to 0.35 and conjectured the sharp value to be $\frac{1}{2}(3 - \sqrt{5})$. This he proved under the additional assumption that f is starlike. Goluzin's method is quite natural, yet apparently incapable of yielding the sharp result. Eventually, Shah [2] verified Goluzin's conjecture by an argument considerably more technical. We propose to follow Shah's proof, which is still the only known path to the sharp result.

All of the methods just indicated use Rogosinski's lemma, the sharpened form of the Schwarz lemma. The function g has the representation $g(z) = f(\omega(z))$, where $|\omega(z)| \leq |z|$ and $\omega'(0) \geq 0$. Thus $\omega(z) \in \Delta_z$, and it is sufficient (in fact, equivalent) to prove that $|f(\zeta)| \leq |f(z)|$ for all $\zeta \in \Delta_z$. If $|\zeta| \leq |z|^2 = r^2$, then

$$|f(\zeta)| \leq \frac{r^2}{(1 - r^2)^2} \leq \frac{r}{(1 + r)^2} \leq |f(z)|,$$

provided $r^2(1 + r)^2 \leq r(1 - r^2)^2$, or $r^2 - 3r + 1 \geq 0$. But this is true precisely for $r \leq \frac{1}{2}(3 - \sqrt{5})$.

The same calculation shows that the radius cannot be increased beyond $\frac{1}{2}(3 - \sqrt{5})$. Simply take f to be the Koebe function $k(z) = z(1 - z)^{-2}$, and let $g(z) = k(z^2)$. Then $g \prec k$, but $|g(z)| > |k(z)|$ for $-1 < z < -\frac{1}{2}(3 - \sqrt{5})$.

Before embarking upon the proof of Theorem 6.7, let us make one final observation. The theorem is totally false without the assumption that f is univalent, at least if we replace subordination by the weaker condition that the range of g be contained in the range of f. For example, let $f(z) = z + 2z^2$ and $g(z) = z$. Note that f is not univalent, since $f'(-\frac{1}{4}) = 0$. Rouché's theorem shows that for each number α with $|\alpha| < 1$,

$$f(z) - \alpha = 2z^2 + (z - \alpha)$$

has the same number of zeros in $|z| < 1$ as does $2z^2$. In particular, the range of f contains the entire unit disk, the range of g. Also, $f'(0) = g'(0) = 1$. Nevertheless, $|g(z)| > |f(z)|$ on the full segment $-1 < z < 0$, since $|1 + 2z| < 1$ there.

Proof of Theorem. For convenience, let $b = \frac{1}{2}(3 - \sqrt{5})$. As we have just remarked, Rogosinski's lemma reduces the proof to showing that if $|z_0| \leq b$, then $|f(\zeta)| \leq |f(z_0)|$ for all $\zeta \in \Delta_{z_0}$. Suppose, on the contrary, that for some $f \in S$ and for some point z_0 with $0 < |z_0| < b$, $|f(\zeta)| > |f(z_0)|$ for some $\zeta \in \Delta_{z_0}$. By the maximum modulus theorem, we may assume that ζ lies on the boundary of Δ_{z_0}. It is clear that $|z_0|^2 < |\zeta| < |z_0|$, since $|f(z)| \leq |f(z_0)|$ for

$|z| \le |z_0|^2$, as we have seen above. Thus we may assume without loss of generality that ζ lies in the interior of the arc γ_{z_0}. It follows that $|f(z_1)| = |f(z_0)|$ for some $z_1 \in \zeta_{z_0}$, $|z_0|^2 < |z_1| < |z_0|$. It is clear geometrically that $0 < \arg\{z_1/z_0\} < \pi/2$.

We now appeal to the inequality (Theorem 4.3, Corollary 8)

$$\left|\log\left(\frac{G(\zeta_1) + G(\zeta_2)}{G(\zeta_1) - G(\zeta_2)} \cdot \frac{\zeta_1 - \zeta_2}{\zeta_1 + \zeta_2}\right)\right|^2 \le \log\frac{|\zeta_1|^2 + 1}{|\zeta_1|^2 - 1} \log\frac{|\zeta_2|^2 + 1}{|\zeta_2|^2 - 1}$$

for odd functions $G \in \Sigma$ and for distinct points ζ_1 and ζ_2 in $|\zeta| > 1$. We take $h(z) = \sqrt{f(z^2)}$ and $G(\zeta) = 1/h(1/\zeta)$; and we choose $\zeta_1 = 1/w_1$ and $\zeta_2 = 1/w_0$, where $w_1 = \sqrt{z_1}$ and $w_0 = \sqrt{z_0}$. The above inequality then takes the form

$$\left|\log\left(\frac{h(w_1) + h(w_0)}{h(w_1) - h(w_0)} \cdot \frac{w_1 - w_0}{w_1 + w_0}\right)\right|^2 \le \log\frac{1 + |z_0|}{1 - |z_0|} \log\frac{1 + |z_1|}{1 - |z_1|}. \tag{3}$$

Since $|f(z_1)| = |f(z_0)|$, we can write $h(w_1)/h(w_0) = e^{i\psi}$, so that

$$\frac{h(w_1) + h(w_0)}{h(w_1) - h(w_0)} = -i \cot\frac{\psi}{2}.$$

Since $z_1 \in \gamma_{z_0}$, it is easy to see that $z_1/z_0 = \rho e^{i\theta}$ lies on the circular arc

$$\rho^2 + \frac{1 - r_0^2}{r_0}\rho \sin\theta = 1, \qquad 0 < \theta < \frac{\pi}{2}, \tag{4}$$

where $r_0 = |z_0|$. Note that $r_0 < \rho < 1$. The left-hand side of (3) is no smaller than

$$\left|\arg\left\{-i\frac{\sqrt{\rho}\,e^{i\theta/2} + 1}{\sqrt{\rho}\,e^{i\theta/2} - 1}\right\}\right|^2 = \left\{\tan^{-1}\left(\frac{1 - \rho}{2\sqrt{\rho}\sin\theta/2}\right)\right\}^2.$$

Thus the inequality (3) implies

$$\tan^{-1}\left(\frac{1 - \rho}{2\sqrt{\rho}\sin\theta/2}\right) \le \left\{\log\frac{1 + r_0}{1 - r_0}\log\frac{1 + \rho r_0}{1 - \rho r_0}\right\}^{1/2}, \tag{5}$$

where ρ and θ are related by (4). We shall obtain a contradiction by showing that (5) cannot hold.

First we eliminate θ by combining (4) with the identity

$$\sin\frac{\theta}{2} = \frac{1}{\sqrt{2}}\sin\theta\,(1 + \cos\theta)^{-1/2}.$$

The result of a tedious calculation is

$$\frac{1-\rho}{2\sqrt{\rho}\sin\theta/2} = \varphi(\rho, R), \qquad R = \frac{r_0}{1-r_0^2},$$

where

$$\varphi(\rho, R) = \frac{1}{\sqrt{2}R}\cdot\frac{\sqrt{\rho}}{1+\rho}\left\{1+\left[1-R^2\left(\frac{1-\rho^2}{\rho}\right)^2\right]\right\}^{1/2}.$$

A calculation shows that $(\partial/\partial\rho)[\varphi(\rho, R)]^2 > 0$ for $r_0 < \rho < 1$. Thus

$$\varphi(\rho, R) \geq \varphi(r_0, R) = \frac{1-r_0}{\sqrt{2r_0}} \geq \frac{1-b}{\sqrt{2b}} = \frac{1}{\sqrt{2}},$$

since $r_0 < b$ and $b^2 - 3b + 1 = 0$. In other words, the left-hand side of (5) is bounded below by $\tan^{-1}(1/\sqrt{2})$.

On the other hand, for $r_0 \leq \rho \leq b$, the right-hand side of (5) is bounded above by

$$\left\{\log\frac{1+b}{1-b}\log\frac{1+b^2}{1-b^2}\right\}^{1/2} = \left\{\log\sqrt{5}\log\left(\frac{3}{\sqrt{5}}\right)\right\}^{1/2} < \tfrac{1}{2}.$$

Since $\tan^{-1}(1/\sqrt{2}) > \frac{1}{2}$, this shows that (5) is false for $r_0 \leq \rho \leq b$.

The case $b < \rho < 1$ is more delicate. First observe that $\varphi(\rho, R)$ is a decreasing function of R, so

$$\varphi(\rho, R) \geq \varphi\left(\rho, \frac{b}{1-b^2}\right) = \varphi\left(\rho, \frac{1}{\sqrt{5}}\right), \qquad r_0 < b < \rho < 1.$$

Thus the inequality (5) implies

$$\tan^{-1}\left(\varphi\left(\rho, \frac{1}{\sqrt{5}}\right)\right) \leq \left\{\log\sqrt{5}\log\frac{1+b\rho}{1-b\rho}\right\}^{1/2}, \tag{6}$$

since $(1+b)/(1-b) = \sqrt{5}$. For convenience, write (6) in the form $u(\rho) \leq v(\rho)$. Both $u(\rho)$ and $v(\rho)$ are increasing functions of ρ. Thus for $b < \rho \leq \frac{3}{5}$, the inequality (6) gives

$$u(b) \leq u(\rho) \leq v(\rho) \leq v(\tfrac{3}{5});$$

while for $\frac{3}{5} \leq \rho < 1$ it gives

$$u(\tfrac{3}{5}) \leq u(\rho) \leq v(\rho) \leq v(1).$$

But these conclusions contradict the numerical facts

$$u(b) > 0.615; \qquad v(\tfrac{3}{5}) < 0.613;$$

$$u(\tfrac{3}{5}) > 0.807; \qquad v(1) < 0.805.$$

Hence the inequality (5) is false also for $b < \rho < 1$, and the supposition that $|f(\zeta)| > |f(z_0)|$ somewhere in Δ_{z_0} has led to a contradiction. This completes the proof.

There is a corresponding theorem for derivatives.

Theorem 6.8. *If $f \in S$, $g \prec f$, and $g'(0) \geq 0$, then $|g'(z)| \leq |f'(z)|$ for all z in the disk $|z| \leq 3 - \sqrt{8} = 0.171\ldots$. This radius is best possible.*

This result is also due to Shah [3]. Goluzin[22] had proved it for $|z| \leq 0.12$ and had conjectured the stronger form of the theorem. Shah's proof is again heavily computational, and not very enlightening. As may be expected, it is based on Dieudonné's lemma. Rather than give the details, we shall content ourselves with a proof (in the next section) for the case in which g is univalent. The following elementary construction, due to Goluzin [23], shows that the radius $3 - \sqrt{8}$ is best possible.

Let $f(z) = z(1 + z)^{-2}$, let

$$\omega(z) = z\,\frac{z + \alpha}{1 + \alpha z}, \qquad 0 \leq \alpha \leq 1,$$

and let $g(z) = f(\omega(z))$. A calculation gives

$$g'(r) = \frac{(1 - r^2)(1 + \alpha\rho)}{4r^2(\alpha + \rho)^3}, \qquad \rho = \frac{1 + r^2}{2r}.$$

For $\alpha = 1$, $g'(r) = f'(r)$. But for $\alpha = 1$,

$$\frac{\partial}{\partial\alpha}\{g'(r)\} = \frac{1 - r^2}{4r^2}\,\frac{\rho - 3}{(\rho + 1)^3} < 0$$

for $\rho < 3$; *i.e.*, for $r > 3 - \sqrt{8}$. Thus for each $r > 3 - \sqrt{8}$, $g'(r) > f'(r)$ for all $\alpha < 1$ sufficiently near 1.

Lewandowski [3] has proved a converse theorem, to the effect that majorization implies subordination. Specifically, he showed that if $f \in S$

and g is an analytic function with $g'(0) \geq 0$ for which $|g(z)| \leq |f(z)|$ in $|z| < 1$, then g is subordinate to f in the disk $|z| \leq 0.21$; that is, $g(D_r) \subset f(D_r)$ for all $r \leq 0.21$, where D_r is the disk $|z| < r$. The best possible radius is unknown, but Lewandowski gave a simple example which shows it is no larger than 0.3. Lewandowski [5] and MacGregor [2] have also shown that under the same hypotheses, $|g'(z)| \leq |f'(z)|$ in $|z| \leq 2 - \sqrt{3}$, and this radius is best possible.

Campbell [1, 2, 3] generalized the subordination–majorization theory to locally univalent families of functions. However, this approach does not simplify the proofs of Shah's theorems (Theorems 6.7 and 6.8) for the class S.

§6.5. Univalent Subordinate Functions

The basic theorem on majorization (Theorem 6.7) can be slightly improved under the additional hypothesis that the subordinate function g is univalent. Biernacki [1, 2] first obtained the result by a variational method, and Goluzin [4, 23] later based a proof on a variant of Loewner's method. We shall follow Goluzin's approach, which is particularly well suited to problems of this type.

Theorem 6.9. *Suppose $f \in S$, and let g be analytic and univalent in the unit disk, with $g \prec f$ and $g'(0) > 0$. Then $|g(z)| \leq |f(z)|$ for all z in the disk $|z| \leq r_0 = 0.390\ldots$, where r_0 is the unique solution to the transcendental equation*

$$\log \frac{1+r}{1-r} + 2 \tan^{-1} r = \frac{\pi}{2}.$$

The radius r_0 is best possible. If equality holds for any z with $|z| < r_0$, then $g = f$.

Proof. Since $f(z)$ can be approximated uniformly on compact subsets by $f(\rho z)/\rho$, $0 < \rho < 1$, there is no loss of generality in supposing that f maps the disk onto a domain D bounded by an analytic Jordan curve C. In view of the Carathéodory convergence theorem, it is sufficient to assume that g maps the disk onto a domain $\tilde{D}$ consisting of D slit along some analytic Jordan arc Γ terminating at a point $w_0 \in C$ and not passing through the origin (*cf.* §3.2). Let Γ be represented by $w = \psi(t)$, $0 \leq t \leq T$, with $\psi(T) = w_0$; and let D_t be the domain D slit along Γ from w_0 to $\psi(t)$. The parametrization may be chosen in such a way that

$$w = g(z, t) = e^t\{z + b_2(t)z^2 + \cdots\}$$

maps the disk onto D_t. Thus $g(z, 0) = g(z)$, while $g(z, T) = f(z)$. Referring to the derivation of the Loewner differential equation (§3.2), we find that $g(z, t)$ satisfies the differential equation (Chapter 3, Exercise 8)

$$\frac{\partial g}{\partial t} = \frac{\partial g}{\partial z} z \frac{1 + \kappa z}{1 - \kappa z}, \tag{7}$$

where $\kappa = \kappa(t)$ is a continuous complex-valued function with $|\kappa(t)| \equiv 1$ for $1 \leq t \leq T$. Transforming this differential equation and extracting real parts, we find

$$\frac{\partial}{\partial t} \log|g(z, t)| = \operatorname{Re}\left\{\frac{zg'(z, t)}{g(z, t)} \cdot \frac{1 + \kappa z}{1 - \kappa z}\right\}, \tag{8}$$

where $g'(z, t) = (\partial/\partial z)g(z, t)$. But since $g(z, t)$ is univalent and normalized by $g(0, t) = 0$ and $g'(0, t) > 0$, a general inequality for the class S (see §3.6) gives

$$\left|\arg \frac{zg'(z, t)}{g(z, t)}\right| \leq \log \frac{1 + r}{1 - r}, \qquad r = |z|.$$

On the other hand,

$$\begin{aligned}\left|\arg \frac{1 + \kappa z}{1 - \kappa z}\right| &= |\arg\{(1 + \kappa z)(1 - \bar{\kappa}\bar{z})\}| \\ &= \left|\tan^{-1}\left(\frac{2 \operatorname{Im}\{\kappa z\}}{1 - r^2}\right)\right| \leq \tan^{-1}\left(\frac{2r}{1 - r^2}\right) = 2 \tan^{-1} r.\end{aligned}$$

In view of these last two inequalities, the equation (8) shows that

$$\frac{\partial}{\partial t} \log|g(z, t)| > 0$$

if

$$\log \frac{1 + r}{1 - r} + 2 \tan^{-1} r < \frac{\pi}{2}.$$

In other words, $|g(z, t)|$ increases from $|g(z)|$ to $|f(z)|$ if $|z| < r_0$. Hence $|g(z)| \leq |f(z)|$ if $|z| \leq r_0$.

In order to show that the radius r_0 is best possible, suppose $r_0 < r < r_1$, where r_1 is the solution to

$$\log \frac{1+r}{1-r} + 2\tan^{-1} r = \frac{3\pi}{2}.$$

Consider the mapping

$$\zeta = \varphi(z) = k^{-1}((1-\varepsilon)k(z)), \qquad 0 < \varepsilon < 1,$$

where $k(z) = z(1-z)^{-2}$ is the Koebe function. The function φ maps the disk $|z| < 1$ onto the disk $|\zeta| < 1$ minus a short radial slit inward from $\zeta = -1$. Note that $\varphi(0) = 0$ and $\varphi'(0) > 0$. For small $\varepsilon > 0$, a simple calculation leads to the formula

$$\varphi(z) = z - \varepsilon z \frac{1-z}{1+z} + O(\varepsilon^2), \qquad |z| < 1.$$

Now let $g(z) = f(\varphi(z))$ and note that all hypotheses of the theorem are satisfied if $f \in S$. The asymptotic formula for φ gives

$$g(z) = f(z) - \varepsilon z f'(z) \frac{1-z}{1+z} + O(\varepsilon^2), \tag{9}$$

or

$$|g(z)| = |f(z)| \left[1 - \varepsilon \operatorname{Re}\left\{ \frac{zf'(z)}{f(z)} \cdot \frac{1-z}{1+z} \right\} + O(\varepsilon^2) \right]. \tag{10}$$

Now choose z with $|z| = r$ so that

$$\arg \frac{1-z}{1+z} = 2\tan^{-1} r,$$

and then choose $f \in S$ so that

$$\arg \frac{zf'(z)}{f(z)} = \log \frac{1+r}{1-r}.$$

Since $r_0 < r < r_1$, it is clear from (10) that $|g(z)| > |f(z)|$ for sufficiently small $\varepsilon > 0$. In view of the rotational symmetry, this demonstrates the failure of the theorem throughout the annulus $r_0 < |z| < r_1$. In fact, by a trivial modification of the construction (at the stage where z is chosen with the given modulus), it can be seen that the theorem fails whenever $r_0 < |z| < 1$.

Alternatively, it is easy to see, by rotational symmetry and the maximum modulus theorem applied to g/f, that the set of points where the theorem holds is a closed disk centered at the origin.

Finally, if $|g(z)| = |f(z)|$ somewhere in the disk $|z| < r_0$, the maximum modulus theorem shows that $g = f$. This completes the proof.

The preceding theorem shows that the radius of majorization increases under the additional hypothesis that g is univalent. One would expect the same hypothesis to give a corresponding improvement in the majorization theorem for derivatives (Theorem 6.8). It seems astonishing that in this case the univalence assumption on g is quite irrelevant and does not increase the radius of majorization beyond $3 - \sqrt{8}$. This unexpected result is due to Goluzin [23].

Theorem 6.10. *Suppose $f \in S$, and let g be analytic and univalent in the unit disk, with $g \prec f$ and $g'(0) > 0$. Then $|g'(z)| \le |f'(z)|$ for all z in the disk $|z| \le 3 - \sqrt{8} = 0.171\ldots$, and this radius is best possible. If equality holds for any z with $|z| < 3 - \sqrt{8}$, then $g = f$.*

Proof. As in the proof of the preceding theorem, we assume that $f(z)$ and $g(z)$ are related by a function $g(z, t)$ satisfying the differential equation (7) in an interval $0 \le t \le T$, with $g(z, 0) = g(z)$ and $g(z, T) = f(z)$. Differentiating (7) with respect to z and extracting real parts, we find

$$\frac{\partial}{\partial t} \log|g'(z, t)| = \operatorname{Re}\{AB + C\}, \tag{11}$$

where

$$A = 1 + \frac{zg''(z, t)}{g'(z, t)}, \qquad B = \frac{1 + \kappa z}{1 - \kappa z}, \qquad C = \frac{2\kappa z}{(1 - \kappa z)^2}.$$

Since $g(z, t)$ satisfies the elementary inequality (Theorem 2.4)

$$\left| \frac{zg''(z, t)}{g'(z, t)} - \frac{2r^2}{1 - r^2} \right| \le \frac{4r}{1 - r^2}, \qquad r = |z|,$$

it follows that A lies in the disk with center $(1 + r^2)/(1 - r^2)$ and radius $4r/(1 - r^2)$. Hence

$$\operatorname{Re}\left\{\left(A - \frac{1 + r^2}{1 - r^2}\right)B\right\} \ge -\frac{4r}{1 - r^2}|B|,$$

so that

$$\begin{aligned}\operatorname{Re}\{AB + C\} &\geq \frac{1 + r^2}{1 - r^2}\operatorname{Re}\{B\} - \frac{4r}{1 - r^2}|B| + \operatorname{Re}\{C\}\\ &= \frac{1 + r^2}{|1 - \kappa z|^2} - \frac{4r}{1 - r^2}\left|\frac{1 + \kappa z}{1 - \kappa z}\right| + \frac{2(1 + r^2)\operatorname{Re}\{\kappa z\} - 4r^2}{|1 - \kappa z|^4}\\ &= \frac{(1 - r^2)^3 - 4r|1 + \kappa z||1 - \kappa z|^3}{(1 - r^2)|1 - \kappa z|^4}.\end{aligned}$$

But

$$|1 + \kappa z|^2|1 - \kappa z|^6 = 16r^4(a + x)(a - x)^3,$$

where $a = (1 + r^2)/2r > 1$ and $x = \operatorname{Re}\{\kappa z\}/r$; $-1 \leq x \leq 1$. For $a > 2$, the derivative of this expression with respect to x is

$$-32r^4(a - x)^2(a + 2x) < 0, \qquad -1 \leq x \leq 1.$$

Thus

$$|1 + \kappa z|^2|1 - \kappa z|^6 \leq 16r^4(a - 1)(a + 1)^3 = (1 - r)^2(1 + r)^6$$

if $a > 2$; *i.e.*, if $r^2 - 4r + 1 > 0$, or $r < 2 - \sqrt{3} = 0.267\ldots$. Consequently,

$$\operatorname{Re}\{AB + C\} \geq |1 - \kappa z|^{-4}(1 + r)^2(r^2 - 6r + 1) > 0$$

for $r < 3 - \sqrt{8} < 2 - \sqrt{3}$. In view of (11), this shows that for $|z| < 3 - \sqrt{8}$, $|g'(z, t)|$ increases from $|g'(z)|$ to $|f'(z)|$ as t increases from 0 to T. Thus $|g'(z)| \leq |f'(z)|$ for $|z| \leq 3 - \sqrt{8}$. An application of the maximum modulus theorem to g'/f' shows that equality cannot occur for $|z| < 3 - \sqrt{8}$ unless $g = f$.

To show that the radius $3 - \sqrt{8}$ cannot be increased, consider again the function φ introduced in the proof of the previous theorem. Let $f(z) = z(1 + z)^{-2}$, and let $g(z) = f(\varphi(z))$. Differentiating the asymptotic formula (9), we find

$$g'(z) = f'(z) - \varepsilon\left[zf'(z)\frac{1 - z}{1 + z}\right]' + O(\varepsilon^2);$$

whereupon a calculation gives

$$|g'(z)| = |f'(z)|\left[1 - \varepsilon \operatorname{Re}\left\{\frac{z^2 - 6z + 1}{(1 + z)^2}\right\} + O(\varepsilon^2)\right].$$

It follows that if $r > 3 - \sqrt{8}$, then $|g'(r)| > |f'(r)|$ for sufficiently small $\varepsilon > 0$, since $r^2 - 6r + 1 < 0$ in this case. Taking rotations into account, we conclude that for any given point z with $3 - \sqrt{8} < |z| < 1$, there are functions f and g satisfying the hypotheses of the theorem for which $|g'(z)| > |f'(z)|$.

Exercises

1. Show that if $f \in S$, $g \in S$, and $g \prec f$, then $f = g$.

2. Prove that each function $f \in S$ other than the identity omits some point in $\mathbb{D}$.

3. Use Littlewood's subordination theorem to show that for $0 < p < \infty$ and for each function f analytic in $\mathbb{D}$, the integral means $M_p(r, f)$ are nondecreasing functions of r.

4. Let $\varphi(t)$ be an arbitrary convex nondecreasing function on $0 \le t < \infty$. Show that if $g \prec f$ in $\mathbb{D}$, then

$$\int_0^{2\pi} \varphi(|g(re^{i\theta})|)\, d\theta \le \int_0^{2\pi} \varphi(|f(re^{i\theta})|)\, d\theta, \qquad 0 \le r < 1.$$

5. Show that Rogosinski's theorem (Theorem 6.2) fails for each p in the range $2 < p < \infty$. More specifically, construct $g \prec f$ with

$$\sum_{k=1}^{n} |b_k|^p \ne O\left(\sum_{k=1}^{n} |a_k|^p\right), \qquad n \to \infty.$$

(*Suggestion*: Appeal to Rogosinski's construction (cited in §6.2) of $g \prec f$ with $a_n = O(1)$ and $b_n \ne o(\sqrt{n})$.)

6. Show that Rogosinski's theorem fails for each p in the range $0 < p < 2$. (*Suggestion*: Produce a bounded function g for which $\sum_{k=1}^{\infty} |b_k|^p = \infty$.)

7. Show that if g is univalently subordinate to f (*i.e.*, if $g(z) = f(\omega(z))$ with ω univalent in $\mathbb{D}$), then

$$\sum_{k=1}^{n} k|b_k|^2 \le \sum_{k=1}^{n} k|a_k|^2, \qquad n = 1, 2, \ldots.$$

(Rogosinski [7].)

8. Prove Rogosinski's conjecture for the first two coefficients: if $g \prec f \in S$, then $|b_n| \le n$ for $n = 1, 2$. (Littlewood [1].)

9. Prove that if $g \prec f$ and the coefficients a_n of f are nonnegative, nondecreasing, and form a convex sequence, then $|b_n| \le |a_n|$, $n = 1, 2, \ldots$. (Rogosinski [7].)

10. Show that if $g \prec f$ in $\mathbb{D}$ and $0 < p \le \infty$, then $M_p(r, g') \le M_p(r, f')$ for $0 < r \le \sqrt{2} - 1$, with equality only for $g(z) = f(\alpha z)$, $|\alpha| = 1$. Show also that the radius $\sqrt{2} - 1$ is best possible for $p = \infty$. (However, for $p = 2$ the radius can be improved to $\frac{1}{2}$, as asserted by Corollary 2 to Theorem 6.3. The best possible radius for other indices p is not known.)

11. Carry out the proof of Corollary 2 to Theorem 6.3.

12. A function g is said to be *quasisubordinate* to f if g is majorized by some function subordinate to f. Generalize Rogosinski's theorem (Theorem 6.2) to quasisubordination. (Robertson [11].)

13. Show that $\log[f(z)/z] \prec \log[k(z)/z]$ for each $f \in S^*$, where k is the Koebe function. (*Hint*: Use the Herglotz representation of starlike functions. See Chapter 2, Exercises 10 and 11.) (Marx [1]. Marx conjectured that $\log f'(z) \prec \log k'(z)$ for each $f \in S^*$, but Hummel [8] disproved this. For other information on the Marx conjecture, see Robinson [2,3], Hummel [3, 10], Duren [5], McLaughlin [1], Duren and McLaughlin [1], and Pfaltzgraff [1].)

14. Let $f \in S$ map the disk onto a convex domain D, and suppose that each analytic function $g \prec f$ has an integral representation of the form

$$g(z) = \int_0^{2\pi} f(e^{it})\, d\mu(t), \qquad |z| < 1,$$

where $d\mu$ is a positive unit measure. Prove that D must be a half-plane. Thus the Herglotz formula (§1.9) cannot be generalized in this manner. (A. W. Goodman [4].)

Chapter 7

Integral Means

The main object of this chapter is to show that among all functions in S, the integral mean $M_p(r, f)$ is largest for the Koebe function. This basic theorem had long been conjectured, but it was proved only in 1973 by Albert Baernstein [1]. His proof uses the surprising fact that a certain maximal function (now known as the Baernstein star-function) is subharmonic. The latter result is of independent interest and has important applications to other branches of function theory, but we shall discuss only its connection with univalent functions.

§7.1. Baernstein's Theorem

An important problem in the theory of univalent functions is to find the sharp upper bounds for the integral means

$$M_p(r, f) = \left\{\frac{1}{2\pi}\int_0^{2\pi} |f(re^{i\theta})|^p \, d\theta\right\}^{1/p}, \qquad 0 < r < 1,$$

for $0 < p < \infty$. In the case $p = 1$, the problem is closely related to the Bieberbach conjecture. The main step in Littlewood's proof (§2.4) that $|a_n| < en$ is to obtain the estimate $M_1(r, f) \leq r/(1 - r)$ for all $f \in S$. Once this estimate is improved to

$$M_1(r, f) \leq M_1(r, k) = r/(1 - r^2),$$

where k is the Koebe function, the proof gives $|a_n| < (e/2)n$. The last estimate is now eclipsed by the results of I. M. Milin (§5.3) and FitzGerald (§5.12), but it is still a central problem to find the sharp estimate for $M_1(r, f)$ and more generally for $M_p(r, f)$, $0 < p < \infty$.

As early as 1951, Bazilevich [4] approached the solution for $p = 1$ and $p = 2$ with a proof that

$$M_p(r, f) < M_p(r, k) + C_p, \qquad p = 1, 2,$$

where C_p is a constant (given numerically) independent of f. There the problem stood until Baernstein [1] introduced new methods to show that $M_p(r, f) \leq M_p(r, k)$ for every p. In fact, he established a more general inequality for the integral means defined in terms of an arbitrary convex function.

A function $\Phi(x)$ continuous on $-\infty < x < \infty$ is said to be *convex* if $\Phi(\frac{1}{2}(x + y)) \leq \frac{1}{2}[\Phi(x) + \Phi(y)]$. It is said to be *strictly convex* if strict inequality holds unless $x = y$.

Theorem 7.1 (Baernstein's Theorem). *Let $\Phi(x)$ be a convex nondecreasing function on $-\infty < x < \infty$. Then for each $f \in S$,*

$$\int_0^{2\pi} \Phi(\log|f(re^{i\theta})|)\, d\theta \leq \int_0^{2\pi} \Phi(\log|k(re^{i\theta})|)\, d\theta, \qquad 0 < r < 1,$$

where k is the Koebe function. If Φ is strictly convex, then equality holds for some r only if f is a rotation of k.

The choice $\Phi(x) = e^{px}$ gives the result already mentioned:

Corollary. *For $0 < p < \infty$ and $f \in S$,*

$$M_p(r, f) \leq M_p(r, k), \qquad 0 < r < 1,$$

with equality only if f is a rotation of k.

The proof of Baernstein's theorem involves a curious maximal function, which we now proceed to define. Let $u(z)$ be a real-valued function defined on the annulus $r_1 < |z| < r_2$. For each $r \in (r_1, r_2)$, suppose $u(re^{i\theta}) \in L^1(0, 2\pi)$. The *Baernstein star-function* of u is

$$u^*(re^{i\theta}) = \sup_{|E|=2\theta} \int_E u(re^{it})\, dt, \qquad 0 \leq \theta \leq \pi,$$

where $|E|$ denotes the Lebesgue measure of the set $E \subset [-\pi, \pi]$. Baernstein [1] showed that the star-function has the following remarkable property.

Lemma 1. *If u is continuous and subharmonic in the annulus $r_1 < |z| < r_2$, then u^* is continuous in the semiannulus*

$$\{re^{i\theta}: r_1 < r < r_2, 0 \leq \theta \leq \pi\}$$

and subharmonic in the interior.

The proof of this lemma is deferred to §7.4. In §7.2 we develop some of the more elementary properties of the star operation. These are purely "real-variable" results which make no reference to complex function theory. In §7.3 we use these results and Lemma 1 to prove Baernstein's theorem.

§7.2. The Star-Function

It is convenient to begin with a simple representation formula for convex functions. For any real-valued function $g(x)$, let $[g(x)]^+ = \max\{g(x), 0\}$.

Lemma 2. *Let $\Phi(s)$ be a convex function on $-\infty < s < \infty$, with $\Phi(s) \equiv 0$ on some interval $(-\infty, s_0)$. Then*

$$\Phi(s) = \int_{-\infty}^{\infty} [s - t]^+ \, d\mu(t)$$

for some nonnegative measure $d\mu$.

Proof. A convex function satisfies a Lipschitz condition on each compact subinterval, and so is absolutely continuous there. Thus

$$\Phi(s) = \int_{-\infty}^{s} \Phi'(t) \, dt = -\int_{-\infty}^{s} \Phi'(t) \, d(s - t),$$

and integration by parts gives

$$\Phi(s) = \int_{-\infty}^{s} (s - t) \, d\Phi'(t) = \int_{-\infty}^{\infty} [s - t]^+ \, d\Phi'(t).$$

Since $d\Phi'(t) \geq 0$ because Φ is convex, this is the desired representation.

Let $g(x)$ be a real-valued function integrable over $(-\pi, \pi)$. The *distribution function* of g is

$$\lambda(t) = |\{x : g(x) > t\}|.$$

It is clear that $\lambda(t)$ is nonincreasing and right-continuous: $\lambda(t) = \lambda(t+)$. By the definition of the Lebesgue integral,

$$\int_{-\pi}^{\pi} g(x) \, dx = -\int_{-\infty}^{\infty} t \, d\lambda(t).$$

Two functions defined on the same set are said to be *equimeasurable* if they have the same distribution function. Thus two equimeasurable functions have equal integrals.

One particular function equimeasurable with g is of special importance. If $\lambda(t)$ is continuous and strictly decreasing, the *symmetric decreasing rearrangement* of g is the function $G(x)$ defined for $0 \le x \le \pi$ as the inverse of $\frac{1}{2}\lambda(t)$, then extended to $[-\pi, 0)$ as an even function: $G(-x) = G(x)$. In the general case, we must resort to the more technical definition

$$G(x) = \min\{t: \lambda(t) \le 2x\}, \qquad 0 < x < \pi.$$

$G(0)$ is taken to be the essential supremum and $G(\pi)$ the essential infimum of g; and again $G(-x) = G(x)$. It is not difficult to see that g and G are equimeasurable.

Now consider the star-function

$$g^*(\theta) = \sup_{|E| = 2\theta} \int_E g(x)\,dx, \qquad 0 \le \theta \le \pi.$$

It is useful to note that "sup" may be replaced by "max"; that is, the supremum is always attained. This is the content of the following lemma.

Lemma 3. *For each θ, $0 \le \theta \le \pi$, there exists a set $E \subset [-\pi, \pi]$ of measure $|E| = 2\theta$ for which $g^*(\theta) = \int_E g(x)\,dx$.*

Proof. For $\theta = 0$ and for $\theta = \pi$, the assertion is obviously true. For $0 < \theta < \pi$, choose t such that $\lambda(t) \le 2\theta \le \lambda(t-)$. Let $A = \{x: g(x) > t\}$ and $B = \{x: g(x) \ge t\}$. Then $|A| = \lambda(t)$ and $|B| = \lambda(t-)$. Choose a measurable set E with $A \subset E \subset B$ and $|E| = 2\theta$. Then for any set F of measure $|F| = 2\theta$,

$$\begin{aligned}\int_F g(x)\,dx &= \int_F [g(x) - t]\,dx + 2\theta t \le \int_{-\pi}^{\pi} [g(x) - t]^+\,dx + 2\theta t \\ &= \int_E [g(x) - t]\,dx + 2\theta t = \int_E g(x)\,dx.\end{aligned}$$

This proves the lemma.

The star-function g^* and the symmetric decreasing rearrangement G are closely related, as the following lemma shows.

Lemma 4. *For each θ, $0 \le \theta \le \pi$,*

$$g^*(\theta) = \int_{-\theta}^{\theta} G(x)\,dx.$$

Proof. For $\theta = 0$, both sides vanish. For $\theta = \pi$, both sides are equal to the integral of g over $[-\pi, \pi]$. For $0 < \theta < \pi$, let E be the set of Lemma 3, and let t be determined by $\lambda(t) \leq 2\theta \leq \lambda(t-)$. Then since $[g(x) - t]^+$ and $[G(x) - t]^+$ are equimeasurable,

$$g^*(\theta) = \int_E g(x)\,dx = \int_{-\pi}^{\pi} [g(x) - t]^+ + 2\theta t$$

$$= \int_{-\pi}^{\pi} [G(x) - t]^+\,dx + 2\theta t.$$

But it follows from the definition of G that

$$\{x\colon G(x) > t\} \subset (-\theta, \theta) \subset \{x\colon G(x) \geq t\}.$$

Thus

$$\int_{-\pi}^{\pi} [G(x) - t]^+\,dx + 2\theta t = \int_{-\theta}^{\theta} [G(x) - t]\,dx + 2\theta t = \int_{-\theta}^{\theta} G(x)\,dx.$$

The next lemma reveals the role of the star-function in the proof of Baernstein's theorem.

Lemma 5. For $g, h \in L^1(-\pi, \pi)$, *the following three statements are equivalent.*

(a) *For each function $\Phi(s)$ convex and nondecreasing on $-\infty < s < \infty$,*

$$\int_{-\pi}^{\pi} \Phi(g(x))\,dx \leq \int_{-\pi}^{\pi} \Phi(h(x))\,dx.$$

(b) *For each $t \in \mathbb{R}$*

$$\int_{-\pi}^{\pi} [g(x) - t]^+\,dx \leq \int_{-\pi}^{\pi} [h(x) - t]^+\,dx.$$

(c) $g^*(\theta) \leq h^*(\theta), \qquad 0 \leq \theta \leq \pi.$

Proof. (a) $\Rightarrow$ (b). This is trivial, since $\Phi(s) = [s - t]^+$ is convex and nondecreasing.

(b) $\Rightarrow$ (a). Since Φ may be approximated by a monotonic sequence of lower truncations, there is no loss of generality in assuming that $\Phi(s) \equiv \alpha$ for all $s \leq s_0$, where α and s_0 are constants. Furthermore, since $\Phi(s) = [\Phi(s) - \alpha] + \alpha$, we may assume $\alpha = 0$. Thus we may assume that Φ has the integral representation of Lemma 2, whereupon an interchange of the order of integration shows that (b) implies (a).

(b) ⇒ (c). Since (b) clearly implies that $\int g \leq \int h$, it is enough to consider $0 < \theta < \pi$. Let $\nu(t)$ be the distribution function of h, and choose t so that $\nu(t) \leq 2\theta \leq \nu(t-)$. Then as in the proof of Lemma 3, there is a set E of measure 2θ such that $h(x) \geq t$ for all $x \in E$ and $h(x) \leq t$ for all $x \notin E$. Hence if F is any set of measure $|F| = 2\theta$,

$$\begin{aligned}\int_F g(x)\,dx &= \int_F [g(x) - t]\,dx + 2\theta t \\ &\leq \int_{-\pi}^{\pi} [g(x) - t]^+\,dx + 2\theta t \leq \int_{-\pi}^{\pi} [h(x) - t]^+\,dx + 2\theta t \\ &= \int_E [h(x) - t]\,dx + 2\theta t = \int_E h(x)\,dx \leq h^*(\theta).\end{aligned}$$

Since F is arbitrary, it follows that $g^*(\theta) \leq h^*(\theta)$.

(c) ⇒ (b). Let $\lambda(t)$ be the distribution function of g. Given $t \in \mathbb{R}$, choose $\theta \in [0, \pi]$ so that $\lambda(t) \leq 2\theta \leq \lambda(t-)$; and let E be a set of measure 2θ such that $g(x) \geq t$ on E and $g(x) \leq t$ elsewhere. Appealing to Lemma 3, choose a set F with $|F| = 2\theta$ so that $h^*(\theta) = \int_F h(x)\,dx$. Then

$$\begin{aligned}\int_{-\pi}^{\pi} [g(x) - t]^+\,dx &= \int_E [g(x) - t]\,dx \leq g^*(\theta) - 2\theta t \\ &\leq h^*(\theta) - 2\theta t = \int_F [h(x) - t]\,dx \leq \int_{-\pi}^{\pi} [h(x) - t]^+\,dx.\end{aligned}$$

§7.3. Proof of Baernstein's Theorem

In view of Lemma 5 ((b) ⇒ (a)), the inequality of Baernstein's theorem will be established if we can show that

$$\int_{-\pi}^{\pi} \log^+\left(\frac{|f(re^{i\theta})|}{\rho}\right) d\theta \leq \int_{-\pi}^{\pi} \log^+\left(\frac{|k(re^{i\theta})|}{\rho}\right) d\theta, \qquad 0 < r < 1, \tag{1}$$

for each $\rho > 0$ and for all $f \in S$.

The first step in the proof is to apply Jensen's theorem to obtain another expression for the left-hand side of (1). Let f be an arbitrary analytic function, let α be a complex number, and let $n(r, \alpha)$ be the number of points (counted according to multiplicity) in $|z| \leq r$ at which $f(z) = \alpha$. Assume $f(0) \neq \alpha$, and let

$$N(r, \alpha) = \int_0^r \frac{n(t, \alpha)}{t}\,dt.$$

Then Jensen's theorem takes the form

$$\frac{1}{2\pi}\int_{-\pi}^{\pi} \log|f(re^{i\theta}) - \alpha|\, d\theta = N(r, \alpha) + \log|f(0) - \alpha|.$$

If $\alpha = e^{i\varphi}$ has modulus one and $f(0) = 0$, this reduces to

$$\frac{1}{2\pi}\int_{-\pi}^{\pi} \log|f(re^{i\theta}) - e^{i\varphi}|\, d\theta = N(r, e^{i\varphi}).$$

Now integrate with respect to φ, using the simple identity

$$\frac{1}{2\pi}\int_{-\pi}^{\pi} \log|\beta - e^{i\varphi}|\, d\varphi = \log^+|\beta|,$$

to obtain

$$\int_{-\pi}^{\pi} \log^+|f(re^{i\theta})|\, d\theta = \int_{-\pi}^{\pi} N(r, e^{i\varphi})\, d\varphi.$$

If f is replaced by f/ρ, this becomes

$$\int_{-\pi}^{\pi} \log^+\left(\frac{|f(re^{i\theta})|}{\rho}\right) d\theta = \int_{-\pi}^{\pi} N(r, \rho e^{i\varphi})\, d\varphi. \tag{2}$$

But it is easy to see that if $f \in S$ and $\alpha \neq 0$ is in the range D of f, then

$$N(r, \alpha) = \log^+\left(\frac{r}{|f^{-1}(\alpha)|}\right), \qquad 0 < r < 1. \tag{3}$$

Now let $u(\zeta) = -\log|f^{-1}(\zeta)|$ be Green's function of D with pole at 0. Extend it to a continuous function in the punctured plane by setting $u(\zeta) = 0$, $\zeta \notin D$. Then the formula (3) takes the form

$$N(r, \zeta) = [u(\zeta) + \log r]^+, \qquad 0 < r < 1,$$

for arbitrary ζ, and equation (2) becomes

$$\int_{-\pi}^{\pi} \log^+\left(\frac{|f(re^{i\theta})|}{\rho}\right) d\theta = \int_{-\pi}^{\pi} [u(\rho e^{i\varphi}) + \log r]^+\, d\varphi. \tag{4}$$

Next let $v(\zeta) = -\log|k^{-1}(\zeta)|$ for ζ in the range of the Koebe function k, and let $v(\zeta) = 0$ elsewhere (*i.e.*, for $-\infty < \zeta \leq -\frac{1}{4}$). In view of (4), the inequality (1) can be recast in the form

$$\int_{-\pi}^{\pi} [u(\rho e^{i\varphi}) + \log r]^+ \, d\varphi \leq \int_{-\pi}^{\pi} [v(\rho e^{i\varphi}) + \log r]^+ \, d\varphi,$$
$$0 < r < 1, \quad 0 < \rho < \infty.$$

But by Lemma 5 ((c) $\Rightarrow$ (b)), this is implied by the inequality

$$u^*(\rho e^{i\varphi}) \leq v^*(\rho e^{i\varphi}), \qquad 0 < \rho < \infty, \quad 0 \leq \varphi \leq \pi. \tag{5}$$

The proof of (5) will make use of Lemma 1 (to be proved in §7.4). The function $u(\zeta)$ is continuous in $0 < |\zeta| < \infty$. In the domain D it is positive and harmonic, and $u(\zeta) \equiv 0$ outside D. In particular, u has the local sub-mean-value property at each point $\zeta \notin D$. This shows that $u(\zeta)$ is subharmonic in $0 < |\zeta| < \infty$. Hence it follows from Lemma 1 that u^* is subharmonic in the (open) upper half-plane.

The next step is to observe that v^* is harmonic in the upper half-plane. First note that $v(\rho e^{i\varphi}) = v(\rho e^{-i\varphi})$, since $k(\bar{z}) = \overline{k(z)}$. Also, $v(\rho e^{i\varphi})$ is a decreasing function of φ in the interval $(0, \pi)$. To see this, let $z = k^{-1}(\zeta)$ and $\zeta = \rho e^{i\varphi}$, and compute

$$\frac{\partial}{\partial \varphi} v(\rho e^{i\varphi}) = \operatorname{Im}\left\{\frac{1 - z}{1 + z}\right\} < 0 \quad \text{for } \operatorname{Im}\{z\} > 0.$$

It is now evident that

$$v^*(\rho e^{i\varphi}) = \int_{-\varphi}^{\varphi} v(\rho e^{i\psi}) \, d\psi, \qquad 0 < \varphi < \pi. \tag{6}$$

This formula allows the direct calculation of the Laplacian

$$\frac{1}{\rho^2}\left\{\frac{\partial^2 v^*}{\partial(\log \rho)^2} + \frac{\partial^2 v^*}{\partial \varphi^2}\right\}.$$

Since $v(\rho e^{i\psi})$ is harmonic for $-\pi < \psi < \pi$,

$$\frac{\partial^2 v^*}{\partial\,(\log \rho)^2}(\rho e^{i\varphi}) = \int_{-\varphi}^{\varphi} \frac{\partial^2 v}{\partial(\log \rho)^2} \, d\psi = -\int_{-\varphi}^{\varphi} \frac{\partial^2 v}{\partial \varphi^2} \, d\psi$$
$$= \frac{\partial v}{\partial \varphi}(\rho e^{-i\varphi}) - \frac{\partial v}{\partial \varphi}(\rho e^{i\varphi}) = -\frac{\partial^2 v^*}{\partial \varphi^2}(\rho e^{i\varphi}).$$

Thus the Laplacian vanishes, and v^* is harmonic in the upper half-plane.

It is also clear from (6) that v^* is continuous in the closed upper half-plane, except at the origin. By Lemma 1, the same is true for u^*. Near the origin, u has the form

$$u(\zeta) = -\log|\zeta| + u_1(\zeta), \tag{7}$$

where u_1 is harmonic and $u_1(0) = 0$. Thus

$$u^*(\rho e^{i\varphi}) + 2\varphi \log \rho \to 0$$

as $\rho \to 0$, uniformly in φ $(0 \leq \varphi \leq \pi)$. Since the same is true of v^*, it follows that

$$[u^*(\rho e^{i\varphi}) - v^*(\rho e^{i\varphi})] \to 0 \quad \text{as } \rho \to 0,$$

uniformly for $\varphi \in [0, \pi]$. As $\zeta \to \infty$, it is geometrically obvious that $u(\zeta) \to 0$; thus $u^*(\rho e^{i\varphi}) \to 0$ as $\rho \to \infty$, uniformly in φ.

Since $(u^* - v^*)$ is subharmonic in the upper half-plane and continuous in its closure, the maximum principle reduces the proof of (5) to showing that $u^*(\zeta) \leq v^*(\zeta)$ on the real axis. On the positive real axis this is trivial, since by definition, $u^*(\zeta) = v^*(\zeta) = 0$ for $\zeta > 0$. Next let d be the distance from 0 to the complement of D. By the Koebe theorem, $d \geq \frac{1}{4}$. In the disk $|\zeta| < d$, $u(\zeta)$ has the form (7), where $u_1(\zeta)$ is harmonic in $|\zeta| < d$ and $u_1(0) = 0$. Thus

$$u^*(\rho e^{i\pi}) = \int_{-\pi}^{\pi} u(\rho e^{i\varphi})\, d\varphi = -2\pi \log \rho, \qquad 0 < \rho \leq d. \tag{8}$$

In fact, since $u(\zeta)$ is subharmonic in $0 < |\zeta| < \infty$, it is clear that u_1 is subharmonic in the whole plane. Applying this remark to

$$v(\zeta) = -\log|\zeta| + v_1(\zeta),$$

we see that

$$v^*(\rho e^{i\pi}) = -2\pi \log \rho + \int_{-\pi}^{\pi} v_1(\rho e^{i\varphi})\, d\varphi \geq -2\pi \log \rho, \qquad 0 < \rho < \infty. \tag{9}$$

In particular, $u^*(\zeta) \leq v^*(\zeta)$ for $-d \leq \zeta < 0$.

The inequality is more difficult to establish on the interval $-\infty < \zeta < -d$. For this purpose, we fix $\varepsilon > 0$ and consider the function

$$Q(\zeta) = u^*(\zeta) - v^*(\zeta) - \varepsilon\varphi, \qquad \zeta = \rho e^{i\varphi},$$

which is subharmonic in the upper half-plane and continuous in its closure, except at $\zeta = 0$. Since $u^*(\zeta) - v^*(\zeta) \to 0$ as $\zeta \to 0$ and as $\zeta \to \infty$, it is clear that

$$\limsup_{\zeta \to 0} Q(\zeta) \leq 0; \qquad \limsup_{\zeta \to \infty} Q(\zeta) \leq 0.$$

Let M be the maximum of $Q(\zeta)$ in the closed upper half-plane. Then $M \geq 0$, and the maximum is attained somewhere on the real axis.

Suppose now that $M > 0$. Then since $u^*(\zeta) \leq v^*(\zeta)$ on the interval $-d \leq \zeta < \infty$, there is some point $\zeta_0 = -\rho_0$ for which $-\infty < \zeta_0 < -d$ and $Q(\zeta_0) = M$. Let $G(\varphi)$ denote the symmetric decreasing rearrangement of $u(\rho_0 e^{i\varphi})$. In view of Lemma 4,

$$\frac{\partial u^*}{\partial \varphi}(\rho_0 e^{i\varphi}) = 2G(\varphi), \qquad 0 \leq \varphi \leq \pi.$$

But because $\rho_0 > d$, there is some point on the circle $|\zeta| = \rho_0$ which lies outside D, so

$$G(\pi) = \inf_{0 \leq \varphi \leq \pi} u(\rho_0 e^{i\varphi}) = 0.$$

Applying the same argument to v^*, we conclude that

$$\frac{\partial Q}{\partial \varphi}(\zeta_0) = \frac{\partial Q}{\partial \varphi}(\rho_0 e^{i\pi}) = -\varepsilon < 0.$$

But this is impossible, since $Q(\zeta)$ has a relative maximum at ζ_0. This contradiction shows that $M = 0$; that is,

$$u^*(\zeta) \leq v^*(\zeta) + \varepsilon\varphi \leq v^*(\zeta) + \varepsilon\pi, \qquad \operatorname{Im}\{\zeta\} \geq 0.$$

Letting ε tend to 0, we obtain (5). This completes the proof of the inequality in Baernstein's theorem.

It now remains only to investigate the case of equality. Under the assumption that Φ is strictly convex, we will show that if f is not a rotation of the Koebe function, then strict inequality holds.

Continuing with the same notation, we first note that if f is not a rotation of k, then $u^*(\zeta) < v^*(\zeta)$ throughout the upper half-plane. To see this, observe that $v(\zeta)$ fails to be harmonic in any annulus $\frac{1}{4} < |\zeta| < \rho$, since it is nonnegative there and equal to zero at interior points of the annulus on the segment $-\rho < \zeta < -\frac{1}{4}$. Thus $v_1(\zeta) = \log|\zeta| + v(\zeta)$ cannot be harmonic in the disk $|\zeta| < \rho$ if $\rho > \frac{1}{4}$. If $h(\zeta)$ is the function harmonic in $|\zeta| < \rho$ and equal to $v_1(\zeta)$ on $|\zeta| = \rho$, it follows that $v_1(\zeta) < h(\zeta)$ in $|\zeta| < \rho$. In particular,

$$0 = v_1(0) < h(0) = \frac{1}{2\pi}\int_{-\pi}^{\pi} v_1(\rho e^{i\varphi})\,d\varphi, \qquad \rho > \tfrac{1}{4}.$$

Comparing this with (8) and (9), and bearing in mind that $d > \frac{1}{4}$ if f is not a rotation of k, we conclude that $u^*(\zeta) < v^*(\zeta)$ for $-d < \zeta < -\frac{1}{4}$. Hence $(u^* - v^*)$ is a nonpositive subharmonic function in the upper half-plane, not identically zero. But by the maximum principle, this implies $u^*(\zeta) < v^*(\zeta)$ everywhere in the half-plane $\operatorname{Im}\{\zeta\} > 0$.

We now claim that

$$\int_{-\pi}^{\pi} [u(\rho e^{i\varphi}) - t]^+\,d\varphi < \int_{-\pi}^{\pi} [v(\rho e^{i\varphi}) - t]^+\,d\varphi$$

if $0 < \lambda_\rho(t) < 2\pi$, where

$$\lambda_\rho(t) = |\{\varphi : u(\rho e^{i\varphi}) > t\}|.$$

Indeed, since $u^*(\rho e^{i\varphi}) < v^*(\rho e^{i\varphi})$ for $0 < \rho < \infty$ and $0 < \varphi < \pi$, this conclusion follows easily from the proof of Lemma 5 ((c) $\Rightarrow$ (b)). On the other hand, it is geometrically clear that unless $f(z) \equiv z$ there will correspond to each $t > 0$ an open interval $I \subset (0, \infty)$ such that $0 < \lambda_\rho(t) < 2\pi$ for all $\rho \in I$. Thus if $f(z) \not\equiv z$, there corresponds to each $r \in (0, 1)$ an open interval I_r such that

$$\int_{-\pi}^{\pi} [u(\rho e^{i\varphi}) + \log r]^+\,d\varphi < \int_{-\pi}^{\pi} [v(\rho e^{i\varphi}) + \log r]^+\,d\varphi, \qquad \rho \in I_r;$$

or equivalently, in view of (4),

$$\int_{-\pi}^{\pi} \log^+\left(\frac{|f(re^{i\theta})|}{\rho}\right) d\theta < \int_{-\pi}^{\pi} \log^+\left(\frac{|k(re^{i\theta})|}{\rho}\right) d\theta, \qquad \rho \in I_r. \tag{10}$$

But this inequality (10) obviously remains true even for $f(z) \equiv z$, with $I_r = (r, r + \varepsilon)$ for sufficiently small $\varepsilon > 0$.

Now let $\Phi(s)$ be an arbitrary nondecreasing, strictly convex function. Fix $r \in (0, 1)$, let I_r be the interval for which (10) holds, and let J_r be the interval $\log I_r$. Let s_0 be a point to the left of J_r at which Φ is differentiable. Decompose Φ in the form

$$\Phi(s) = \Phi_1(s) + \Phi_2(s),$$

where

$$\Phi_1(s) = \begin{cases} \Phi(s), & s \le s_0 \\ \Phi(s_0) + \Phi'(s_0)(s - s_0), & s \ge s_0. \end{cases}$$

Then Φ_1 and Φ_2 are nondecreasing convex functions on $(-\infty, \infty)$, and Φ_2 is strictly increasing on (s_0, ∞). By Lemma 2, Φ_2 has the form

$$\Phi_2(s) = \int_{-\infty}^{s} [s - t]^+ \, d\mu(t), \qquad d\mu(t) \ge 0. \tag{11}$$

Since Φ_2 is strictly increasing on (s_0, ∞), $\mu(J_r) > 0$. Rewriting (10) in the form

$$\int_{-\pi}^{\pi} [\log|f(re^{i\theta})| - t]^+ \, d\theta < \int_{-\pi}^{\pi} [\log|k(re^{i\theta})| - t]^+ \, d\theta, \qquad t \in J_r,$$

using the representation (11), and interchanging the order of integration, we obtain

$$\int_{-\pi}^{\pi} \Phi_2(\log|f(re^{i\theta})|) \, d\theta < \int_{-\pi}^{\pi} \Phi_2(\log|k(re^{i\theta})|) \, d\theta.$$

But by Baernstein's theorem,

$$\int_{-\pi}^{\pi} \Phi_1(\log|f(re^{i\theta})|) \, d\theta \le \int_{-\pi}^{\pi} \Phi_1(\log|k(re^{i\theta})|) \, d\theta.$$

Adding these two inequalities, we conclude that strict inequality holds in Baernstein's theorem for the function Φ. This completes the proof.

§7.4. Subharmonic Property of the Star-Function

We now turn to the proof of Lemma 1, the main tool in the proof of Baernstein's theorem.

First we consider the assertion that u^* is continuous in the given semi-annulus. Choose an arbitrary pair of points $z = re^{i\theta}$ and $z' = r'e^{i\theta'}$ with $r, r' \in (r_1, r_2)$ and $\theta, \theta' \in [0, \pi]$. By Lemma 3, there is a set $E \subset [-\pi, \pi]$ of measure $|E| = 2\theta$ for which

$$u^*(re^{i\theta}) = \int_E u(re^{it}) \, dt.$$

Let $E' \subset [-\pi, \pi]$ be an arbitrary set of measure $|E'| = 2\theta'$, chosen so that $E' \subset E$ if $\theta' \leq \theta$ and $E \subset E'$ if $\theta \leq \theta'$. Then

$$\begin{aligned} u^*(z) - u^*(z') &\leq \int_E u(re^{it})\,dt - \int_{E'} u(r'e^{it})\,dt \\ &= \int_E u(re^{it})\,dt - \int_{E'} u(re^{it})\,dt + \int_{E'} [u(re^{it}) - u(r'e^{it})]\,dt \\ &\leq \int_F |u(re^{it})|\,dt + \int_{-\pi}^{\pi} |u(re^{it}) - u(r'e^{it})|\,dt, \end{aligned}$$

where $F = (E - E') \cup (E' - E)$ has measure $|F| = 2|\theta - \theta'|$. Interchanging the roles of z and z', and recalling that u is continuous, we see that $|u^*(z) - u^*(z')| < \varepsilon$ if $|z - z'| < \delta$. Thus u^* is continuous.

The subharmonicity of u^* lies deeper. It is convenient to view the function $u(re^{it})$ as defined (for fixed r) on the unit circle $\mathbb{T}$ rather than on the interval $[-\pi, \pi]$. Let n be a positive integer, and let

$$u_n^*(re^{i\theta}) = \sup_E \int_E u(re^{it})\,dt, \qquad 0 \leq \theta \leq \pi,$$

where the supremum is extended over all sets $E \subset \mathbb{T}$ of measure $|E| = 2\theta$ which are the union of n or fewer disjoint closed arcs. Clearly,

$$u_n^*(re^{i\theta}) \leq u_{n+1}^*(re^{i\theta}) \leq u^*(re^{i\theta}), \qquad n = 1, 2, \ldots.$$

A simple measure-theoretic argument shows that $u_n^*(re^{i\theta}) \to u^*(re^{i\theta})$ as $n \to \infty$. (Surround the complement of the set E of Lemma 3 by an open set of approximately equal measure, express this open set as a countable union of disjoint open arcs, pass to a finite union, and take complements.) Therefore, it is sufficient to show that each function u_n^* ($n = 1, 2, \ldots$) is subharmonic. The preceding argument may be adapted to show that u_n^* is continuous; we need only verify that u_n^* also has the local sub-mean-value property.

The proof will require some additional notation. For $0 < \rho < r$, let

$$r + \rho e^{i\psi} = r(\psi)e^{i\alpha(\psi)}, \qquad |\alpha(\psi)| < \frac{\pi}{2}.$$

Note that $r(-\psi) = r(\psi)$ and $\alpha(-\psi) = -\alpha(\psi)$. For $r_1 < r < r_2$, $0 \leq \theta \leq \pi$, and arbitrary real φ, define

$$v(r, \theta, \varphi) = \int_{-\theta}^{\theta} u(re^{i(t+\varphi)})\,dt.$$

We will need the identity

$$\int_{-\pi}^{\pi} v(r(\psi), \theta + \alpha(\psi), \varphi)\, d\psi = \int_{-\pi}^{\pi} v(r(\psi), \theta, \varphi + \alpha(\psi))\, d\psi, \qquad 0 < \theta < \pi, \tag{12}$$

valid for ρ so small that $r_1 < r(\psi) < r_2$ and $0 < \theta + \alpha(\psi) < \pi$. To prove (12), write

$$\int_{-\pi}^{\pi} v(r(\psi), \theta + \alpha(\psi), \varphi)\, d\psi = \int_{-\pi}^{\pi} [J_1(\psi) + J_2(\psi)]\, d\psi,$$

where

$$J_1(\psi) = \int_{-\theta - \alpha(\psi)}^{-\theta + \alpha(\psi)} u(r(\psi)e^{i(t+\varphi)})\, dt;$$

$$J_2(\psi) = \int_{-\theta + \alpha(\psi)}^{\theta + \alpha(\psi)} u(r(\psi)e^{i(t+\varphi)})\, dt.$$

But $\int_{-\pi}^{\pi} J_1(\psi)\, d\psi = 0$, since $J_1(-\psi) = -J_1(\psi)$. On the other hand, a simple transformation gives

$$J_2(\psi) = v(r(\psi), \theta, \varphi + \alpha(\psi)),$$

which completes the proof of (12).

If $I(\theta, \varphi)$ denotes the closed arc of the unit circle described counterclockwise from $e^{i(\varphi - \theta)}$ to $e^{i(\varphi + \theta)}$, we may write

$$v(r, \theta, \varphi) = \int_{I(\theta, \varphi)} u(re^{it})\, dt.$$

We are now ready to show that u_n^* has the local sub-mean-value property. Fix $re^{i\theta}$ with $r_1 < r < r_2$ and $0 < \theta < \pi$. The supremum in the definition of u_n^* is attained, simply because a continuous function on a compact subset of the torus $\mathbb{T}^{2n}$ has a maximum there. Thus there exists a set

$$E = \bigcup_{j=1}^{m} I(\theta_j, \varphi_j), \qquad \sum_{j=1}^{m} \theta_j = \theta, \qquad m \le n,$$

composed of disjoint arcs $I(\theta_j, \varphi_j)$, for which

$$u_n^*(re^{i\theta}) = \int_E u(re^{it})\, dt.$$

For $0 < \rho < r$ and $-\pi \le \psi \le \pi$, define the set

$$E(\psi) = I(\theta_1 + \alpha(\psi), \varphi_1) \cup \left\{ \bigcup_{j=2}^{m} I(\theta_j, \varphi_j + \alpha(\psi)) \right\}.$$

Let ρ be chosen small enough to keep the arcs in $E(\psi)$ disjoint for all ψ. Then $E(\psi)$ has measure

$$|E(\psi)| = 2\theta + 2\alpha(\psi),$$

so by the definition of u_n^*,

$$\begin{aligned} u_n^*(r(\psi)e^{i(\theta+\alpha(\psi))}) &\ge \int_{E(\psi)} u(r(\psi)e^{it})\,dt \\ &= v(r(\psi), \theta_1 + \alpha(\psi), \varphi_1) + \sum_{j=2}^{m} v(r(\psi), \theta_j, \varphi_j + \alpha(\psi)). \end{aligned}$$

Now integrate with respect to ψ and use (12) to obtain

$$\int_{-\pi}^{\pi} u_n^*(re^{i\theta} + \rho e^{i\psi})\,d\psi \ge \sum_{j=1}^{m} \int_{-\pi}^{\pi} v(r(\psi), \theta_j, \varphi_j + \alpha(\psi))\,d\psi.$$

But since u is assumed to be subharmonic,

$$\begin{aligned} \int_{-\pi}^{\pi} v(r(\psi), \theta_j, \varphi_j + \alpha(\psi))\,d\psi &= \int_{-\theta_j}^{\theta_j} \int_{-\pi}^{\pi} u(r(\psi)e^{i(t+\varphi_j+\alpha(\psi))})\,d\psi\,dt \\ &= \int_{-\theta_j}^{\theta_j} \int_{-\pi}^{\pi} u(re^{i(t+\varphi_j)} + \rho e^{i\psi})\,d\psi\,dt \\ &\ge 2\pi \int_{-\theta_j}^{\theta_j} u(re^{i(t+\varphi_j)})\,dt. \end{aligned}$$

Thus for sufficiently small ρ,

$$\begin{aligned} \frac{1}{2\pi} \int_{-\pi}^{\pi} u_n^*(re^{i\theta} + \rho e^{i\psi})\,d\psi &\ge \sum_{j=1}^{m} \int_{-\theta_j}^{\theta_j} u(re^{i(t+\varphi_j)})\,dt \\ &= \int_E u(re^{it})\,dt = u_n^*(re^{i\theta}). \end{aligned}$$

This shows that each function u_n^* has the local sub-mean-value property at each point of the open semiannulus. Hence u_n^* is subharmonic for each n, which implies that u^* is subharmonic in the semiannulus. This completes the proof of Lemma 1.

§7.5. Integral Means of Derivatives

The question naturally arises whether Baernstein's theorem extends to derivatives. More specifically, is it true that $M_p(r, f') \le M_p(r, k')$ for all $f \in S$?

For $p < \frac{1}{3}$ this is certainly false. It can be seen by direct calculation that k' belongs to the Hardy space H^p for all $p < \frac{1}{3}$, so that $M_p(r, k')$ remains bounded as r tends to 1. On the other hand, there exist functions $f \in S$ whose derivatives f' have a radial limit on no set of positive measure. (An example is sketched in Chapter 8, Exercise 17. In fact, f' can have much worse behavior. See Lohwater, Piranian, and Rudin [1] for the ultimate example.) In particular, $f' \notin H^p$ for any $p > 0$; that is, $M_p(r, f')$ is unbounded. (See, for instance, Duren [6], Chapter 2.) For such a function f, and for each $p < \frac{1}{3}$, the inequality $M_p(r, f') \le M_p(r, k')$ must fail for all r near 1.

In general, the sharp upper bound of $M_p(r, f')$ for $f \in S$ is unknown. For $p > \frac{1}{3}$ it seems a reasonable conjecture that $M_p(r, f') \le M_p(r, k')$.

There are two kinds of evidence in support of this conjecture. First, it is asymptotically correct, at least for $p > \frac{2}{5}$. Feng and MacGregor [1] have shown that for each $p > \frac{2}{5}$, and for each positive integer n,

$$M_p(r, f^{(n)}) = O(M_p(r, k^{(n)})), \qquad r \to 1,$$

for all $f \in S$. In particular, for $p > \frac{2}{5}$,

$$M_p(r, f') = O((1 - r)^{1/p - 3}), \qquad r \to 1.$$

(See Exercises 4, 5, and 6.) Second, the conjecture is true for certain subclasses of S. For instance, it is true for the subclass K of close-to-convex functions, as the following theorem asserts.

Theorem 7.2. *For* $0 < p < \infty$,

$$M_p(r, f') \le M_p(r, k'), \qquad 0 < r < 1,$$

for all $f \in K$.

For $p = 1$ this theorem has the geometric interpretation that among all close-to-convex functions, the arclength of the image of the circle $|z| = r$ is greatest for the Koebe function. In this context it was first proved by Duren [4] and Clunie and Duren [1], whose proof can be extended to $p > 1$. MacGregor [5] gave another proof for $p \ge 1$, using the theory of extreme points. The extension to $0 < p < 1$ was obtained by Leung [3] as an application of Baernstein's theory.

The proof depends on two further lemmas concerning the star-function.

Lemma 6. *Let g and h be real-valued functions integrable over $[-\pi, \pi]$. Then*

$$(g + h)^*(\theta) \leq g^*(\theta) + h^*(\theta), \qquad 0 \leq \theta \leq \pi.$$

Equality occurs if both g and h are symmetrically decreasing.

Proof. This follows immediately from the definition of the star-function.

Lemma 7. *Let u and v be subharmonic in $\mathbb{D}$, and suppose that v is subordinate to u. Then for each r, $0 \leq r < 1$,*

$$v^*(re^{i\theta}) \leq u^*(re^{i\theta}), \qquad 0 \leq \theta \leq \pi.$$

Proof. Let Φ be an arbitrary convex nondecreasing function on $(-\infty, \infty)$. Then the composite functions $\Phi \circ u$ and $\Phi \circ v$ are subharmonic, and $\Phi \circ v \prec \Phi \circ u$. By Littlewood's subordination theorem (Theorem 6.1), or rather by its proof, we conclude that

$$\int_{-\pi}^{\pi} \Phi(v(re^{i\theta}))\, d\theta \leq \int_{-\pi}^{\pi} \Phi(u(re^{i\theta}))\, d\theta, \qquad 0 \leq r \leq 1.$$

In view of Lemma 5 (§7.2), this is equivalent to the asserted inequality $v^* \leq u^*$.

Proof of Theorem. By the definition of close-to-convexity, the derivative of a function $f \in K$ has the representation

$$zf'(z) = e^{-i\alpha} g(z)p(z), \qquad -\frac{\pi}{2} < \alpha < \frac{\pi}{2},$$

where $g \in S^*$ and

$$p(z) = e^{i\alpha} + p_1 z + p_2 z^2 + \cdots$$

has positive real part in $\mathbb{D}$. Thus with obvious notation

$$\begin{aligned} \log|f'(z)| &= \log\left|\frac{g(z)}{z}\right| + \log|p(z)| \\ &= G(z) + P(z). \end{aligned}$$

Either from the proof of Baernstein's theorem or directly from the Herglotz representation of a starlike function (see Chapter 2, Exercise 11), we have

$$G^*(re^{i\theta}) \leq K^*(re^{i\theta}), \qquad 0 \leq \theta \leq \pi,$$

where $K(z) = \log|k(z)/z|$ and k is the Koebe function. By Lemma 7,

$$P^*(re^{i\theta}) \le H^*(re^{i\theta}), \qquad 0 \le \theta \le \pi,$$

where

$$H(z) = \log\left|\frac{1+z}{1-z}\right|.$$

Since the functions H and K are both symmetrically decreasing, Lemma 6 now gives

$$\begin{aligned}(\log|f'(z)|)^* &\le G^*(z) + P^*(z) \le K^*(z) + H^*(z)\\ &= (K+H)^*(z) = (\log|k'(z)|)^*.\end{aligned}$$

In view of Lemma 5, it follows that

$$\int_{-\pi}^{\pi} \Phi(\log|f'(re^{i\theta})|)\,d\theta \le \int_{-\pi}^{\pi} \Phi(\log|k'(re^{i\theta})|)\,d\theta$$

for every convex nondecreasing function Φ. The theorem results from the choice $\Phi(x) = e^{px}$.

It may be remarked that the key to the proof is a "lucky accident," the identity

$$k'(z) = \frac{1+z}{(1-z)^3} = \frac{k(z)}{z}\cdot\frac{1+z}{1-z}.$$

Leung's proof generalizes to certain Bazilevich functions. He shows that equality can occur only if f is a rotation of the Koebe function.

Brown [3] has found another proof of the theorem which has various ramifications.

Baernstein and Brown [1] have proved that $M_p(r, f') \le C_\beta M_p(r, k')$ for $0 < p < \infty$ and for all functions $f \in S$ which map the disk onto the complement of an arc with radial angle bounded by β, where $0 < \beta < \pi/2$ and C_β is a constant depending only on β. The radial angle is defined as the angle between the tangent and radial vectors to the arc. The condition $\beta < \pi/2$ ensures that the omitted arc has monotonic modulus. We shall see in Chapter 10 that every support point of S, or every solution to a linear extremal problem, maps the disk onto the complement of an analytic arc with radial angle at most $\pi/4$.

Exercises

1. Use Baernstein's inequality $M_1(r, f) \le M_1(r, k)$ to prove Rogosinski's conjecture with constant $e/2$. In other words, show that if $g(z) = \sum_{n=1}^{\infty} b_n z^n$ is subordinate to some function $f \in S$, then $|b_n| < (e/2)n$, $n = 1, 2, \ldots$. (Compare Chapter 2, Exercise 17.)

2. For a function $f \in S$, let $L_r(f)$ denote the arclength of the Jordan curve which is the image under f of the circle $|z| = r$. Show that $L_r(f) = 2\pi r M_1(r, f')$.

3. For the Koebe function k, show that

$$L_r(k) = 2\rho\left\{\frac{2E(\rho)}{1-\rho^2} - K(\rho)\right\} > \frac{\pi r(1+r)}{2(1-r)^2},$$

where $\rho = 2r/(1 + r^2)$, and K and E denote the standard elliptic integrals of the first and second kinds, with modulus ρ. Show also that $L_r(k) \sim 4(1 - r)^{-2}$ as $r \to 1$. (Duren [4].) (These results may be compared with the estimate $L_r(f) \le 2\pi r(1 - r)^{-2}$ for arbitrary $f \in S$, obtained in §2.4 with the help of Baernstein's inequality $M_1(r, f) \le M_1(r, k)$. It is not known whether $L_r(f) \le L_r(k)$ for all $f \in S$.)

4. Show that for each index p, $0 < p \le \infty$, the Koebe function has the property $M_p(r, k') \sim C(1 - r)^{1/p-3}$ as $r \to 1$, for some constant C depending on p.

5. It is known (see Duren [6], Chapter 5) that for $p > 0$ and $\alpha > 0$, the conditions $M_p(r, f) = O((1 - r)^{-\alpha})$ and $M_p(r, f') = O((1 - r)^{-\alpha-1})$ are equivalent for arbitrary functions f analytic in $\mathbb{D}$. Use this to show that $M_p(r, f') = O((1 - r)^{1/p-3})$ for each $p > \frac{1}{2}$ and for every $f \in S$.

6. Show that $M_p(r, f') = O((1 - r)^{1/p-3})$ for each $p > \frac{2}{5}$ and for every $f \in S$. (Feng and MacGregor [1].)

7. Show that for each function $f \in S$, the inequality

$$\int_{-\pi}^{\pi} \Phi\left(\log\frac{\rho}{|f(re^{i\theta})|}\right) d\theta \le \int_{-\pi}^{\pi} \Phi\left(\log\frac{\rho}{|k(re^{i\theta})|}\right) d\theta$$

holds for $0 < r < 1$, $\rho > 0$, and for every convex nondecreasing function Φ. Conclude that $M_p(r, 1/f) \le M_p(r, 1/k)$ for $0 < r < 1$ and $0 < p < \infty$. (Baernstein [1].)

8. For functions $f \in S$, let the logarithmic coefficients γ_n be defined as usual by

$$\log\frac{f(z)}{z} = 2\sum_{n=1}^{\infty} \gamma_n z^n.$$

Show that $\sum_{n=1}^{\infty} |\gamma_n|^2 \leq \pi^2/6$, with equality for rotations of the Koebe function. (*Hint*: Work with the convex function

$$\Phi(u) = \begin{cases} u^2, & u \geq 0 \\ 0, & u < 0.) \end{cases}$$

(Duren and Leung [1].)

Chapter 8

Some Special Topics

This chapter is devoted to three unrelated topics. The first topic is the asymptotic coefficient problem for bounded functions in S and for functions in Σ. The second broad topic is the problem of coefficient multipliers for S and certain subclasses. This includes diverse questions such as the univalence of sections and the univalence of integrals, but the main focus is the proof of the Pólya–Schoenberg conjecture on the multipliers of convex functions. The third topic is an important criterion for univalence involving the Schwarzian derivative, with various ramifications. The chapter concludes with brief surveys of three additional topics: Bieberbach–Eilenberg functions, univalent polynomials, and functions of bounded boundary rotation.

§8.1. Bounded Univalent Functions

The Cauchy integral formula shows that every bounded analytic function in the unit disk has bounded coefficients. A deeper argument based on the Riemann–Lebesgue lemma shows that $a_n \to 0$ if $f(z) = \sum a_n z^n$ is bounded in $\mathbb{D}$ (and more generally if $f \in H^1$), and nothing stronger can be said in general. (See Duren [6], pp. 38, 100.) However, a bounded *univalent* function has finite Dirichlet integral:

$$\iint_{|z|<1} |f'(z)|^2 \, dx \, dy = \pi \sum_{n=1}^{\infty} n|a_n|^2 < \infty,$$

and so $a_n = o(n^{-1/2})$. This simple estimate was suspected to be best possible until 1966, when Clunie and Pommerenke [2] discovered an essential improvement. Here is a weakened version of their result.

Theorem 8.1. *There is an absolute constant $\alpha > 0$ such that $a_n = O(n^{-1/2-\alpha})$ for every bounded function $f \in S$.*

Proof. For each $\delta > 0$, the Schwarz inequality gives

$$\int_0^{2\pi} |f'(re^{i\theta})|^{1+\delta}\, d\theta \le \{I(r)J(r)\}^{1/2}, \tag{1}$$

where

$$I(r) = \int_0^{2\pi} |f'(re^{i\theta})|^{2\delta}\, d\theta; \qquad J(r) = \int_0^{2\pi} |f'(re^{i\theta})|^2\, d\theta.$$

Since $J(r)$ increases with r,

$$r(1-r)J(r) \le \int_r^1 tJ(t)\, dt \le \int_0^1 tJ(t)\, dt = \pi \sum_{n=1}^{\infty} n|a_n|^2 < \infty.$$

Thus

$$J(r) = O((1-r)^{-1}), \qquad r \to 1. \tag{2}$$

The estimation of $I(r)$ is more difficult. Since f is locally univalent, the function

$$F(z) = [f'(z)]^{\delta} = \sum_{n=0}^{\infty} c_n z^n, \qquad c_0 = 1,$$

is analytic in the unit disk, and

$$I(r) = \int_0^{2\pi} |F(re^{i\theta})|^2\, d\theta = 2\pi \sum_{n=0}^{\infty} |c_n|^2 r^{2n}.$$

It follows that

$$I''(r) \le 8\pi \sum_{n=1}^{\infty} n^2 |c_n|^2 r^{2n-2}.$$

On the other hand, the elementary inequality

$$\left|\frac{f''(z)}{f'(z)}\right| \le \frac{6}{1-r}, \qquad z = re^{i\theta},$$

holds for every $f \in S$ (*cf.* Theorem 2.4), and so

$$\begin{aligned} 2\pi \sum_{n=1}^{\infty} n^2 |c_n|^2 r^{2n-2} &= \int_0^{2\pi} |F'(z)|^2\, d\theta \\ &= \delta^2 \int_0^{2\pi} \left|\frac{f''(z)}{f'(z)}\right|^2 |f'(z)|^{2\delta}\, d\theta \le \frac{36\delta^2}{(1-r)^2} I(r). \end{aligned}$$

Combining these inequalities, we see that

$$[\log I(r)]'' = \frac{I''(r)}{I(r)} - \left[\frac{I'(r)}{I(r)}\right]^2 \leq \frac{I''(r)}{I(r)} \leq \frac{144\delta^2}{(1-r)^2},$$

whereupon two integrations from 0 to r yield

$$\log I(r) \leq \log 2\pi - 144\delta^2 \log(1-r),$$

or

$$I(r) \leq 2\pi(1-r)^{-144\delta^2}. \tag{3}$$

The estimates (1), (2), and (3) give

$$\int_0^{2\pi} |f'(re^{i\theta})|^{1+\delta}\, d\theta = O((1-r)^{-1/2-72\delta^2}). \tag{4}$$

For fixed γ, $0 < \gamma < \frac{1}{2}$, let

$$E_1 = \{\theta: |f'(re^{i\theta})| \leq (1-r)^{-\gamma}\};$$

$$E_2 = \{\theta: |f'(re^{i\theta})| > (1-r)^{-\gamma}\}.$$

Then by (4),

$$\begin{aligned}\int_0^{2\pi} |f'(re^{i\theta})|\, d\theta &= \int_{E_1} |f'(re^{i\theta})|\, d\theta + \int_{E_2} |f'(re^{i\theta})|\, d\theta \\ &\leq 2\pi(1-r)^{-\gamma} + (1-r)^{\gamma\delta} \int_0^{2\pi} |f'(re^{i\theta})|^{1+\delta}\, d\theta \\ &\leq 2\pi(1-r)^{-\gamma} + C(1-r)^{\gamma\delta-1/2-72\delta^2}.\end{aligned}$$

Now choose $\delta = \gamma/144$ to give the exponent $\gamma\delta - \frac{1}{2} - 72\delta^2$ its maximum value $\gamma^2/288 - \frac{1}{2}$, which is larger than $-\gamma$ for γ sufficiently near $\frac{1}{2}$. Thus for some $\gamma < \frac{1}{2}$ we have

$$\int_0^{2\pi} |f'(re^{i\theta})|\, d\theta \leq C(1-r)^{-\gamma}.$$

It follows from the Cauchy formula that

$$n|a_n| \leq \frac{1}{2\pi} r^{-n+1} \int_0^{2\pi} |f'(re^{i\theta})|\, d\theta \leq Cr^{-n}(1-r)^{-\gamma}.$$

The choice $r = 1 - 1/n$ therefore gives $a_n = O(n^{\gamma-1})$, which completes the proof of the theorem.

This version of the proof provides the relatively poor estimate $\alpha > 0.00086\ldots$. The more refined argument of Clunie and Pommerenke [2] gives $\alpha > 1/300$. The best value of α is unknown.

The problem has a surprisingly close connection with the asymptotic coefficient problem for the class Σ of univalent functions

$$g(z) = z + \sum_{n=0}^{\infty} b_n z^{-n}, \qquad 1 < |z| < \infty.$$

The area theorem gives $\sum_{n=1}^{\infty} n|b_n|^2 \leq 1$, so again $b_n = o(n^{-1/2})$. By a method parallel to the above, Clunie and Pommerenke [2] improved this to $b_n = O(n^{-1/2-\beta})$, where $\beta = 1/300$. The best value of β is again unknown.

This latter result actually implies Theorem 8.1 with $\alpha = 1/300$. In fact, it can be shown that $\alpha \geq \beta$. More precisely, one can establish the following theorem, due to Pommerenke [7].

Theorem 8.2. *If the coefficients b_n of each function $g \in \Sigma$ satisfy $b_n = O(n^{-1/2-\beta+\varepsilon})$ for every $\varepsilon > 0$, then the same is true for the coefficients a_n of each bounded function $f \in S$.*

Proof. Suppose, on the contrary, that $a_n \neq O(n^{-1/2-\beta+2\varepsilon})$ for some bounded $f \in S$ and some $\varepsilon > 0$. Consider the cube-root transform

$$h(z) = [f(z^3)]^{1/3} = \sum_{n=0}^{\infty} c_n z^{3n+1}, \qquad c_0 = 1.$$

We claim that $c_n \neq O(n^{-1/2-\beta+\varepsilon})$. Otherwise, we could deduce

$$\begin{aligned}|a_{n+1}| &\leq \sum\sum_{i+j+k=n}\sum |c_i c_j c_k| \leq O(n^{-1/2-\beta+\varepsilon}) \sum_{j+k\leq n}\sum |c_j c_k| \\ &\leq O(n^{-1/2-\beta+\varepsilon}) \left\{\sum_{k=0}^{n} |c_k|\right\}^2 = O(n^{-1/2-\beta+2\varepsilon}),\end{aligned}$$

since

$$\sum_{k=1}^{n} |c_k| \leq \left\{\sum_{k=1}^{n} \frac{1}{k}\right\}^{1/2} \left\{\sum_{k=1}^{n} k|c_k|^2\right\}^{1/2} = O((\log n)^{1/2}).$$

Now observe that $|h(z)| < M$ for some constant M, and note that $\psi(w) = w/M^2 + 1/w$ is univalent in $|w| < M$. Thus $g(\zeta) = \psi(h(1/\zeta))$ belongs to Σ, and

$$g(\zeta) = \zeta + M^{-2} \sum_{\nu=0}^{\infty} c_\nu \zeta^{-3\nu-1} + \sum_{\nu=1}^{\infty} d_\nu \zeta^{-3\nu+1}.$$

In particular, $b_{3\nu+1} = c_\nu/M^2$, so $b_n \neq O(n^{-1/2-\beta+\varepsilon})$. This completes the proof.

For bounded *close-to-convex* functions, Clunie and Pommerenke [1] improved the estimate of Theorem 8.1 to $a_n = O(1/n)$, a best possible result. In general, however, the coefficients of bounded univalent functions need not be $O(1/n)$, as Littlewood [2] demonstrated in 1938. Clunie [1] later used a similar construction to show that the coefficients b_n of functions in Σ need not be $O(1/n)$. (In particular, this thoroughly disproves the conjecture $|b_n| \leq 2/(n+1)$ discussed in §4.7.) Littlewood's construction uses lacunary power series, a natural device for producing bounded analytic functions whose coefficients are occasionally large. The difficulty, however, lies in showing that a given power series is univalent. Pommerenke [7, 8, 14] managed to construct examples by the more delicate method of successive composition of univalent functions. This approach is somewhat easier than Littlewood's, and the results are sharper. In the proof of the next theorem we shall follow Pommerenke [8].

Theorem 8.3. *For each integer $m \geq 1$, there exists a bounded univalent function*

$$f(z) = \sum_{\nu=0}^{\infty} a_{m\nu+1} z^{m\nu+1} \tag{5}$$

with $|f(z)| < 1$ in $\mathbb{D}$, all $a_n \geq 0$, and $a_n \neq O(n^{-0.83})$.

Proof. It will be convenient to use the notation

$$\sum_{n=0}^{\infty} \alpha_n z^n \ll \sum_{n=0}^{\infty} \beta_n z^n$$

to indicate that $\alpha_n \leq \beta_n$ for all n. Let

$$p(z) = 1 + \sum_{n=1}^{\infty} c_n z^n$$

be analytic in $|z| < 1$ and have the properties $\operatorname{Re}\{p(z)\} > 0$, $c_n \geq 0$ for all n, and

$$\lambda = \sum_{n=1}^{\infty} \frac{c_n}{n} < \infty.$$

Choose an integer $q \geq 2$ and form the functions

$$\varphi_k(z) = ze^{-(\lambda/m)q^{-k}} e^{\psi_k(z)},$$

where

$$\psi_k(z) = \frac{1}{mq^k} \sum_{n=1}^{\infty} \frac{c_n}{n} z^{nmq^k}, \qquad k = 1, 2, \ldots.$$

Trivial estimates give $|\psi_k(z)| < (\lambda/m)q^{-k}$ and $|\varphi_k(z)| < 1$. It is also obvious that

$$\psi_k(z) \gg (c_1/m)q^{-k} z^{mq^k},$$

and so

$$\varphi_k(z) \gg ze^{-(\lambda/m)q^{-k}}\{1 + (c_1/m)q^{-k} z^{mq^k}\}.$$

A simple calculation gives

$$\frac{z\varphi_k'(z)}{\varphi_k(z)} = 1 + \sum_{n=1}^{\infty} c_n z^{nmq^k} = p(z^{mq^k}),$$

which shows that φ_k is starlike and therefore univalent in $|z| < 1$.

Now define $f_1(z) = z$ and $f_{k+1}(z) = f_k(\varphi_k(z))$ for $k = 1, 2, \ldots$. Thus $f_{k+1} = \varphi_1 \circ \varphi_2 \circ \cdots \circ \varphi_k$. Let

$$f_k(z) = \sum_{n=1}^{\infty} a_{kn} z^n.$$

It is clear from the corresponding properties of the functions φ_k that $|f_k(z)| < 1$ in $|z| < 1$, that $a_{kn} \geq 0$, and that $a_{kn} = 0$ unless $n = m\nu + 1$; that is, each f_k has m-fold symmetry. Observe that

$$\begin{aligned} f_{k+1}(z) = f_k(\varphi_k(z)) &\gg a_{kn}[\varphi_k(z)]^n \\ &\gg a_{kn} e^{-(\lambda n/m)q^{-k}} z^n \{1 + (nc_1/m)q^{-k} z^{mq^k}\}. \end{aligned} \tag{6}$$

Define

$$n_k = 1 + m(1 + q + \cdots + q^{k-1}) = 1 + m(q^k - 1)/(q - 1),$$

so that $n_{k+1} = n_k + mq^k$. Then by (6),

$$a_{k+1, n_{k+1}} \geq a_{kn_k}(n_k c_1/m)q^{-k} \exp\{-\lambda n_k q^{-k}/m\}.$$

With the notation $A_k = n_k a_{kn_k}$, it follows that

$$A_{k+1} \geq A_k \frac{c_1 q}{q-1}(1 - q^{-k-1}) \exp\left\{\frac{-\lambda q^{-k}}{m} - \frac{\lambda}{q-1}\right\}. \tag{7}$$

Now define β by

$$q^\beta = \frac{c_1 q}{q-1} \exp\left\{-\frac{\lambda}{q-1}\right\} \tag{8}$$

and conclude from (7) that

$$A_k \geq Bq^{\beta(k-1)}, \qquad k = 1, 2, \ldots,$$

where

$$B = A_1 \prod_{j=1}^{\infty}(1 - q^{-j}) \exp\left\{-\frac{\lambda}{m(q-1)}\right\} > 0.$$

Since $n_k \sim mq^k/(q-1)$, this implies

$$a_{kn_k} \geq Cn_k^{\beta-1}, \qquad k = 1, 2, \ldots, \tag{9}$$

for some constant $C > 0$. It also follows from (6) that

$$a_{k+1, n_j} \geq a_{kn_j} \exp\{-\lambda n_j q^{-k}/m\},$$

which may be iterated and combined with (9) to give

$$a_{kn_j} \geq Dn_j^{\beta-1}, \qquad 1 \leq j \leq k, \tag{10}$$

for some constant $D > 0$.

Because $|f_k(z)| < 1$, some subsequence converges uniformly on compact subsets to an analytic function f of the form (5). The limit function is not constant because

$$f_k'(0) = a_{k1} \geq e^{-\lambda/m(q-1)} > 0.$$

Thus f is univalent, and it is clear that $|f(z)| < 1$. Since $a_{kn} \to a_n$ for each n, it follows from (10) that

$$a_{n_j} \geq D n_j^{\beta - 1}, \qquad j = 1, 2, \ldots. \tag{11}$$

The final step is to make an appropriate choice of the function p upon which the construction is based. Fix a number τ in the interval $0 < \tau < \pi$ and let

$$p(z) = 1 + \frac{4}{\tau^2} \sum_{n=1}^{\infty} \frac{1 - \cos n\tau}{n^2} z^n, \qquad |z| \leq 1.$$

Then by the elementary theory of Fourier series, one verifies that

$$\operatorname{Re}\{p(e^{i\theta})\} = \begin{cases} 2\pi\tau^{-2}(\tau - |\theta|), & |\theta| \leq \tau \\ 0, & \tau \leq |\theta| \leq \pi. \end{cases}$$

In particular, $\operatorname{Re}\{p(z)\} > 0$ in $|z| < 1$. Now choose $\tau = \pi/3$ and compute

$$c_1 = 4\tau^{-2}(1 - \cos \tau) = 18\pi^{-2}$$

and

$$\begin{aligned} \lambda &= 4\tau^{-2} \sum_{n=1}^{\infty} n^{-3}(1 - \cos n\tau) \\ &= 36\pi^{-2}\zeta(3) - 36\pi^{-2} \sum_{n=1}^{\infty} n^{-3} \cos(n\pi/3) \\ &= 36\pi^{-2}\Big\{\zeta(3) + \sum_{\nu=1}^{\infty} (-1)^{\nu-1}(3\nu)^{-3}\Big\} \\ &\quad - 18\pi^{-2} \sum_{\nu=0}^{\infty} (-1)^{\nu}\{(3\nu + 1)^{-3} - (3\nu + 2)^{-3}\} \\ &= 37\pi^{-2}\zeta(3) - 18\pi^{-2} \sum_{\nu=0}^{\infty} (-1)^{\nu}\{(3\nu + 1)^{-3} - (3\nu + 2)^{-3}\} \\ &< 37\pi^{-2}\zeta(3) - 18\pi^{-2}\{(1 - 2^{-3}) - (4^{-3} - 5^{-3})\} < 2.93, \end{aligned}$$

where $\zeta(n) = \sum_{\nu=1}^{\infty} \nu^{-n}$ is the Riemann zeta function. Now choose $q = 14$ to obtain from (8) that $\beta > 0.17$. Hence the estimate (11) shows that $a_n \neq O(n^{-0.83})$.

Corollary. *There is a function*

$$g(z) = z + b_0 + b_1 z^{-1} + b_2 z^{-2} + \cdots$$

of class Σ *for which* $b_n \neq O(n^{-0.83})$.

This result is obtained by combining Theorems 8.2 and 8.3. It shows that the conjecture $|b_n| \leq 2/(n+1)$, which was discussed in Chapter 4, is not even asymptotically correct.

Theorem 8.3 also disproves an old conjecture of Szegö concerning the asymptotic behavior of functions of class S with m-fold symmetry. Recall that $S^{(m)}$ is the subclass of all functions $f \in S$ with power series expansions of the form

$$f(z) = z + \sum_{\nu=1}^{\infty} a_{m\nu+1} z^{m\nu+1}.$$

Szegö conjectured that $a_n = O(n^{2/m-1})$ as $n \to \infty$ for each $f \in S^{(m)}$, $m = 1, 2, \ldots$. For $m = 1$ this is certainly true, and for $m = 2$ it is the theorem of Littlewood and Paley (Theorem 2.23 in §2.11). Beyond this, Levin [2] proved it for $m = 3$ and showed for $m = 4$ that $a_n = O(n^{-1/2} \log n)$. The conjecture seems highly plausible because the mth-root transform of the Koebe function, the function $f(z) = \{k(z^m)\}^{1/m}$, has coefficients satisfying

$$a_{m\nu+1} \sim [\Gamma(2/m)]^{-1}(m\nu + 1)^{2/m-1}, \qquad \nu \to \infty.$$

However, the example constructed by Littlewood [2] disproves the conjecture for sufficiently large m, and Pommerenke's construction (Theorem 8.3) disproves it for all $m \geq 12$. For $4 \leq m \leq 11$ the conjecture remains unsettled.

Pommerenke's construction can also be applied without essential change to show that the logarithmic coefficients γ_n of a function $f \in S$ need not be $O(1/n)$. This problem arose in connection with Milin's theory (see §5.4). Recall the definition

$$\log \frac{f(z)}{z} = 2 \sum_{n=1}^{\infty} \gamma_n z^n.$$

Theorem 8.4. *There is a bounded function* $f \in S$ *with logarithmic coefficients* $\gamma_n \neq O(n^{-0.83})$.

Proof. Let $f_{k+1} = f_k \circ \varphi_k$ as in the proof of Theorem 8.3, assuming for simplicity that $m = 1$. Write

$$\log \frac{f_k(z)}{z} = 2 \sum_{n=0}^{\infty} \gamma_{kn} z^n$$

and observe that the identity

$$\log \frac{f_{k+1}(z)}{z} = \log \frac{f_k(\varphi_k(z))}{\varphi_k(z)} + \log \frac{\varphi_k(z)}{z}$$

shows inductively that all $\gamma_{kn} \geq 0$. The same identity gives

$$2 \sum_{n=0}^{\infty} \gamma_{k+1,n} z^n = 2 \sum_{n=0}^{\infty} \gamma_{kn}[\varphi_k(z)]^n + \log \frac{\varphi_k(z)}{z},$$

which yields a precise analogue of the relation (6) and therefore leads in the same manner to an inequality

$$\gamma_{kn_j} \geq Dn_j^{\beta-1}, \qquad 1 \leq j \leq k,$$

analogous to (10). Again passing to the limit through a suitable subsequence, we reach the desired conclusion, with $\beta > 0.17$ as in the proof of Theorem 8.3.

§8.2. Sections of Univalent Functions

Let $f(z) = z + a_2 z^2 + a_3 z^3 + \cdots$ be analytic in the unit disk. The nth partial sum

$$s_n(z) = s_n(z; f) = z + a_2 z^2 + \cdots + a_n z^n$$

of the power series is called the nth *section* (German: *Abschnitt*) of f. It is a consequence of Rouché's theorem that for each $f \in S$ and for each given radius $\rho < 1$, the nth sections s_n are univalent in the disk $|z| < \rho$ for all n sufficiently large. (See Chapter 1, Exercise 5.) A theorem of Szegö [3] sharpens this result in a remarkable way.

Theorem 8.5 (Szegö's Theorem). *Every section of a function of class S is univalent in the disk $|z| < \frac{1}{4}$. The radius $\frac{1}{4}$ is best possible.*

Proof. Observe first that the second section

$$s_2(z) = s_2(z; k) = z + 2z^2$$

of the Koebe function is not univalent in any larger disk, since $s_2'(-\frac{1}{4}) = 0$. Thus the radius $\frac{1}{4}$ is best possible.

It is easy to prove the theorem for $n = 2$. For arbitrary $f \in S$,

$$s_2(z_1; f) - s_2(z_2; f) = (z_1 - z_2)[1 + a_2(z_1 + z_2)]$$

for all points z_1 and z_2 in the unit disk. But for $|z_1| < \frac{1}{4}$ and $|z_2| < \frac{1}{4}$,

$$|1 + a_2(z_1 + z_2)| \geq 1 - 2(|z_1| + |z_2|) > 0,$$

since $|a_2| \leq 2$. Thus s_2 is univalent in $|z| < \frac{1}{4}$.

The case $n = 3$ is the most difficult part of the proof and will be deferred to the end. We consider next the general case $n \geq 4$. It is enough to show that $s_n(z_1) \neq s_n(z_2)$ for $|z_1| = |z_2| = \frac{1}{4}$, $z_1 \neq z_2$. Since

$$f(z) = s_n(z) + \sum_{k=n+1}^{\infty} a_k z^k,$$

this will follow from the inequality

$$\left| \frac{f(z_1) - f(z_2)}{z_1 - z_2} \right| > \left| \sum_{k=n+1}^{\infty} a_k \left(\frac{z_1^k - z_2^k}{z_1 - z_2} \right) \right|, \qquad |z_1| = |z_2| = \tfrac{1}{4}. \tag{12}$$

A proof of (12) will be based on the following consequence of the Lebedev inequalities, already noted in §4.4 (Theorem 4.3, Corollary 7). For each $f \in S$ and for $|z_1| = |z_2| = r < 1$,

$$\left| \frac{f(z_1) - f(z_2)}{z_1 - z_2} \right| \geq |f(z_1)f(z_2)| \left(\frac{1 - r^2}{r^2} \right) \geq \frac{1 - r}{(1 + r)^3}, \tag{13}$$

where the growth theorem (§2.3) has been used. For $r = \frac{1}{4}$ the right-hand member of (13) reduces to $\frac{48}{125}$. Thus we need only show that the right-hand member of (12) is less than $\frac{48}{125}$. But in view of FitzGerald's estimate $|a_k| < \sqrt{\frac{7}{6}}k$ (see §5.12), we find for $n \geq 4$ and for $|z_1| = |z_2| = \frac{1}{4}$ that

$$\left| \sum_{k=n+1}^{\infty} a_k \left(\frac{z_1^k - z_2^k}{z_1 - z_2} \right) \right| \leq \sum_{k=n+1}^{\infty} k |a_k| (\tfrac{1}{4})^{k-1}$$

$$< \sqrt{\tfrac{7}{6}} \sum_{k=5}^{\infty} k^2 (\tfrac{1}{4})^{k-1} < 0.17 < \tfrac{48}{125}.$$

This proves (12) and establishes the theorem for $n \geq 4$. It may be remarked that FitzGerald's estimate is stronger than necessary, but the above proof fails for $n = 3$ even if the Bieberbach conjecture is assumed.

The case $n = 3$ can be handled by appeal to the Loewner formulas for a_2 and a_3 (see §3.5):

$$a_2 = 2 \int_0^{\infty} e^{-t} \kappa(t)\, dt;$$

$$a_3 = a_2^2 - 2 \int_0^{\infty} e^{-2t} [\kappa(t)]^2\, dt,$$

where κ is a continuous function with $|\kappa(t)| = 1$. In order to show that $s_3(z_1) \neq s_3(z_2)$ for distinct points z_1 and z_2 with $|z_1| = |z_2| = \frac{1}{4}$, it will be sufficient to prove

$$1 + \operatorname{Re}\{a_2(z_1 + z_2) + a_3(z_1^2 + z_1 z_2 + z_2^2)\} > 0. \tag{14}$$

There is no loss of generality in supposing that z_1 and z_2 are complex conjugates:

$$z_1 = \tfrac{1}{4}e^{i\psi} \quad \text{and} \quad z_2 = \tfrac{1}{4}e^{-i\psi}, \qquad 0 < \psi \leq \frac{\pi}{2}.$$

(The expression in (14) is unchanged if z_1 and z_2 are replaced by αz_1 and αz_2, where $|\alpha| = 1$; while f is replaced by the rotation $\alpha f(\bar{\alpha} z)$.) With the notation $u = \cos\psi, a_2 = x + iy$, and $a_3 = \xi + i\eta$, the inequality (14) to be proved then becomes

$$G = 1 + \tfrac{1}{2}ux + \tfrac{1}{16}(4u^2 - 1)\xi > 0, \qquad 0 \leq u < 1. \tag{15}$$

Suppose first that $\xi \geq 0$. Then (15) is obviously true if $u \geq \frac{1}{2}$, since $x \geq -2$. If $0 \leq u < \frac{1}{2}$, then since $\xi \leq 3$,

$$G > 1 - \tfrac{1}{2} - \tfrac{3}{16} > 0.$$

Next suppose $\xi < 0$. Then the graph of G as a function of u is a parabola opening downward, so the inequality $G > 0$ need be verified only at the endpoints of the interval $0 \leq u \leq 1$. For $u = 0$ it is obviously true. For $u = 1$ the assertion reduces to

$$G = 1 + \tfrac{1}{2}x + \tfrac{3}{16}\xi > 0.$$

But by the Loewner formula for a_3,

$$\xi = x^2 - y^2 - 2\int_0^\infty e^{-2t}[1 - 2\sin^2\theta(t)]\,dt,$$

where $\kappa(t) = e^{i\theta(t)}$. Furthermore, an application of the Schwarz inequality to the imaginary part of the Loewner formula for a_2 gives

$$y^2 \leq 4\int_0^\infty e^{-t}\sin^2\theta(t)\int_0^\infty e^{-t}\,dt = 4\int_0^\infty e^{-t}\sin^2\theta(t).$$

Therefore, since $x + \frac{3}{8}x^2 \geq -\frac{2}{3}$, we have

$$G = 1 + \tfrac{1}{2}x + \tfrac{3}{16}x^2 - \tfrac{3}{16}y^2 - \tfrac{3}{16} + \tfrac{3}{4}\int_0^\infty e^{-2t}\sin^2\theta(t)\,dt$$

$$\geq 1 - \tfrac{1}{3} - \tfrac{3}{16} - \tfrac{3}{4}\int_0^\infty (e^{-t} - e^{-2t})\,dt = \tfrac{7}{16} - \tfrac{1}{3} > 0.$$

This proves (15) and establishes the univalence of s_3 in $|z| < \frac{1}{4}$, which completes the proof of the theorem.

The proof for $n \geq 4$ is a modification of Szegö's argument due to Jenkins [1]. It is clear that for $n \geq 4$ the method will prove the univalence of s_n in a disk of radius greater than $\frac{1}{4}$. The exact (largest) radius of univalence ρ_n is unknown. Jenkins observed that the above method of proof shows $\rho_n \geq 1 - (4 + \varepsilon)n^{-1}\log n$ for each $\varepsilon > 0$ and for all large n, an improvement on an earlier result of Levin [1].

If f is convex, starlike, or close-to-convex, then all of its sections have the same property in $|z| < \frac{1}{4}$. The first two of these assertions were established by Szegö [3]. All three are very special cases of general theorems on convolutions due to Ruscheweyh and Sheil-Small [1], to be developed in the next section. (See Exercise 6.) Robertson [5] proved that the nth section of an arbitrary function $f \in S^*$ is starlike in a disk of radius $1 - 4n^{-1}\log n$. For the Koebe function he showed that 4 can be replaced by 3, but by no smaller constant. The general theorems on convolutions allow the immediate inference that 4 can be replaced by 3 for all $f \in S^*$, and that the corresponding result holds, with the same radius, for the classes of convex and close-to-convex functions, respectively. (See Exercise 7.)

§8.3. Convolutions of Convex Functions

The *convolution*, or *Hadamard product*, of two power series

$$f(z) = \sum_{n=1}^{\infty} a_n z^n \quad \text{and} \quad g(z) = \sum_{n=1}^{\infty} b_n z^n$$

convergent in $|z| < 1$ is the function $h = f * g$ with power series

$$h(z) = \sum_{n=1}^{\infty} a_n b_n z^n, \qquad |z| < 1.$$

The term "convolution" arises from the formula

$$h(r^2 e^{i\theta}) = \frac{1}{2\pi}\int_0^{2\pi} f(re^{i(\theta - t)})g(re^{it})\,dt, \qquad r < 1.$$

Convolution has the algebraic properties of ordinary multiplication. The geometric series

$$\ell(z) = \sum_{n=1}^{\infty} z^n = \frac{z}{1-z}$$

acts as the identity element under convolution: $f * \ell = f$ for all f.

The *integral convolution* $H = f \circledast g$ is defined by

$$H(z) = \sum_{n=1}^{\infty} \frac{a_n b_n}{n} z^n = \int_0^z \frac{h(\zeta)}{\zeta}\, d\zeta.$$

An old conjecture attributed to Mandelbrojt and Schiffer asserted that univalence is preserved under integral convolution: $f \circledast g \in S$ if $f \in S$ and $g \in S$. This implies the Bieberbach conjecture, because if any $f \in S$ had $|a_n| > n$ for some n, the integral convolution of f with itself sufficiently many times would produce a function with nth coefficient larger than en. However, the Mandelbrojt–Schiffer conjecture is false. Counterexamples were constructed by Hayman [4], by Epstein and Schoenberg [1], and by Loewner and Netanyanu [1]. In fact, these constructions show that $f \circledast g$ need not be locally univalent. The more modest conjecture that $f \circledast \ell \in S$ for each $f \in S$ is also false. This latter conjecture was proposed by Biernacki [7], who gave an erroneous proof. A counterexample will be presented in the next section. Bshouty [2] showed that $f \circledast g$ need not be univalent even if f and g are univalent functions with real coefficients.

The classes T and S^* of typically real and starlike functions are preserved under integral convolution. Robertson [6] showed that $f \in T$ and $g \in T$ imply $f \circledast g \in T$. Pólya and Schoenberg [1], in connection with their work on de la Vallée Poussin means, conjectured that $f \in S^*$ and $g \in S^*$ imply $f \circledast g \in S^*$. Since $zf'(z)$ is starlike if and only if f is convex, an equivalent statement is that $f \in C$ and $g \in C$ imply $f * g \in C$; in other words, that the class of convex functions is preserved under convolution. Pólya and Schoenberg verified several special cases. Robertson [6] had shown that if f and g are convex and have real coefficients, then $f * g$ is univalent and is convex in the direction of the imaginary axis: each vertical line intersects its range in a segment. Suffridge [1] proved that the convolution of every pair of convex functions is univalent and close-to-convex.

In 1973, Ruscheweyh and Sheil-Small [1] finally succeeded in proving the Pólya–Schoenberg conjecture. Their argument is ingenious and very intricate. It leads to a wealth of results, including both the Pólya–Schoenberg conjecture and its analogue for the space K of close-to-convex functions. The main theorems will now be stated.

Theorem 8.6. *If $f \in C$ and $g \in C$, then $f * g \in C$.*

Theorem 8.7. *If $f \in C$ and $g \in K$, then $f * g \in K$.*

An equivalent formulation of Theorem 8.6 is:

Theorem 8.6′. *If $f \in C$ and $g \in S^*$, then $f * g \in S^*$.*

These results are immediate consequences of another theorem which may be viewed as a common generalization. This more general theorem has other applications and plays a central role in the theory of convolutions.

Theorem 8.8. *If $f \in C$, $g \in S^*$, and φ is an analytic function with positive real part in $\mathbb{D}$, then $[f * (\varphi g)]/[f * g]$ is also an analytic function with positive real part.*

Momentarily assuming the truth of Theorem 8.8, we shall deduce Theorems 8.6′ and 8.7.

Deduction of Theorem 8.6′. Let $\varphi(z) = zg'(z)/g(z)$ and observe that

$$f(z) * [\varphi(z)g(z)] = f(z) * [zg'(z)] = z(f * g)'(z).$$

Since $\text{Re}\{\varphi(z)\} > 0$ by the hypothesis that $g \in S^*$, it follows from Theorem 8.8 that $f * g \in S^*$.

Deduction of Theorem 8.7. Since $g \in K$, there is a starlike function h with $h(0) = 0$ (not necessarily normalized so that $h \in S^*$) such that $\varphi(z) = zg'(z)/h(z)$ has positive real part. Apply Theorem 8.8 to $[f * (\varphi h)]/[f * h]$, observing that

$$f(z) * [\varphi(z)h(z)] = f(z) * [zg'(z)] = z(f * g)'(z)$$

and that $f * h$ is starlike, by Theorem 8.6′. Hence $f * g \in K$.

Theorem 8.8 is a straightforward consequence of the following three lemmas, applied in succession. Recall that $\mathbb{T}$ denotes the unit circle, the set of complex numbers α with $|\alpha| = 1$.

Lemma 1. *For each $g \in S^*$ and $\alpha \in \mathbb{T}$, there exist $\beta \in \mathbb{T}$ and $\gamma \in \mathbb{T}$ such that*

$$\text{Re}\{\gamma(1 - \alpha z)(1 - \beta z)g(z)/z\} > 0, \qquad |z| < 1.$$

Lemma 2. *Let h be analytic in the unit disk, with $h(0) = 0$, and suppose*

$$\text{Re}\{(1 - \alpha z)(1 - \beta z)h(z)/z\} > 0, \qquad |z| < 1,$$

for some $\alpha, \beta \in \mathbb{T}$. Then for each $f \in C$, $(f * h)(z) \neq 0$ in $0 < |z| < 1$.

Lemma 3. *Let f and g be analytic in the unit disk, with $f(0) = g(0) = 0$, and suppose*

$$f(z) * \left[\frac{1 + \alpha z}{1 - \beta z} g(z)\right] \neq 0, \qquad 0 < |z| < 1,$$

for all $\alpha, \beta \in \mathbb{T}$. Then for each function φ analytic and with positive real part in the disk, $[f * (\varphi g)]/[f * g]$ is also an analytic function with positive real part.

Proof of Lemma 1. It is sufficient to assume that g is continuous in the closed disk, since an arbitrary $g \in S^*$ may be dilated to such a function, and an obvious compactness argument may then be applied to deduce the general result. Since the argument of a starlike function increases monotonically, we may assume (after a rotation of the image) that $u(e^{it}) = \operatorname{Re}\{g(e^{it})\}$ is positive for $a < t < b$ and negative for $b < t < a + 2\pi$, where $\alpha = e^{ia}$. Let $\beta = e^{ib}$. Since $g(0) = 0$, we may set $\omega = e^{it}$ and write

$$g(z) = \frac{1}{2\pi}\int_0^{2\pi} \frac{\omega + z}{\omega - z} u(\omega)\,dt = \frac{1}{2\pi}\int_0^{2\pi} \left\{\frac{\omega + z}{\omega - z} - \frac{\alpha + z}{\alpha - z}\right\} u(\omega)\,dt;$$

or

$$\pi(\alpha - z)\frac{g(z)}{z} = \int_0^{2\pi} \frac{\alpha - \omega}{\omega - z} u(\omega)\,dt = \int_0^{2\pi} (\alpha - \omega)\left\{\frac{1}{\omega - z} - \frac{1}{\beta - z}\right\} u(\omega)\,dt.$$

Thus $h(z) = \pi(\alpha - z)(\beta - z)g(z)/z$ has the form

$$h(z) = \int_0^{2\pi} \frac{(\alpha - \omega)(\beta - \omega)}{\omega - z} u(\omega)\,dt,$$

so that

$$e^{-(1/2)i(a+b)}h(z) = 4\int_a^{a+2\pi} \sin\left(\frac{t - a}{2}\right)\sin\left(\frac{b - t}{2}\right)\frac{u(e^{it})}{1 - ze^{-it}}\,dt.$$

But $\operatorname{Re}\{(1 - \zeta)^{-1}\} > 0$ in $|\zeta| < 1$, and so the integrand in the last formula has positive real part both for $a < t < b$ and for $b < t < a + 2\pi$. This completes the proof.

The proof of Lemma 2 requires some preparation, and will be postponed.

Proof of Lemma 3. The choice $\beta = -\alpha$ shows $(f * g)(z) \neq 0$ in $0 < |z| < 1$. The hypothesis $f(0) = g(0) = 0$ implies that $f * g$ vanishes at the origin, but

$f * (\varphi g)$ has a zero there of at least the same order. Thus $f * (\varphi g)/f * g$ is analytic in the unit disk. In view of the decomposition

$$\frac{1+\alpha z}{1-\beta z} = \tfrac{1}{2}(1+\gamma)\frac{1+\beta z}{1-\beta z} + \tfrac{1}{2}(1-\gamma), \qquad \gamma = \alpha/\beta,$$

the hypothesis implies that

$$h(z) = \frac{f(z) * \left[\dfrac{1+\beta z}{1-\beta z}\, g(z)\right]}{f(z) * g(z)} \neq \frac{\gamma-1}{\gamma+1}, \qquad 0 < |z| < 1.$$

In other words, h fails to take any value on the imaginary axis. Since $h(0) = 1$, it follows that $\operatorname{Re}\{h(z)\} > 0$. The function φ can now be represented by the Herglotz formula

$$\varphi(z) = \int_{\mathbb{T}} \frac{1+\beta z}{1-\beta z}\, d\mu(\beta) + ic,$$

and the convolution $f * (\varphi g)$ can be expressed in integral form, whereupon an appeal to Fubini's theorem gives the desired conclusion.

Lemma 2 is best understood in a more general context. We shall therefore digress briefly to establish some global properties of convex functions. Everything is a consequence of the following proposition, essentially due to Ruscheweyh and Sheil-Small [1].

Proposition. *If $f \in C$, then for all z, ζ, and w in $\mathbb{D}$,*

$$\operatorname{Re}\left\{\frac{z}{z-\zeta}\,\frac{\zeta-w}{z-w}\,\frac{f(z)-f(w)}{f(\zeta)-f(w)} - \frac{\zeta}{z-\zeta}\right\} > \tfrac{1}{2}.$$

Proof (*cf.* Schober [1]). The function

$$F(z,\zeta,w) = \frac{2z}{z-\zeta}\,\frac{\zeta-w}{z-w}\,\frac{f(z)-f(w)}{f(\zeta)-f(w)} - \frac{2\zeta}{z-\zeta} - 1$$

is holomorphic in the polydisk

$$\mathbb{D}^3 = \{(z,\zeta,w) : |z| < 1, |\zeta| < 1, |w| < 1\}.$$

For distinct constants $\alpha = e^{ia}$ and $\beta = e^{ib}$, $0 < a, b < 2\pi$,

$$\operatorname{Re}\{F(\alpha w, \beta w, w)\} = -\frac{\sin\left(\frac{b}{2}\right)}{\sin\left(\frac{a-b}{2}\right)\sin\left(\frac{a}{2}\right)}\operatorname{Im}\left\{\frac{f(\alpha w) - f(w)}{f(\beta w) - f(w)}\right\}.$$

But since the image under f of each disk $|w| \le \rho < 1$ is convex,

$$\arg\frac{f(\alpha w) - f(w)}{f(\beta w) - f(w)} \in \begin{cases}(0, \pi), & 0 < a < b < 2\pi;\\ (-\pi, 0), & 0 < b < a < 2\pi.\end{cases}$$

Thus $\sin[(a - b)/2]$ and $\operatorname{Im}\{\cdot\}$ have opposite signs, and so

$$\operatorname{Re}\{F(\alpha w, \beta w, w)\} > 0.$$

But $\operatorname{Re}\{F(z, \zeta, w)\}$ attains its minimum on the distinguished boundary

$$\mathbb{T}^3 = \{(z, \zeta, w): |z| = |\zeta| = |w| = 1\}$$

of $\mathbb{D}^3$, so it follows that $\operatorname{Re}\{F(z, \zeta, w)\} > 0$ throughout $\mathbb{D}^3$.

Corollary 1. *If $f \in C$, then*

(a) $\operatorname{Re}\left\{\frac{zf'(z)}{f(z)}\right\} > \frac{1}{2}, \qquad |z| < 1;$

(b) $\operatorname{Re}\left\{\frac{f(z)}{z}\right\} > \frac{1}{2}, \qquad |z| < 1.$

Part (a) was first discovered by Strohhäcker [1] and part (b) by Marx [1].

Corollary 2. *If $f \in C$, then for each fixed $z_0 \in \mathbb{D}$, the function*

$$F(z) = z\left(\frac{f(z) - f(z_0)}{z - z_0}\right)^2$$

is starlike.

This is due to Suffridge [3] and Sheil-Small [1].

Corollary 3. *If $f \in C$, then*

$$\operatorname{Re}\left\{\frac{f(z) - f(\zeta)}{z - \zeta}\,\frac{z}{f(z)}\right\} > \tfrac{1}{2}.$$

Corollary 4. *If $f \in C$, then for $|\alpha| \le 1$ and $|\beta| \le 1$,*

$$\operatorname{Re}\left\{\frac{f(z) * [z(1-\alpha z)^{-1}(1-\beta z)^{-1}]}{f(z) * [z(1-\alpha z)^{-1}]}\right\} > \tfrac{1}{2}.$$

Proof. This is a restatement of Corollary 3, since the denominator is

$$f(z) * \frac{1}{\alpha}\ell(\alpha z) = \frac{1}{\alpha} f(\alpha z)$$

and the numerator can be written as

$$f(z) * \frac{1}{\alpha-\beta}\left\{\frac{1}{1-\alpha z} - \frac{1}{1-\beta z}\right\} = \frac{f(\alpha z) - f(\beta z)}{\alpha - \beta}.$$

Corollary 5. *If $f \in C$, then for $|\alpha| \le 1$, $|\beta| \le 1$, and $|\gamma| \le 1$,*

$$\operatorname{Re}\left\{\frac{f(z) * [z(1-\alpha z)^{-1}(1-\beta z)^{-1}(1-\gamma z)^{-1}]}{f(z) * [z(1-\alpha z)^{-1}(1-\beta z)^{-1}]}\right\} > \tfrac{1}{2}.$$

Proof. This is a restatement of the proposition itself. Replace z, ζ, and w by αz, βz, and γz, and reduce the given expression to

$$\begin{aligned}
&\frac{\alpha}{\alpha-\beta}\cdot\frac{f(z) * [z(1-\alpha z)^{-1}(1-\gamma z)^{-1}]}{f(z) * [z(1-\beta z)^{-1}(1-\gamma z)^{-1}]} - \frac{\beta}{\alpha-\beta} \\
&\quad = \frac{1}{\alpha-\beta}\,\frac{f(z) * \{\alpha z(1-\alpha z)^{-1}(1-\gamma z)^{-1} - \beta z(1-\beta z)^{-1}(1-\gamma z)^{-1}\}}{f(z) * [z(1-\beta z)^{-1}(1-\gamma z)^{-1}]} \\
&\quad = \frac{f(z) * [z(1-\alpha z)^{-1}(1-\beta z)^{-1}(1-\gamma z)^{-1}]}{f(z) * [z(1-\beta z)^{-1}(1-\gamma z)^{-1}]}.
\end{aligned}$$

Proof of Lemma 2. Replacing z by γz for suitable $\gamma \in \mathbb{T}$, we may rotate $h(z)$ to $\bar{\gamma}h(\gamma z)$; thus we may assume $h'(0) = 1$. The Herglotz formula then gives

$$h(z) = \int_{\mathbb{T}} \frac{z(1+\omega z)}{(1-\alpha z)(1-\beta z)(1-\omega z)}\, d\mu(\omega)$$

for some unit measure $d\mu$. If the kernel is put in the form

$$2z(1-\alpha z)^{-1}(1-\beta z)^{-1}(1-\omega z)^{-1} - z(1-\alpha z)^{-1}(1-\beta z)^{-1},$$

the convolution of h with $f \in C$ reduces to

$$(f * h)(z) = F(z) \int_{\mathbb{T}} G(z, \omega)\, d\mu(\omega),$$

where

$$F(z) = f(z) * [z(1 - \alpha z)^{-1}(1 - \beta z)^{-1}]$$

and

$$G(z, \omega) = 2\frac{f(z) * [z(1 - \alpha z)^{-1}(1 - \beta z)^{-1}(1 - \omega z)^{-1}]}{F(z)} - 1.$$

But $F(z) \neq 0$ for $z \neq 0$, by Corollary 4, and $\operatorname{Re}\{G(z, \omega)\} > 0$ by Corollary 5. Thus $\int G(z, \omega)\, d\mu(\omega) \neq 0$, and it follows that $(f * h)(z) \neq 0$ in $0 < |z| < 1$. This establishes Lemma 2, and therefore concludes the proof of Theorem 8.8, hence of Theorems 8.6 and 8.7.

A different proof of the Pólya–Schoenberg conjecture was found by Suffridge [5].

Theorem 8.8 has another interesting application which involves the concept of subordination, already explored in Chapter 6. Recall that a function g is said to be *subordinate* to f if $g(z) = f(\omega(z))$ for some analytic function ω with $|\omega(z)| \leq |z|$ in $|z| < 1$. This is denoted by $g \prec f$. If ω is not merely a rotation of the disk (that is, if $|\omega(z)| < |z|$), then g is said to be *properly subordinate* to f.

Lemma 4. *A function h analytic in the unit disk is properly subordinate to a convex function f if and only if*

$$\operatorname{Re}\left\{\frac{f(z) - h(\alpha z)}{zf'(z)}\right\} > 0, \qquad |z| < 1,$$

for each $\alpha \in \mathbb{T}$.

Proof. A domain is convex if and only if it is starlike with respect to each of its points. Thus if f is convex, then

$$\operatorname{Re}\left\{\frac{zf'(z)}{f(z) - f(z_0)}\right\} > 0, \qquad |z_0| < |z| < 1.$$

Under the assumption that $h(z) = f(\omega(z))$, choose $z_0 = \omega(\alpha z)$ to reach the desired conclusion. Conversely, if the condition of the lemma holds for all $\alpha \in \mathbb{T}$, the maximum principle extends it to $|\alpha| \leq 1$. In particular, $f(0) = h(0)$.

If h is not subordinate to f, then f omits some value $w_0 = h(z_0)$. But for $|z_0| < |z|$,

$$\text{Re}\left\{\frac{f(z) - w_0}{zf'(z)}\right\} = \text{Re}\left\{\frac{f(z) - h(\alpha z)}{zf'(z)}\right\} > 0,$$

where $\alpha = z_0/z$. The maximum principle extends this inequality to $0 < |z| < 1$, but it leads to a contradiction near $z = 0$. Hence h is properly subordinate to f, since the condition of the lemma rules out the possibility that $f(z) \equiv h(\beta z)$ for some $\beta \in \mathbb{T}$.

Theorem 8.9. *If $h \prec f \in C$, then $g * h \prec g * f$ for all $g \in C$.*

Proof. The theorem is trivial unless h is properly subordinate to f. In this case, Lemma 4 says that for each $\alpha \in \mathbb{T}$,

$$\varphi(z) = \frac{f(z) - h(\alpha z)}{zf'(z)}$$

has positive real part. In view of Lemma 4, it is sufficient to show that

$$\psi(z) = \frac{(g * f)(z) - (g * h)(\alpha z)}{z(g * f)'(z)}$$

has positive real part for each $\alpha \in \mathbb{T}$. But ψ has the form

$$\psi(z) = \frac{g(z) * [\varphi(z)zf'(z)]}{g(z) * [zf'(z)]}.$$

Thus Theorem 8.8 may be applied to conclude that $\text{Re}\{\psi(z)\} > 0$.

§8.4. Coefficient Multipliers

A complex sequence $\{\lambda_n\}$ is said to be a *coefficient multiplier*, or simply a *multiplier*, of a family $\mathscr{F}$ of analytic functions into a family $\mathscr{G}$ if $\sum \lambda_n a_n z^n$ belongs to $\mathscr{G}$ for each function $\sum a_n z^n$ in $\mathscr{F}$. The general problem is to characterize the set of multipliers of $\mathscr{F}$ into $\mathscr{G}$. This may be viewed as a converse to the problem of describing the convolution of two functions belonging to given families.

According to Theorem 8.6, a sequence $\{\lambda_n\}$ is a multiplier of the class C into itself if and only if the function $g(z) = \sum \lambda_n z^n$ belongs to C. (For the "only if" part, simply note that if $\{\lambda_n\}$ is a multiplier of C into C, then $g = \ell * g \in C$, since $\ell \in C$.) Equivalently, $\{\lambda_n\}$ multiplies S^* into itself if and only if $g \in C$. The multipliers of K into itself are also characterized by the condition $g \in C$. (The necessity will be proved shortly. The sufficiency is the content of Theorem 8.7 in the preceding section.)

The multipliers of S into itself are unknown. Their full description is an open problem, but it is easy to give some examples:

(i) Let $\lambda_n = \zeta^{n-1}$, $n = 1, 2, \ldots$, where $0 < |\zeta| < 1$. Then $g(z) = \sum \lambda_n z^n = z/(1 - \zeta z)$, and for each function $f(z) = \sum a_n z^n$ in S,

$$(f * g)(z) = \sum_{n=1}^{\infty} \lambda_n a_n z^n = \zeta^{-1} f(\zeta z)$$

also belongs to S.

(ii) Let $\lambda_k = r^{k-1}$ for $1 \le k \le n$, where $0 < r < 1$; and let $\lambda_k = 0$ for $k > n$. Then

$$g(z) = \sum_{k=1}^{\infty} \lambda_k z^k = z(1 - r^n z^n)/(1 - rz),$$

and

$$(f * g)(z) = \sum_{k=1}^{n} r^{k-1} a_k z^k = \frac{1}{r} s_n(rz; f),$$

where $s_n(z; f) = z + \cdots + a_n z^n$ is the nth section of f. Thus if r is chosen no larger than the radius of univalence of the nth section of functions in S (see §8.2), the sequence $\{\lambda_k\}$ will be a multiplier of S into itself. In particular, $\{\lambda_k\}$ is a multiplier for $r \le \frac{1}{4}$, by Szegö's theorem.

(iii) If $\lambda_1 = 1$ and

$$\sum_{n=2}^{\infty} n^2 |\lambda_n| \le 1/c,$$

where c is a constant such that $|a_n| \le cn$ for all n and for every $f \in S$ (for instance, $c = \sqrt{\frac{7}{6}}$), then $\{\lambda_n\}$ is a multiplier of S into S. Indeed, for each function $f(z) = \sum a_n z^n$ in S, this condition implies that $h(z) = \sum \lambda_n a_n z^n$ is suitably normalized and

$$\sum_{n=2}^{\infty} n |\lambda_n a_n| \le c \sum_{n=2}^{\infty} n^2 |\lambda_n| \le 1,$$

so that $h \in S$. In fact, $h \in S^*$. (See Chapter 2, Exercise 24.)

In connection with this last example, it may be remarked that any condition on the moduli $|\lambda_n|$ alone which is sufficient to ensure that $\{\lambda_n\}$ be a multiplier of S into S, must actually imply that $\sum n^2|\lambda_n|^2 < \infty$. Suppose, on the contrary, that for some sequence $\{\mu_n\}$ with $\sum n^2|\mu_n|^2 = \infty$, $\{\lambda_n\} = \{\varepsilon_n \mu_n\}$ is a multiplier for each choice of signs $\varepsilon_n = \pm 1$. Then $\{\lambda_n\}$ multiplies the Koebe function into $h(z) = \sum \varepsilon_n \mu_n n z^n$, a function which for almost every choice of signs fails to have a radial limit on any set of positive measure, since its coefficients are not square-summable. (See, for instance, Duren [6], Appendix A.) In particular, $h \notin H^p$ for all $p > 0$, and so $h \notin S$ for almost every choice of signs. (Recall that $S \subset H^p$ for each $p < \frac{1}{2}$, as proved in §2.10.)

An obvious necessary condition for $\{\lambda_n\}$ to be a multiplier is that $g(z) = \sum \lambda_n z^n$ be univalent. For if $\{\lambda_n\}$ multiplies S into itself, then $g = \ell * g \in S$. It is also necessary that g be *convex*. This surprising result is contained in the following theorem.

Theorem 8.10. *Let $g(z) = z + \sum_{n=2}^{\infty} \lambda_n z^n$ be analytic in $\mathbb{D}$, and suppose $f * g$ is locally univalent there for each $f \in K$. Then $g \in C$.*

Corollary 1. *If $f * g \in K$ for each $f \in K$, then $g \in C$.*

Corollary 2. *If $f * g \in S$ for each $f \in S$, then $g \in C$.*

Corollary 3. *For each integer n, the radius of univalence of the family of sections $s_n(z; f)$ of functions $f \in S$ is no larger than the radius of convexity of $s_n(z; \ell) = z + z^2 + \cdots + z^n$.*

Proof of Theorem. Let $G(z) = zg'(z)$ and $h = f * g$. Then

$$zh'(z) = (f * G)(z) \neq 0, \qquad 0 < |z| < 1,$$

for each $f \in K$. We are to show that this implies $G \in S^*$, or $\mathrm{Re}\{zG'(z)/G(z)\} > 0$. Equivalently, we must show that

$$zG'(z) - iaG(z) \neq 0, \qquad 0 < |z| < 1,$$

for each real number a. Observe that

$$zG'(z) - iaG(z) = (f_a * G)(z),$$

where $f_a = k - ia\ell$ and k is the Koebe function. But $f_a \in K$ for each real a, since

$$\mathrm{Re}\{f_a'(z)/\ell'(z)\} = \mathrm{Re}\{k'(z)/\ell'(z)\} = \mathrm{Re}\{(1 + z)/(1 - z)\} > 0.$$

Thus

$$(f_a * G)(z) = z(f_a * g)'(z) \neq 0, \qquad 0 < |z| < 1,$$

by the hypothesis of local univalence, and we have proved that $G \in S^*$, or $g \in C$.

Note that the proof uses only the weaker hypothesis that $f_a * g$ is locally univalent for each function f_a. This elegant argument was communicated to the author by T. Sheil-Small.

The converse of Corollary 2 is false: there are convex functions which fail to multiply S into itself. One example is the function

$$g(z) = \log \frac{1}{1-z} = \sum_{n=1}^{\infty} \frac{1}{n} z^n.$$

It is clear that g is convex; in fact, $zg'(z) = \ell(z)$ belongs not only to S^*, but to C. For this function g, the assertion that $f * g \in S$ for each $f \in S$ is equivalent to the Biernacki conjecture that $f \circledast \ell \in S$ for each $f \in S$. This conjecture was mentioned at the beginning of §8.3. It will now be disproved by a counterexample due to Krzyż and Lewandowski [1]. In particular, this will provide a counterexample to the Mandelbrojt–Schiffer conjecture that $f \in S$ and $g \in S$ imply $f \circledast g \in S$.

Theorem 8.11. *There exists a function $f \in S$ for which $h = f \circledast \ell \notin S$; that is, for which the integral*

$$h(z) = \int_0^z f(\zeta)/\zeta \, d\zeta$$

is not univalent in $|z| < 1$.

Corollary. *There exist functions $f \in S$ and $g \in C$ such that $f * g \notin S$.*

Proof of Theorem. Consider the function

$$f(z) = z(1 - iz)^{i-1}, \qquad |z| < 1.$$

It is easily verified that f is spirallike (see §2.7), hence univalent. Indeed, a simple calculation gives

$$\frac{zf'(z)}{f(z)} = \frac{1+z}{1-iz},$$

and so

$$\text{Re}\left\{e^{i\pi/4} \frac{zf'(z)}{f(z)}\right\} = \frac{1}{\sqrt{2}} \frac{1-|z|^2}{|1-iz|^2} > 0, \qquad |z| < 1.$$

Thus $f \in S$. On the other hand,

$$h(z) = \int_0^z f(\zeta)/\zeta \, d\zeta = (1 - iz)^i - 1.$$

Now set $z_0 = i(e^{2\pi} - 1)/(e^{2\pi} + 1)$ and observe that

$$\begin{aligned} h(z_0) + 1 &= \exp\left\{i \log \frac{2e^{2\pi}}{e^{2\pi} + 1}\right\} \\ &= \exp\left\{i \log \frac{2}{e^{2\pi} + 1}\right\} = h(-z_0) + 1. \end{aligned}$$

Hence $h(z_0) = h(-z_0)$, so $h \notin S$.

The proof shows that among all functions $f \in S$ the radius of univalence of the family of integrals h is no larger than

$$(e^{2\pi} - 1)/(e^{2\pi} + 1) = 0.996 \ldots.$$

The exact radius of univalence is unknown. Lewandowski [4] found the lower bound 0.91.

Campbell and Singh [1] have found another simple construction which gives the corollary.

§8.5. Criteria for Univalence

The simplest analytic criteria for univalence have limited application because they imply special properties such as starlikeness. These conditions are sufficient for univalence but far from necessary. We shall now develop a general criterion due to Nehari [1] which involves the Schwarzian derivative. This is a very useful sufficient condition which is almost necessary in a certain sense.

The *Schwarzian derivative* of an analytic, locally univalent function f is defined by

$$\{f, z\} = \left[\frac{f''(z)}{f'(z)}\right]' - \frac{1}{2}\left[\frac{f''(z)}{f'(z)}\right]^2.$$

It is important primarily because of its invariance under linear fractional transformations. More precisely, if $\mathscr{S}$ denotes the mapping from f to its

Schwarzian derivative, a direct calculation confirms that $\mathscr{S}(T \circ f) = \mathscr{S}(f)$ for every nondegenerate linear fractional transformation

$$T(w) = \frac{aw + b}{cw + d}, \qquad ad - bc \neq 0.$$

This is actually a special case of a much more general transformation property. If $z = \varphi(\zeta)$ is an analytic function for which the composition $F = f \circ \varphi$ is defined, then

$$\{F, \zeta\} = \{f, \varphi(\zeta)\}[\varphi'(\zeta)]^2 + \{\varphi, \zeta\}.$$

Since $\mathscr{S}(T) = 0$ for every linear fractional mapping T, this implies that $\mathscr{S}(T \circ \varphi) = \mathscr{S}(\varphi)$. It also implies

$$\{f \circ T, \zeta\} = \{f, T(\zeta)\}[T'(\zeta)]^2 \tag{16}$$

for linear fractional mappings T.

The problem of finding the most general function with prescribed Schwarzian derivative amounts to solving a nonlinear differential equation of third order. It has a remarkably simple solution which may be expressed as follows.

Lemma 1. *For a given analytic function p, the general function f with Schwarzian derivative $\mathscr{S}(f) = 2p$ has the form $f = g_1/g_2$, where g_1 and g_2 are arbitrary linearly independent solutions of the linear differential equation*

$$g'' + pg = 0. \tag{17}$$

Proof. A straightforward calculation, facilitated by the fact that the Wronskian of g_1 and g_2 is constant, shows that every such function f has Schwarzian derivative $2p$. Conversely, suppose $\mathscr{S}(f) = 2p$ and let g satisfy the differential equation

$$2f'g' + f''g = 0.$$

Divide by f' and differentiate to obtain

$$2g'' + \left(\frac{f''}{f'}\right)' g + \frac{f''}{f'} g' = 0,$$

or

$$2g'' + \mathscr{S}(f)g = 0,$$

which shows that g satisfies (17). Furthermore, the function $h = fg$ satisfies

$$h'' + ph = f''g + 2f'g' = 0,$$

by the construction of g. Thus $f = h/g$ has the required form.

A general fact about linear fractional transformations will also be needed.

Lemma 2. *For each pair of distinct points z_1 and z_2 in the unit disk there is a linear fractional mapping*

$$\varphi(z) = e^{i\sigma}\frac{z - \alpha}{1 - \bar{\alpha}z}, \qquad |\alpha| < 1,$$

of the disk onto itself for which $0 < \varphi(z_1) = -\varphi(z_2)$.

Proof. We first observe that $|\varphi(z_1)| = |\varphi(z_2)|$ for some choice of α. To see this, let

$$\psi_j(\alpha) = \frac{z_j - \alpha}{1 - \bar{\alpha}z_j}, \qquad j = 1, 2,$$

and observe that $0 = |\psi_1(z_1)| < |\psi_2(z_1)|$ and $0 = |\psi_2(z_2)| < |\psi_1(z_2)|$. Thus by continuity, $|\psi_1(\alpha)| = |\psi_2(\alpha)|$ for some α on any given continuous curve joining z_1 and z_2. The lemma therefore reduces to the special case where $|z_1| = |z_2|$. After a suitable rotation, we may assume $z_1 = \overline{z_2} = re^{i\theta}$ and $\mathrm{Re}\{z_1\} > 0$. Hence we need only show that $\varphi(z) = -\varphi(\bar{z})$ for some α satisfying $|\alpha| < 1$, where $z = z_1 = re^{i\theta}$. This equation is equivalent to

$$(z - \alpha)(1 - \bar{\alpha}\bar{z}) + (\bar{z} - \alpha)(1 - \bar{\alpha}z) = 0,$$

or

$$(1 + |\alpha|^2)r\cos\theta = \alpha + \bar{\alpha}r^2.$$

Since $0 < \cos\theta < 1$, this last equation has a solution α with $0 < \alpha < 1$. (See Exercise 16 for another proof.)

We shall also make use of the following lemma.

Lemma 3. *Let u be a real-valued, continuously differentiable function in the interval $[-1, 1]$, not the zero function. Suppose $u(x) = O(1 - x)$ as $x \to 1$ and $u(x) = O(1 + x)$ as $x \to -1$. Then*

$$\int_{-1}^{1}(1 - x^2)^{-2}[u(x)]^2\,dx < \int_{-1}^{1}[u'(x)]^2\,dx.$$

Proof. Observe that

$$\begin{aligned}0 &< \int_{-1}^{1} \{(1-x^2)^{-1}xu + u'\}^2\,dx\\ &= \int_{-1}^{1} (1-x^2)^{-2}x^2u^2\,dx + 2\int_{-1}^{1}(1-x^2)^{-1}xuu'\,dx + \int_{-1}^{1} u'^2\,dx.\end{aligned}$$

Integrate the second term by parts to obtain

$$\begin{aligned}&2\int_{-1}^{1}(1-x^2)^{-1}xuu'\,dx\\ &\qquad = [(1-x^2)^{-1}xu^2]_{-1}^{1} - \int_{-1}^{1}(1+x^2)(1-x^2)^{-2}u^2\,dx.\end{aligned}$$

The integrated terms vanish because of the assumed behavior of u at ± 1, and the desired inequality follows.

We are now prepared to consider Nehari's criterion for univalence.

Theorem 8.12 (Nehari's Theorem). *Let f be analytic in $\mathbb{D}$ and suppose its Schwarzian derivative satisfies*

$$|\{f, z\}| \le 2(1-|z|^2)^{-2}, \qquad |z| < 1. \tag{18}$$

Then f is univalent in $\mathbb{D}$.

Proof. We first observe that it is sufficient to prove that $f(r) \ne f(-r)$ for $0 < r < 1$ under the hypothesis (18). Indeed, if $f(z_1) = f(z_2)$, then by Lemma 2 some conformal automorphism T of the disk produces a function $F = f \circ T$ with $F(r) = F(-r)$ and with Schwarzian derivative

$$\{F, \zeta\} = \{f, T(\zeta)\}[T'(\zeta)]^2,$$

according to the formula (16). The inequality (18) therefore gives

$$\begin{aligned}|\{F, \zeta\}| &\le 2(1-|T(\zeta)|^2)^{-2}|T'(\zeta)|^2\\ &\le 2(1-|\zeta|^2)^{-2},\end{aligned}$$

as a direct calculation shows. Thus F also satisfies the hypothesis (18), which shows that it suffices to prove $f(r) \ne f(-r)$.

In view of Lemma 1, it is equivalent to prove that if p is an analytic function satisfying

$$|p(z)| \leq (1 - |z|^2)^{-2}, \qquad |z| < 1, \tag{19}$$

then the ratio g_1/g_2 of two linearly independent solutions to (17) takes different values at $\pm r$ for each r, $0 < r < 1$. If, on the contrary,

$$\frac{g_1(r)}{g_2(r)} = \frac{g_1(-r)}{g_2(-r)} = \alpha$$

for some $\alpha \in \mathbb{C}$ and for some r with $0 < r < 1$, then $g = g_1 - \alpha g_2$ satisfies $g(r) = g(-r) = 0$. Conversely, if some nontrivial solution of (17) vanishes at $\pm r$, the same argument shows that some ratio g_1/g_2 takes equal values at $\pm r$. The theorem is therefore equivalent to the statement that if p satisfies (19), then no nontrivial solution to (17) can vanish at both r and $-r$ for any r in the interval $0 < r < 1$.

Suppose, on the contrary, that $g \not\equiv 0$ satisfies (17) and $g(r) = g(-r) = 0$ for some r, $0 < r < 1$. Multiplying the equation (17) by $\bar{g}$ and integrating by parts, we find

$$\begin{aligned} 0 &= \int_{-r}^{r} p(x)|g(x)|^2\, dx + \int_{-r}^{r} \overline{g(x)} g''(x)\, dx \\ &= \int_{-r}^{r} p(x)|g(x)|^2\, dx - \int_{-r}^{r} |g'(x)|^2\, dx, \end{aligned}$$

because $g(\pm r) = 0$. It now follows from (19) that

$$\int_{-r}^{r} |g'(x)|^2\, dx \leq \int_{-r}^{r} (1 - x^2)^{-2} |g(x)|^2\, dx.$$

Letting $g(rt) = u(t) + iv(t)$ and using the inequality

$$r^2(1 - r^2t^2)^{-2} \leq (1 - t^2)^{-2}, \qquad -1 \leq t \leq 1,$$

we deduce that

$$\int_{-1}^{1} \{u'^2 + v'^2\}\, dt \leq \int_{-1}^{1} (1 - t^2)^{-2} \{u^2 + v^2\}\, dt,$$

which contradicts Lemma 3. This completes the proof.

The constant 2 in Nehari's theorem is best possible and cannot be replaced by any larger number. This is demonstrated by the example

$$f(z) = \left(\frac{1 - z}{1 + z}\right)^{\alpha}, \qquad \alpha \in \mathbb{C},$$

for which

$$\{f, z\} = 2(1 - \alpha^2)(1 - z^2)^{-2}.$$

This function f is univalent in the unit disk or, equivalently, the function w^{α} is univalent in the right half-plane if and only if the exponent $\alpha = a + ib$ satisfies $a^2 + b^2 \leq 2|a|$. The choice $\alpha = ib$ therefore gives a nonunivalent function f with

$$|\{f, z\}| \leq 2(1 + b^2)(1 - |z|^2)^{-2}, \qquad |z| < 1.$$

This example is due to Hille [1].

On the other hand, the inequality (18) becomes a necessary condition for univalence if the constant 2 is replaced by 6. This is proved by inversion of the function

$$F(z) = \frac{f\left(\dfrac{z + \zeta}{1 + \bar{\zeta}z}\right) - f(\zeta)}{(1 - |\zeta|^2)f'(\zeta)} = z + A_2 z^2 + A_3 z^3 + \cdots$$

obtained from f by a disk automorphism. The function

$$G(z) = F(1/z)^{-1} = z + B_0 + B_1 z^{-1} + \cdots$$

belongs to Σ, and so $B_1 = A_2^2 - A_3$ satisfies $|B_1| \leq 1$. A calculation gives

$$B_1 = -\tfrac{1}{6}(1 - |\zeta|^2)^2\{f, \zeta\},$$

which proves that

$$|\{f, \zeta\}| \leq 6(1 - |\zeta|^2)^{-2}, \qquad |\zeta| < 1,$$

for each $f \in S$. This result is due to Kraus [1] and was rediscovered by Nehari [1]. The Koebe function shows that the constant 6 is best possible.

If the condition (18) is strengthened to

$$|\{f, z\}| \leq C(1 - |z|^2)^{-2}, \qquad |z| < 1,$$

for some constant $C < 2$, then f maps the disk onto a Jordan domain on the Riemann sphere. This important result is due to Ahlfors and Weill [1], whose proof invokes the theory of quasiconformal mappings. Duren and Lehto [1] obtained the same conclusion under a weaker hypothesis.

Nehari's condition (18) may be awkward to verify because it requires the calculation of the Schwarzian derivative. It is often simpler to work directly with the ratio f''/f'. As an immediate consequence of Nehari's theorem, one sees that a condition of the form

$$|f''(z)/f'(z)| \le C(1 - |z|^2)^{-1}, \qquad |z| < 1,$$

implies the univalence of f if the constant C is small enough. Duren, Shapiro, and Shields [1] observed that it is sufficient to take $C = 2(\sqrt{5} - 2) = 0.472\ldots$. Becker [1] then applied Loewner's method to increase the constant to $C = 1$. Hille's example shows that the condition is not sufficient if $C > 2$. Pommerenke gave an example to show that $C > 4/e = 1.471\ldots$ is not sufficient. (See Exercises 19 and 20.) The best possible value of C is not known. The condition becomes necessary for univalence if $C = 6$. (See Exercise 21.)

Nehari's theorem can also be applied to show that if f is univalent in $|z| < 1$, then

$$f_\alpha(z) = \int_0^z [f'(\zeta)]^\alpha \, d\zeta$$

is univalent for each complex α of sufficiently small modulus. Duren, Shapiro, and Shields [1] derived this result with $|\alpha| \le (\sqrt{5} - 2)/3 = 0.078\ldots$. Becker [1] improved it to $|\alpha| \le \frac{1}{6}$, and Pfaltzgraff [2] made a further improvement to $|\alpha| \le \frac{1}{4}$. On the other hand, Royster [1] showed that f_α need not be univalent for any α with $|\alpha| > \frac{1}{3}$. The exact region of exponents α which lead to the univalence of f_α is unknown.

Nehari's condition (18) defines a subclass of S which excludes the Koebe function and its rotations. The problem therefore arises to find the sharp growth and distortion theorems, the maximum value of $|a_n|$, etc. for this subclass. Essén and Keogh [1] have solved some of these problems in general form.

In addition to the condition (18), Nehari [1] showed that $|\{f, z\}| \le \pi^2/2$ is sufficient for univalence, and the constant $\pi^2/2$ is best possible. In a similar vein, Pokornyi [1] announced that $|\{f, z\}| \le 4(1 - |z|^2)^{-1}$ is sufficient for univalence, and Nehari [4] supplied a proof. The constant 4 is again best possible. In fact, Nehari [4, 10] established a general criterion which includes all of these as special cases. Other related criteria were found by Ahlfors [4] and by Becker [2, 3]. Nehari [9] also showed that all convex univalent functions satisfy the condition (18), and that the constant 2 is best possible in this case. Various other conditions of this type are discussed in the survey article by Avhadiev and Aksentev [1].

§8.6. Additional Topics

We close this chapter with brief accounts of three special classes of functions: Bieberbach–Eilenberg functions, univalent polynomials, and functions of bounded boundary rotation. Our discussion is confined to a survey of results, with all proofs omitted.

1. Bieberbach–Eilenberg Functions

The *Bieberbach–Eilenberg class B* consists of all functions

$$f(z) = a_1 z + a_2 z^2 + \cdots$$

analytic in $\mathbb{D}$ such that $f(z)f(\zeta) \neq 1$ for all pairs of points $z, \zeta \in \mathbb{D}$. A *univalent* function $f \in B$ is said to belong to the class B_u. These functions may be viewed as generalizations of analytic functions vanishing at the origin and mapping the unit disk into itself. For this special class of bounded functions the coefficient inequality $|a_n| \leq 1$ is trivial and is sharp if univalence is not required, an extremal function being simply $f(z) = z^n$. In 1916, Bieberbach [2] introduced the class B_u and showed that the inequality $|a_1| \leq 1$ extends to this class. Later, Eilenberg [1] showed that univalence is not necessary. Rogosinski [5] proved $|a_2| \leq 1$ and conjectured that $|a_n| \leq 1$ for all n if $f \in B$. His conjecture was confirmed by Lebedev and Milin [1] in 1951. A few years later, Jenkins [5] obtained the sharp bound

$$|f(z)| \leq r(1 - r^2)^{-1/2}, \qquad |z| = r < 1,$$

for functions $f \in B$, with equality occurring only for suitable rotations of the (univalent) functions

$$f_r(z) = (1 - r^2)^{1/2} z(1 + irz)^{-1}. \tag{20}$$

Then in 1970, Aharonov [3] and Nehari [7] independently showed that in fact

$$\sum_{n=1}^{\infty} |a_n|^2 \leq 1 \tag{21}$$

for every $f \in B$, and that the coefficients of functions $f \in B_u$ satisfy

$$|a_n| \leq e^{-\gamma/2}(n-1)^{-1/2}, \qquad n = 2, 3, \ldots, \tag{22}$$

where γ is Euler's constant. The inequality (21) easily implies those of Lebedev–Milin and Jenkins. Furthermore, Jenkins' function (20) shows that the estimate (22) is best possible in order of magnitude, and that the constant $e^{-\gamma/2}$ cannot be replaced by anything smaller than $e^{-1/2}$.

Other results on Bieberbach–Eilenberg functions have been obtained by Alenicyn [1], Jenkins [6, 14, 16], Garabedian and Schiffer [3], Hummel and Schiffer [1, 2], Schiffer and Schmidt [1], Aharonov [2, 7], Kühnau [1], and Hummel [7]. Rogosinski [5] showed that every Bieberbach–Eilenberg function is subordinate to a univalent one.

The class B_u of univalent Bieberbach–Eilenberg functions is closely related to the class S. Indeed, if $f \in B_u$, then the function

$$F(z) = \frac{f(z)}{f'(0)[1 + f(z)]^2}$$

belongs to S. Conversely, if $F \in S$ and $F(z) \neq \omega$, then

$$f(z) = \frac{1 - \sqrt{1 - F(z)/\omega}}{1 + \sqrt{1 - F(z)/\omega}}$$

belongs to B_u.

Essentially equivalent to the Bieberbach–Eilenberg functions are the *Gelfer functions*

$$g(z) = 1 + c_1 z + c_2 z^2 + \cdots, \qquad |z| < 1,$$

satisfying the condition $g(z) + g(\zeta) \neq 0$ for all points $z, \zeta \in \mathbb{D}$. They were introduced by Gelfer [1] in 1946. The transformation

$$g(z) = \frac{1 - f(z)}{1 + f(z)}$$

relates the (univalent) Gelfer functions to the (univalent) Bieberbach–Eilenberg functions. A typical Gelfer function is the half-plane mapping

$$g(z) = \frac{1 - z}{1 + z} = 1 + 2z + 2z^2 + \cdots.$$

This suggests the general conjecture that $|c_n| \leq 2$, at least for univalent Gelfer functions. The inequality $|c_1| \leq 2$ holds without the assumption of univalence and is equivalent to the original Bieberbach–Eilenberg result $|a_1| \leq 1$. Gelfer proved that $|c_n| \leq 2\sqrt{2e}$ for all n. However, Hummel [9] developed a variational method for univalent Gelfer functions and used it to show (with the help of numerical calculation) that the sharp upper bound for $|c_2|$ is not 2 but is 2.00011.... Grinspan [3] then concluded that $\max |c_n| \geq 2.00011\ldots$ for every even index n.

2. Univalent Polynomials

Which polynomials of the form

$$p(z) = z + a_2 z^2 + \cdots + a_n z^n$$

are univalent in the unit disk? This question has not been answered in full generality, but some interesting partial results are available. The problem is to describe the region $\mathbb{W}_n \subset \mathbb{C}^{n-1}$ occupied by the points $(a_2, \ldots, a_n)$ for which the polynomial p is univalent.

For $n = 2$ the problem is easy: $z + a_2 z^2$ is univalent if and only if $|a_2| \leq \frac{1}{2}$. For $n > 2$ the sharp inequality $|a_n| \leq 1/n$ is necessary for univalence (see Chapter 2, Exercise 25) but is far from sufficient. The full description of the region $\mathbb{W}_3$ is already a difficult problem, solved by Kössler [1] in 1951, and later (independently) by Cowling and Royster [1] and by Brannan [1]. It is interesting to note that the three methods begin with different general criteria for univalence of polynomials. Brannan uses Dieudonné's criterion (see Chapter 2, Exercise 34) in conjunction with Cohn's rule (see Marden [1]) for computing the number of zeros a polynomial has in the unit disk. Cowling and Royster observe that p is univalent if and only if the function

$$\varphi(z; \alpha) = z[p(z) - p(\alpha z)]^{-1}$$

is analytic in $\mathbb{D}$ for each $\alpha \in \mathbb{T}$, $\alpha \neq 1$. Kössler shows that the univalence of p in $\overline{\mathbb{D}}$ is equivalent to the nonvanishing of the resultant of a certain pair of associated polynomials of two variables. For the case $n = 3$, all three approaches lead to the extremal univalent polynomial

$$q_3(z) = z + \tfrac{2}{3}\sqrt{2}z^2 + \tfrac{1}{3}z^3,$$

which simultaneously maximizes $|a_2|$ and $|a_3|$. It is noteworthy that this polynomial has real coefficients.

Suffridge [2] investigated polynomials with real coefficients and generalized q_3 to the univalent polynomial

$$q_n(z) = z + \sum_{k=2}^{n} A_{nk} z^k, \qquad A_{nk} = \frac{n-k+1}{n} \frac{\sin \dfrac{k\pi}{n+1}}{\sin \dfrac{\pi}{n+1}}.$$

He found that among all polynomials $p \in S$ with real coefficients and with highest coefficient $a_n = 1/n$, the sharp inequalities $|a_k| \le A_{nk}$ hold. The polynomials q_n have elegant geometric properties, mapping the disk onto the interior of a closed curve with $n - 1$ cusps and with certain self-intersections on the real axis.

Michel [1] investigated the case $n = 4$ and showed that Suffridge's estimates hold for all polynomials

$$p(z) = z + a_2 z^2 + a_3 z^3 + a_4 z^4$$

with real coefficients, without the assumption that $a_4 = \frac{1}{4}$. Thus

$$|a_2| \le \tfrac{3}{8}(1 + \sqrt{5}) = 1.21352\ldots, \qquad |a_3| \le \tfrac{1}{4}(1 + \sqrt{5}) = 0.80901\ldots,$$

with equality only for $p(z) = q_4(z)$ or $-q_4(-z)$. Michel's method is the same as Brannan's, but it makes additional use of an inequality of Fejér on nonnegative trigonometric polynomials. By analogy with the result for $n = 3$, it is natural to suspect that the above bounds extend to the full class of univalent polynomials of degree 4 with complex coefficients. However, Michel found that the sharp bound for the third coefficient in this class is very slightly larger:

$$|a_3| \le A = \frac{\sqrt{3}}{2}\sqrt{\sqrt{15} - 3} = 0.80915\ldots.$$

The extremal function has the form

$$Q_4(z) = z + \tfrac{3}{2}Ae^{it}z^2 + Ae^{-it}z^3 + \tfrac{1}{4}z^4$$

for a certain value of t, showing also that the bound on a_2 increases when the coefficients are allowed to be complex. Suffridge [4] showed that for $n > 5$ none of the sharp bounds for real coefficients remain valid in the full class of univalent polynomials of degree n.

Quine [1] characterized the interior points of $\mathbb{W}_n$ by the condition that the boundary curve $w = p(e^{i\theta})$ have no cusps or self-intersections. (Equivalently, p is univalent in a larger disk.) Quine [2] applied this result to obtain certain information about the boundary of $\mathbb{W}_n$.

The univalent polynomials are dense in the full class S of univalent functions (Chapter 1, Exercises 5 and 6). Thus the Bieberbach conjecture will be settled if it can be proved for polynomials. Suffridge [4] showed that even the subclass of polynomials with highest coefficient $a_n = 1/n$ is dense in S. His proof shows that the polynomials with real coefficients and $a_n = 1/n$ are dense in S_R, so Suffridge's estimate $|a_k| \leq A_{nk}$ can be used to prove the Bieberbach conjecture for functions with real coefficients. (Note that for each fixed k, the bounds A_{nk} increase to k as $n \to \infty$.) Horowitz [2] has verified the Bieberbach conjecture for all univalent polynomials of degree up to 27. His method makes use of a result of Dieudonné [3] on univalent polynomials, FitzGerald's inequality (Theorem 5.14) and a generalization, the truth of the Bieberbach conjecture for $n \leq 6$, and extensive machine calculation.

Starlike, close-to-convex, and typically real polynomials were studied by Brannan [4], Brannan and Brickman [1], and Royster and Suffridge [1].

3. Functions of Bounded Boundary Rotation

For a Jordan domain with smooth (*i.e.*, continuously differentiable) boundary, the *boundary rotation* α is defined as the total variation of the direction angle of the tangent to the boundary curve under a complete circuit. Thus $\alpha \geq 2\pi$, with equality if and only if the domain is convex. If $\alpha \leq 4\pi$, it is intuitively obvious that the domain is close-to-convex, in view of Kaplan's geometric criterion (Theorem 2.18).

For a general simply connected domain D, the boundary rotation is defined as follows. Let $\{D_n\}$ be an exhaustion of D; that is, a sequence of subdomains D_n with $\overline{D_n} \subset D_{n+1} \subset \overline{D_{n+1}} \subset D$ and $\bigcup_{n=1}^{\infty} D_n = D$. Let α_n be the infimum of the boundary rotations of all smoothly bounded Jordan domains G_n with $D_n \subset G_n \subset D$. Then α_n increases to a limit α, finite or infinite, called the boundary rotation of D. It is easily seen that α is independent of the choice of exhaustion $\{D_n\}$. The domain D is said to have *bounded boundary rotation* if $\alpha < \infty$. This concept can be extended to domains on a Riemann surface. The details may be found in Paatero [1].

A function analytic and locally univalent in a given simply connected domain is said to be of bounded boundary rotation if its range (regarded as lying on a Riemann surface) has bounded boundary rotation. It is convenient to consider the class V of normalized functions

$$f(z) = z + a_2 z^2 + a_3 z^3 + \cdots$$

analytic and locally univalent in the unit disk $\mathbb{D}$, and of bounded boundary rotation. For each real number $k \geq 2$, the class V_k consists of all $f \in V$ of boundary rotation $\alpha \leq k\pi$.

Functions of bounded boundary rotation were introduced by Loewner [1], but it was Paatero [1, 2] who systematically developed their properties. Paatero [1] showed that V coincides with the class of functions

$$f(z) = \int_0^z \exp\left\{-\int_0^{2\pi} \log(1 - ze^{-i\theta})\, d\mu(\theta)\right\} dz,$$

where μ is a real-valued function of bounded variation with $\int_0^{2\pi} d\mu(\theta) = 2$. This is the continuous version of the Schwarz–Christoffel formula. Each function $f \in V$ has a unique representing measure $d\mu$. The boundary rotation of f is

$$\alpha = \pi \int_0^{2\pi} |d\mu(\theta)|.$$

It can also be expressed as

$$\alpha = \lim_{r\to 1} \int_0^{2\pi} \left|\text{Re}\left\{1 + \frac{zf''(z)}{f'(z)}\right\}\right| d\theta, \qquad z = re^{i\theta},$$

as is clear from the geometric interpretation of the integrand (see §2.5).

The classes V_k obviously expand as k increases. V_2 is simply the class C of convex mappings. Paatero [1] showed that $V_4 \subset S$, and Rényi [1] proved the Bieberbach conjecture $|a_n| \le n$ for V_4. Much later, Pinchuk [1] and Brannan [2] observed that V_4 is (properly) contained in the class K of close-to-convex functions. However, each class V_k with $k > 4$ contains non-univalent functions. Kirwan [3] showed that the radius of univalence of V_k for $k > 4$ is $\tan(\pi/k)$.

It is interesting to consider extremal problems within the class V_k. The Koebe function belongs to V_4 but not to V_k for any $k < 4$, so even the standard growth and distortion problems will have new solutions for $k < 4$. For arbitrary $k \ge 2$, Loewner [1] obtained the sharp distortion theorem

$$\frac{(1-r)^{k/2-1}}{(1+r)^{k/2+1}} \le |f'(z)| \le \frac{(1+r)^{k/2-1}}{(1-r)^{k/2+1}}, \qquad |z| = r < 1,$$

for all $f \in V_k$, with equality only for certain rotations of the "wedge mapping"

$$F_k(z) = \frac{1}{k}\left[\left(\frac{1+z}{1-z}\right)^{k/2} - 1\right] = z + \sum_{n=2}^{\infty} A_n(k) z^n.$$

This function plays the role of the Koebe function in V_k. In particular, F_4 is the Koebe function and F_2 is the half-plane mapping $\ell(z) = z/(1 - z)$, the typical extremal function for problems involving convex functions.

Thus it is natural to conjecture that for each $k \geq 2$, the coefficient inequality

$$|a_n| \leq A_n(k), \qquad n = 2, 3, \ldots,$$

holds for every $f \in V_k$, where $A_n(k)$ are the coefficients of the wedge mapping F_k. The first few coefficients are easily computed:

$$A_2(k) = k/2, \qquad A_3(k) = (k^2 + 2)/6, \qquad A_4(k) = (k^3 + 8k)/24.$$

In 1952, Lehto [1] verified the coefficient conjecture for $n = 2$ and 3. (He attributed the result for $n = 2$ to Pick.) However, the full conjecture yielded only after an additional 20 years of effort. First it was proved for $n = 4$ by Schiffer and Tammi [1], using a variational method, for $2 \leq k \leq 4$. Then Lonka and Tammi [1], Brannan [3], and Coonce [1] extended the inequality for $n = 4$ to all $k \geq 2$. Noonan [1] then proved the inequality for all n if k is an even integer. For general k he obtained an asymptotic result, the analogue of Hayman's regularity theorem (Theorem 5.6), thus giving additional support to the conjecture.

Then came the decisive step. In 1973, Brannan, Clunie, and Kirwan [1] succeeded in proving the conjecture for all n if $k \geq 4$, and for all $n \leq 14$ if $k < 4$. Their method depends upon the observation (essentially due to Brannan [2]) that V_k is contained in the class $K(\beta)$ of close-to-convex functions of order β, where $\beta = k/2 - 1$. (For $\beta > 1$ the class $K(\beta)$ properly contains the class $K = K(1)$ of close-to-convex functions.) Their proof uses extreme-point theory and the Krein–Milman theorem. (See Chapter 9.) For $k \geq 2$ they reduced the problem to showing that

$$\left(\frac{1 + e^{it}z}{1 - z}\right)^a \ll \left(\frac{1 + z}{1 - z}\right)^a, \qquad a \geq 1, \quad 0 \leq t \leq 2\pi, \tag{23}$$

in the sense of coefficient majorization (*cf.* §8.1), but they were able to verify this, hence the coefficient conjecture, for only the first 14 coefficients. Aharonov and Friedland [1] and Brannan [5] then found general proofs of (23), thus completing the proof of the coefficient conjecture for functions of bounded boundary rotation.

Exercises

1. For each function $f \in S$, show that

$$M_2^2(r, f'/f) = O\left(\frac{1}{1 - r} \log M_\infty(r, f)\right) = O\left(\frac{1}{1 - r} \log \frac{1}{1 - r}\right)$$

as $r \to 1$. (*Hint*: Consider area integrals over annuli.) Deduce that the logarithmic coefficients of f satisfy

$$\sum_{n=1}^{N} n^2 |\gamma_n|^2 = O(N \log N).$$

(Duren and Leung [1]. A construction by Hayman [9] shows that "O" cannot be improved to "o".)

2. According to Theorem 8.3, the logarithmic coefficients of bounded univalent functions need not be $O(1/n)$ and may in fact fail to be $O(n^{-0.83})$. Show, however, that

$$\sum_{n=1}^{N} n^2 |\gamma_n|^2 = O(N)$$

for all bounded functions $f \in S$. More generally, establish this estimate for all functions $f \in S$ whose image has finite area. (Duren and Leung [1].)

3. Show that if a function $f \in S$ maps $\mathbb{D}$ onto a Jordan domain with rectifiable boundary, then $\gamma_n = O(1/n)$. (Duren and Leung [1].)

4. For a function $f \in S$, Milin's lemma (§5.4) asserts that

$$\sum_{k=1}^{n} \left(k|\gamma_k|^2 - \frac{1}{k} \right) \le \delta < 0.312.$$

By Corollary 1 to Theorem 5.5, these sums are also bounded below if f has Hayman index $\alpha > 0$. Using these facts, show that

$$\sum_{n=1}^{N} n^2 |\gamma_n|^2 = O(N),$$

or equivalently that $M_2^2(r, f'/f) = O(1/(1-r))$, for all functions $f \in S$ with $\alpha > 0$. (Duren and Leung [1].)

5. For each n ($n = 2, 3, \ldots$), show that the radius of convexity of the nth section $z + a_2 z^2 + \cdots + a_n z^n$ of a function $f \in C$ is smallest for $f(z) = \ell(z) = z(1-z)^{-1}$.

6. Show that all sections of the Koebe function are starlike in the disk $|z| < \frac{1}{4}$. Apply Theorems 8.6 and 8.7 to conclude that all sections of an arbitrary convex, starlike, or close-to-convex function ($f \in C$, S^*, or K) are respectively convex, starlike, or close-to-convex in $|z| < \frac{1}{4}$.

7. Prove that the nth section of the Koebe function is starlike in the disk $|z| < 1 - 3n^{-1} \log n$, $n \ge 5$, and that 3 can be replaced by no smaller

constant. (Robertson [5].) Conclude that if f belongs to C, S^*, or K, then its nth section is respectively convex, starlike, or close-to-convex in the disk $|z| < 1 - 3n^{-1}\log n$, $n \geq 5$.

8. Prove that every section of each odd univalent function $h \in S^{(2)}$ is univalent in the disk $|z| < 1/\sqrt{3}$, and that this radius is best possible. (Joh [1]; Jenkins [1].)

9. The *de la Vallée Poussin means* of an analytic function $f(z) = \sum_{k=1}^{\infty} a_k z^k$ are the polynomials V_n defined by

$$V_n(z; f) = \binom{2n}{n}^{-1} \sum_{k=1}^{n} \binom{2n}{n+k} a_k z^k.$$

(a) Show that $V_n(z; \ell)$ is convex in $\mathbb{D}$ for $n = 1, 2, \ldots$, where $\ell(z) = z(1 - z)^{-1}$.

(b) For arbitrary $f \in C$, show that $V_n(z; f)$ is convex in $\mathbb{D}$ for $n = 1, 2, \ldots$. (Pólya and Schoenberg [1].)

10. Derive the results of Strohhäcker and Marx (Corollary 1) from the proposition in §8.3.

11. Deduce from the proposition in §8.3 that

$$\operatorname{Re}\left\{\frac{zf'(z)}{f(z) - f(\zeta)} - \frac{\zeta}{z - \zeta}\right\} > \tfrac{1}{2}$$

for each function $f \in C$ and for all points $z, \zeta \in \mathbb{D}$. Use this to prove Corollary 2. Conversely, show that if f is analytic in $\mathbb{D}$ with $f(0) = 0$ and $f'(0) = 1$, and if f satisfies the above inequality, then $f \in C$. (Sheil-Small [1]; Suffridge [3].)

12. Show that if $\varphi(z) = 1 + \sum_{n=1}^{\infty} c_n z^n$ and $\psi(z) = 1 + \sum_{n=1}^{\infty} d_n z^n$ have positive real part in $\mathbb{D}$, then so does their "convolution"

$$\chi(z) = 1 + \tfrac{1}{2} \sum_{n=1}^{\infty} c_n d_n z^n.$$

(Schur [2], p. 125; Komatu [1].)

13. Prove that if f and g belong to the class T of typically real functions, then $f \circledast g \in T$. (Robertson [6].)

14. Prove that for each pair of functions $g(z) = z + \sum_{n=1}^{\infty} b_n z^{-n}$ and $h(z) = z + \sum_{n=1}^{\infty} c_n z^{-n}$ of class Σ, the convolution $z + \sum_{n=1}^{\infty} b_n c_n z^{-n}$ also belongs to Σ. (Robertson [7].)

15. Let f be analytic and locally univalent, and let $\mathscr{S}(f)$ be its Schwarzian derivative. Prove that $\mathscr{S}(f) = 0$ if and only if f is a linear fractional transformation.

16. Prove Lemma 2 of §8.5 in the following more conceptual manner. Let ζ be the midpoint of the hyperbolic geodesic joining z_1 to z_2, and let φ be a mapping of $\mathbb{D}$ onto itself with $\varphi(\zeta) = 0$. (Suggested to the author by Brad Osgood.)

17. Let $\{n_k\}$ be a lacunary sequence of positive integers: $n_{k+1}/n_k \geq q > 1$. Show that for a sufficiently small constant $c > 0$, the function f defined by

$$f'(z) = \exp\left\{c \sum_{k=1}^{\infty} z^{n_k}\right\}$$

is univalent in $\mathbb{D}$, yet $f'(z)$ has a radial limit almost nowhere. In particular, $f' \notin H^p$ for all $p > 0$. (For a more emphatic demonstration of this phenomenon, consult the paper of Lohwater, Piranian, and Rudin [1].)

18. Use the function $f(z) = -\log(1 - z)$ to show that Nehari's condition (18) does not imply that f maps the disk onto a Jordan domain on the Riemann sphere. (Gehring and Pommerenke [1] have shown that up to linear fractional transformation this is the only example.)

19. Show that the function

$$f_n(z) = \int_0^z e^{\lambda\zeta^n}\, d\zeta = z + \frac{\lambda}{n+1} z^{n+1} + \cdots$$

is not univalent in $\mathbb{D}$ if $\lambda > 2(n + 1)/n$. Show, moreover, that

$$\sup_{z \in \mathbb{D}} (1 - |z|^2)|f_n''(z)/f_n'(z)| \to \frac{2\lambda}{e} \quad \text{as } n \to \infty.$$

Thus for no constant $C > 4/e = 1.471\ldots$ is the condition $|f''(z)/f'(z)| \leq C(1 - |z|^2)^{-1}$ sufficient for univalence. (Example due to Ch. Pommerenke.)

20. Becker has shown that the condition (slightly weaker than that cited in §8.5)

$$|zf''(z)/f'(z)| \leq C(1 - |z|^2)^{-1}, \qquad |z| < 1,$$

implies that f is univalent in $\mathbb{D}$ if $C = 1$. Use the function $f(z) = e^{\lambda z}$ with $\lambda > \pi$ to show that this condition is not sufficient for univalence if $C > 2\sqrt{3}\pi/9 = 1.209\ldots$. (Becker [1]. Pommerenke [15] has shown that $C > 1.121$ is not sufficient.)

21. Show that if f is univalent in $\mathbb{D}$, then $|f''(z)/f'(z)| \leq 6(1 - |z|^2)^{-1}$. Show also that the constant 6 in this bound is best possible.

Chapter 9

General Extremal Problems

In previous chapters we have treated extremal problems individually, bringing special methods to bear on each problem. We shall now introduce a more general setting which will permit the simultaneous consideration of a wide class of extremal problems. The main idea is to embed the class S in the *linear* space of all analytic functions in the unit disk, in order to use the familiar theory of functional analysis. Under the topology of uniform convergence on compact subsets of the disk, S becomes a compact set in a locally convex linear space. It is then a natural problem to identify the extreme points of S. By the Krein–Milman theorem, there is an extreme point of S among the support points associated with each linear functional. In this chapter we shall give partial descriptions of the extreme points and support points of S. We shall take advantage of the underlying linear space not only to consider linear functionals, but also to introduce the more general concept of a differentiable functional. Among other things, we shall prove that the extremal functions of S associated with every differentiable functional have dense range. More detailed information comes from variational methods, which are developed in Chapter 10.

§9.1. Functionals on Linear Spaces

The class S of univalent functions has no linear structure. Simple examples show that the sum of two functions in S need not be univalent. In fact, A. W. Goodman [5] has constructed a pair of functions in S whose sum has *infinite* valence.

Nevertheless, it is useful to view S as a subset of the linear space $\mathscr{A} = \mathscr{H}(\mathbb{D})$ of all analytic functions in the unit disk, endowed with the topology of uniform convergence on compact subsets. This topology is *locally convex*: every neighborhood contains a convex neighborhood. Furthermore, the topology is induced by the translation–invariant metric

$$\rho(f, g) = \sum_{n=1}^{\infty} 2^{-n} \min\{p_n(f - g), 1\},$$

where $p_n(f)$ denotes the maximum of $|f(z)|$ over the closed subdisk $|z| \leq 1 - 1/n$. In other words, $f_k(z) \to f(z)$ uniformly on each compact subset of the unit disk if and only if $\rho(f_k, f) \to 0$. It is clear that S is a compact subset of $\mathscr{A}$, since S is a compact normal family.

Although the space $\mathscr{A}$ is metrizable, it is not normable. For some purposes, it is more convenient to view S as a subset of the Banach space $\mathscr{B}$ of all analytic functions for which

$$\|f\| = \sum_{n=1}^{\infty} 2^{-n} p_n(f) < \infty.$$

That S is contained in $\mathscr{B}$ follows from the growth theorem (§2.3), which implies that $p_n(f) < n^2$ for each $f \in S$. If $\|f_k - f\| \to 0$, then obviously $\rho(f_k, f) \to 0$; that is, $f_k(z) \to f(z)$ uniformly on compact subsets. The converse is false.

Now let X be an arbitrary complex linear space with a locally convex topology, and let ϕ be a complex-valued functional defined on a subset of X. Then ϕ is said to have a *Gâteaux differential* at a point $x_0 \in X$ if it is defined in some neighborhood of x_0 and the limit

$$\delta(h; x_0) = \lim_{t \to 0+} t^{-1}[\phi(x_0 + th) - \phi(x_0)]$$

exists for each $h \in X$. (Here $t > 0$.) If X is a Banach space and $\|h\| = 1$, the Gâteaux differential $\delta(h; x_0)$ may be viewed as the directional derivative of ϕ at x_0 in the direction h. Even in a finite-dimensional space, however, the Gâteaux differential need not be linear as a function of h.

If X is a Banach space and ϕ is defined in a neighborhood of $x_0 \in X$, then ϕ is said to have a *Fréchet differential* at x_0 if there is a continuous linear functional $\ell(\cdot\,; x_0)$ such that

$$\phi(x_0 + h) = \phi(x_0) + \ell(h; x_0) + r(h; x_0),$$

where $r(h; x_0) = o(\|h\|)$ as $\|h\| \to 0$. This is equivalent to saying that the difference quotient

$$t^{-1}[\phi(x_0 + th) - \phi(x_0)] \to \delta(h; x_0)$$

uniformly on the unit sphere $\|h\| = 1$, and that the Gâteaux differential $\delta(\cdot\,; x_0)$ is linear. In this case, the Gâteaux differential coincides with the Fréchet differential $\ell(\cdot; x_0)$.

Here are some examples.

EXAMPLE 1. Let ϕ be a continuous linear functional on a Banach space X. Then for each $x_0 \in X$,

$$t^{-1}[\phi(x_0 + th) - \phi(x_0)] = \phi(h), \qquad h \in X, \qquad t > 0.$$

Thus a linear functional is its own Fréchet differential.

EXAMPLE 2. For fixed ζ with $|\zeta| < 1$, let

$$\phi(f) = \log f'(\zeta), \qquad f \in S.$$

(The monodromy theorem guarantees the existence of a single-valued branch of the logarithm which vanishes at $\zeta = 0$.) This functional ϕ has a local extension to the Banach space $\mathscr{B}$; that is, it can be extended to some neighborhood of each point $f \in S$. If $f + h$ lies in such a neighborhood, then

$$\log[f'(\zeta) + h'(\zeta)] = \log f'(\zeta) + \frac{h'(\zeta)}{f'(\zeta)} + O(\|h\|^2),$$

or

$$\phi(f + h) = \phi(f) + \ell(h; f) + O(\|h\|^2), \qquad \|h\| \to 0,$$

where the Fréchet differential $\ell(\cdot; f)$ is the continuous linear functional defined by $\ell(h; f) = h'(\zeta)/f'(\zeta)$.

EXAMPLE 3. Let $f \in \mathscr{B}$, and let $F(z_0, z_1, \ldots, z_n)$ be holomorphic in a neighborhood (in $\mathbb{C}^{n+1}$) of the point $(f(\zeta), f'(\zeta), \ldots, f^{(n)}(\zeta))$. Then the functional

$$\phi(g) = F(g(\zeta), g'(\zeta), \ldots, g^{(n)}(\zeta)), \qquad g \in \mathscr{B},$$

is defined in some neighborhood of f and has the Fréchet differential

$$\ell(h; f) = \sum_{k=0}^{n} F_k(f(\zeta), f'(\zeta), \ldots, f^{(n)}(\zeta))h^{(k)}(\zeta),$$

where $F_k = \partial F/\partial z_k$.

EXAMPLE 4. This will explain our insistence that a functional whose real part is to be maximized in S have a Fréchet differential and not only a Gâteaux differential. The methods of variation within S are *nonlinear*. Near each fixed function $f \in S$, they typically produce a one-parameter family of functions

$$f^* = f + \varepsilon g + O(\varepsilon^2), \qquad \varepsilon > 0,$$

where "$O(\varepsilon^2)$" indicates a term which is bounded, uniformly on each compact subset, by a constant multiple of ε^2. Now let ϕ be a functional defined for all functions $(f + h) \in \mathscr{B}$ in some neighborhood of f, and suppose ϕ has a Fréchet differential $\ell(\cdot\,; f)$ at f. Then $\phi(f^*)$ is defined for sufficiently small ε, and

$$\begin{aligned}\phi(f^*) &= \phi(f) + \ell(\varepsilon g + O(\varepsilon^2); f) + o(\varepsilon)\\ &= \phi(f) + \varepsilon\ell(g; f) + o(\varepsilon).\end{aligned}$$

For further information on Fréchet and Gâteaux differentials in linear spaces, the reader is referred to the book of Hille and Phillips [1] or to the expository article of Nashed [1].

§9.2. Representation of Linear Functionals

In treating general linear extremal problems for univalent functions, it is important to have a structural formula for an arbitrary continuous linear functional on the space $\mathscr{A}$ of all analytic functions on the unit disk. The simplest formula, and the most useful, is an integral representation described as follows.

Theorem 9.1. *Each continuous linear functional L on the space $\mathscr{A}$ has the form*

$$L(h) = \iint_E h(z)\, d\mu(z), \qquad h \in \mathscr{A},$$

where μ is a complex-valued measure supported on a compact subset E of $\mathbb{D}$.

This theorem can be generalized to an arbitrary domain in the complex plane. The representing measure μ is not unique. We shall prove the following more specific version of the theorem, due in more general form to Caccioppoli [1], which has other implications as well (*cf.* Schober [1]).

Theorem 9.2. *Each continuous linear functional L on the space $\mathscr{A}$ has the form*

$$L(h) = \frac{1}{2\pi i}\int_C h(z)g(z)\, dz, \qquad h \in \mathscr{A},$$

where C is the circle $|z| = r$ for some $r < 1$, and g is a function analytic in $|z| \geq r$ with $g(\infty) = 0$.

Proof. Let D_n be the disk $|z| < 1 - 1/n$, $n = 2, 3, \ldots,$ and let $C_n = \partial D_n$. Let

$$\|h\|_n = \sup_{z \in D_n} |h(z)|, \qquad h \in \mathscr{A}.$$

We claim that there exist a constant $M > 0$ and an integer $m \geq 2$ such that $|L(h)| \leq M\|h\|_m$ for all $h \in \mathscr{A}$. Otherwise, there would be a sequence of functions $h_n \in \mathscr{A}$ with $\|h_n\|_n = 1$ and $L(h_n) \to \infty$ as $n \to \infty$. Since $\{h_n\}$ is uniformly bounded on compact subsets, it is a normal family. Thus some subsequence $\{h_{n_k}\}$ converges uniformly on compact subsets to a function $h \in \mathscr{A}$. Since L is continuous, it follows that $L(h_{n_k}) \to L(h) \neq \infty$, a contradiction. This proves that L is a bounded linear functional on the subspace $\mathscr{A}$ of the Banach space $\mathscr{B}$ of all functions analytic on D_m and continuous on $\overline{D_m}$. By the Hahn–Banach theorem, L can be extended to a bounded linear functional $\hat{L}$ on $\mathscr{B}$. Let

$$g(\zeta) = \hat{L}\left(\frac{1}{\zeta - z}\right), \qquad \zeta \notin \overline{D_m}.$$

Clearly, g is continuous and $g(\infty) = 0$. To show g is analytic, choose distinct points ζ and ζ_0 outside $\overline{D_m}$ and observe that

$$\frac{g(\zeta) - g(\zeta_0)}{\zeta - \zeta_0} = -\hat{L}\left(\frac{1}{(\zeta - z)(\zeta_0 - z)}\right) \to -\hat{L}\left(\frac{1}{(\zeta_0 - z)^2}\right)$$

as $\zeta \to \zeta_0$. Now choose $C = \partial D_{m+1}$ and appeal to the Cauchy formula:

$$h(z) = \frac{1}{2\pi i}\int_C \frac{h(\zeta)}{\zeta - z}\, d\zeta, \qquad h \in \mathscr{A}, \quad z \in \overline{D_m}.$$

Since the integral is the uniform limit of Riemann sums, we may now conclude that

$$L(h) = \hat{L}(h) = \frac{1}{2\pi i}\int_C h(\zeta)\hat{L}\left(\frac{1}{\zeta - z}\right) d\zeta = \frac{1}{2\pi i}\int_C h(\zeta)g(\zeta)\, d\zeta.$$

This completes the proof.

Linear functionals can also be represented in terms of Taylor coefficients. Since the Taylor series of a function $h \in \mathscr{A}$ converges uniformly on compact subsets of the disk, a continuous linear functional L has the form

$$L(h) = \sum_{n=0}^{\infty} c_n a_n, \qquad h(z) = \sum_{n=0}^{\infty} a_n z^n,$$

where $c_n = L(z^n)$. The following theorem of Toeplitz [1] makes this representation more precise.

Theorem 9.3. *Each continuous linear functional on $\mathscr{A}$ has the form*

$$L(h) = \sum_{n=0}^{\infty} c_n a_n, \qquad h(z) = \sum_{n=0}^{\infty} a_n z^n,$$

for some sequence of complex constants c_n such that $\limsup_{n\to\infty} |c_n|^{1/n} < 1$. *Conversely, each such sequence $\{c_n\}$ generates a continuous linear functional L in this manner.*

Proof. By Theorem 9.2, the functional L has the form

$$L(h) = \frac{1}{2\pi i} \int_C h(z)g(z)\, dz, \qquad h \in \mathscr{A},$$

for some function $g(z) = \sum_{n=1}^{\infty} b_n z^{-n}$ analytic outside a circle $|z| = \rho < 1$, where C is a circle $|z| = r$, $\rho < r < 1$. Thus

$$L(h) = \sum_{n=0}^{\infty} c_n a_n, \qquad c_n = L(z^n) = b_{n+1}.$$

The convergence of the series expansion for g implies

$$\limsup_{n\to\infty} |c_n|^{1/n} = \limsup_{n\to\infty} |b_n|^{1/n} \le \rho < 1.$$

Conversely, any sequence $\{c_n\}$ with $\limsup |c_n|^{1/n} < 1$ gives rise to a function g analytic in $|z| \ge 1$ which generates a continuous linear functional L as above.

§9.3. Extreme Points and Support Points

Let X be a topological vector space over the field of complex numbers, and let E be a subset of X. A point $x \in E$ is called an *extreme point* of E if it has no representation of the form

$$x = ty + (1 - t)z, \qquad 0 < t < 1,$$

as a proper convex combination of two distinct points y and z in E. A point $x \in E$ is called a *support point* of E if there is a continuous linear functional L, not constant on E, such that $\text{Re}\{L(x)\} \ge \text{Re}\{L(y)\}$ for all $y \in E$. For example, if X is the complex plane and E is the closed interior of a convex polygon, every boundary point is a support point but only the vertices are extreme points.

The *convex hull* of a set $E \in X$ is the smallest convex set containing E. The *closed convex hull* of E is the smallest closed convex set containing E; it is the closure of the convex hull of E. The Krein–Milman theorem asserts that every compact subset of a locally convex topological vector space is contained in the closed convex hull of its extreme points (see, for instance, Dunford and Schwartz [1]). In particular, every compact set has extreme points. Furthermore, there is at least one extreme point among the support points associated with every continuous linear functional.

This last statement may be justified as follows. Let L be a continuous linear functional on a locally convex space X, and let E be a compact subset of X. Let

$$\sigma_L(E) = \{x \in E : \text{Re}\{L(x)\} = \max_{y \in E} \text{Re}\{L(y)\}\}$$

be the associated set of support points. Clearly, $\sigma_L(E)$ is a nonempty compact subset of E. By the Krein–Milman theorem, $\sigma_L(E)$ contains an extreme point. This point $x \in \sigma_L(E)$ is not a proper convex combination of two distinct points in $\sigma_L(E)$. If it had a more general representation $x = ty + (1 - t)z$ for $y, z \in E$, then it would follow at once from the linearity of L that $y, z \in \sigma_L(E)$. Thus x is actually an extreme point of E.

Although a support point need not be an extreme point, it seems intuitively obvious that every extreme point must be a support point. This is false, however, even in a Hilbert space. The following counterexample is due to Klee [1].

Consider the space ℓ^2 of all complex sequences $x = (x_1, x_2, \ldots)$ such that

$$\|x\|^2 = \sum_{n=1}^{\infty} |x_n|^2 < \infty.$$

Let $T : \ell^2 \to \ell^2$ be the compact (or "completely continuous") operator defined by $y = Tx$, where $y_n = 2^{-n}x_n$. Let

$$B = \{x \in \ell^2 : \|x\| \le 1\}$$

be the closed unit ball in ℓ^2, and let $E = TB$ be its image under T. Then E is a compact set, because it is the image under a compact operator of the closed bounded set B (which is weakly compact). Clearly, E is convex. The extreme points of B are simply the points x with $\|x\| = 1$. Since T is linear and one-to-one, it follows that the extreme points of E are the points $y = Tx$ with $\|x\| = 1$.

Now let $y = Tx$ be a support point of E. Then by the Riesz representation theorem there is a point $z \neq 0$ in ℓ^2 such that

$$\mathrm{Re}\{(Tx, z)\} \geq \mathrm{Re}\{(Tw, z)\} = \mathrm{Re}\{(w, Tz)\}$$

for all $w \in B$. Thus $\|Tz\| \leq \mathrm{Re}\{(Tx, z)\}$, which gives

$$\|x\| \, \|Tz\| \leq \mathrm{Re}\{(Tx, z)\} \leq |(Tx, z)| = |(x, Tz)|.$$

Since $|(x, Tz)| \leq \|x\| \, \|Tz\|$, it follows that $(x, Tz) = \|x\| \, \|Tz\|$, which implies $Tz = \lambda x$ for some constant $\lambda > 0$. In particular, $x \in T(\ell^2)$ for every support point Tx.

It now remains only to choose a point $x \in \ell^2$ with $\|x\| = 1$ and $x \notin T(\ell^2)$. For example, let $x_n = 2^{-n/2}$, $n = 1, 2, \ldots$. Then $\|x\| = 1$ but $T^{-1}x = w \notin \ell^2$, since $w_n = 2^{n/2}$, $n = 1, 2, \ldots$. Thus Tx is an extreme point of E but not a support point.

This result can be applied to give an example of the same phenomenon in the topological vector space $\mathscr{A}$ of all analytic functions in the unit disk. The previous construction provides a compact convex set K in the Hardy space H^2 and a point $f \in K$ which is an extreme point but not a support point of K. Here the terms "compact" and "support point" refer to the Hilbert space topology of H^2, which is finer than that of $\mathscr{A}$. Hence K is also a compact subset of $\mathscr{A}$. Furthermore, every continuous linear functional on $\mathscr{A}$ has a restriction to H^2 which is continuous with respect to the Hilbert space topology of H^2. Consequently, f cannot be a support point of K with respect to the topology of $\mathscr{A}$. (The preceding observation was communicated to the author by Louis Brickman.)

We have already observed (§9.1) that the set S of univalent functions is a compact subset of the locally convex space $\mathscr{A}$. In view of the Krein–Milman theorem, it is important to identify the extreme points of S, because the solution to any linear extremal problem over S can be reduced to its solution over the set of extreme points. The solutions to linear extremal problems are the support points of S. Some of their general properties are known, but comparatively little is known about the extreme points of S. It is not known whether every extreme point is a support point, or whether every support point is an extreme point.

One important example of a linear extremal problem over S is the coefficient problem of maximizing $\mathrm{Re}\{a_n\}$. For $n = 2$, it is useful to consider the more general problem of maximizing $\mathrm{Re}\{e^{-i\theta}a_2\}$ for fixed θ, $0 \leq \theta < 2\pi$. It is easily seen (cf. §2.2) that the unique solution is the rotated Koebe function

$$k_\theta(z) = z(1 - e^{i\theta}z)^{-2}.$$

Thus the rotations of the Koebe function are support points of S, and they are clearly extreme points as well. By considering the linear extremal problem of maximizing $\operatorname{Re}\{f(\zeta)\}$, we shall show in §10.9 by means of a variational method that S has other support points and other extreme points. First, however, we shall use elementary methods to establish some general properties of the support points and extreme points of S.

§9.4. Properties of Extremal Functions

In §2.9 we applied an elementary variational method to show that if a function $f \in S$ maximizes $\operatorname{Re}\{a_n\}$ for some n, then its coefficients satisfy the *Marty relation*

$$(n-1)\overline{a_{n-1}} + 2a_2 a_n - (n+1)a_{n+1} = 0.$$

This result will now be generalized to an arbitrary differentiable functional.

Let ϕ be a complex-valued functional defined and continuous in an open subset of $\mathscr{A}$ containing S, and having a Fréchet differential $\ell(\cdot\,; f)$ at each $f \in S$. Since S is a compact set, $\operatorname{Re}\{\phi\}$ attains a maximum value at some point $f \in S$. As in the special case where $\phi = a_n$ (*cf.* §2.9), a small complex parameter $\zeta = \rho e^{i\varphi}$ will be used to vary f to

$$\begin{aligned} f^*(z) &= \frac{f\left(\dfrac{z+\zeta}{1+\bar{\zeta}z}\right) - f(\zeta)}{(1-|\zeta|^2)f'(\zeta)} \\ &= f(z) + \rho g(z) + O(\rho^2), \end{aligned}$$

where

$$g(z) = e^{i\varphi}[f'(z) - f''(0)f(z) - 1] - e^{-i\varphi}z^2 f'(z)$$

and "$O(\rho^2)$" indicates a term bounded on each compact subset of the disk by a constant multiple of ρ^2. Thus (*cf.* §9.1, Example 4)

$$\phi(f^*) = \phi(f) + \rho\ell(g; f) + o(\rho).$$

Since $\operatorname{Re}\{\phi(f^*)\} \le \operatorname{Re}\{\phi(f)\}$, it follows that $\operatorname{Re}\{\ell(g; f)\} \le 0$. But because $\ell(\cdot\,; f)$ is linear, this is equivalent to

$$\operatorname{Re}\{e^{i\varphi}[\ell(f' - f''(0)f - 1; f) - \overline{\ell(h; f)}]\} \le 0,$$

where $h(z) = z^2 f'(z)$. Since $e^{i\varphi}$ is arbitrary, it follows that

$$\ell(f'; f) = f''(0)\ell(f; f) + \ell(1; f) + \overline{\ell(h; f)}.$$

For the case of the nth coefficient functional $\phi = a_n$, this reduces to the Marty relation.

Another elementary variation may be used to show that the extremal function f has dense range. For the special case where $\phi = a_n$, this result is due to Marty [2].

Theorem 9.4. *Let ϕ be a continuous functional with a Fréchet differential on S, and let $\operatorname{Re}\{\phi\}$ attain its maximum value on S at a point $f \in S$. Suppose that the Fréchet differential $\ell(\cdot; f)$ of ϕ at f is not constant on S. Then the range of f is dense in $\mathbb{C}$.*

Proof. Suppose, on the contrary, that f omits some open disk $|w - w_0| < \rho$. Then for each constant $e^{i\varphi}$, the function

$$F(w) = w + \frac{e^{i\varphi}\rho^2}{w - w_0}$$

is univalent on the range of f. Normalize F by forming

$$G(w) = \frac{F(w) - F(0)}{F'(0)} = w + \frac{e^{i\varphi}\rho^2 w^2}{w_0^2(w - w_0)} + O(\rho^4).$$

Then $f^* = G \circ f \in S$, and

$$\phi(f^*) = \phi(f) + \rho^2 e^{i\varphi} w_0^{-2} \ell(f^2(f - w_0)^{-1}; f) + o(\rho^2).$$

Because f is extremal, we conclude by letting $\rho \to 0$ that

$$\operatorname{Re}\{e^{i\varphi} w_0^{-2} \ell(f^2(f - w_0)^{-1}; f)\} \le 0;$$

or, since $e^{i\varphi}$ is arbitrary,

$$\ell(f^2(f - w_0)^{-1}; f) = 0.$$

For notational convenience, let $L = \ell(\cdot\,; f)$. Define the linear functional M by

$$M(h) = L(h) - L(1)h(0) - L(f)h'(0),$$

where f is the extremal function. Then

$$M\left(\frac{1}{f - w}\right) = L\left(\frac{1}{f - w} + \frac{1}{w} + \frac{f}{w^2}\right) = \frac{1}{w^2} L\left(\frac{f^2}{f - w}\right) = 0$$

for all w in an open set Ω omitted by f. But the functional M has an integral representation

$$M(h) = \iint_E h(z)\, d\mu(z),$$

where E is a compact subset of $\mathbb{D}$ and μ is a complex-valued measure supported on E. (See §9.2.) Thus the analytic function

$$H(w) = M\left(\frac{1}{f - w}\right) = \iint_E \frac{d\mu(z)}{f(z) - w}$$

vanishes identically on Ω. Changing variables, we find

$$H(w) = \iint_{\tilde{E}} \frac{d\tilde{\mu}(\omega)}{\omega - w} = 0, \qquad w \in \Omega,$$

where $d\tilde{\mu}(w) = (d\mu \circ f^{-1})(w)$ and $\tilde{E} = f(E)$. Since $\tilde{E}$ does not separate the omitted set Ω from ∞, we may conclude that $H(w) \equiv 0$ in a neighborhood of ∞. Consequently,

$$\iint_{\tilde{E}} \omega^k\, d\tilde{\mu}(\omega) = 0, \qquad k = 0, 1, 2, \ldots,$$

which shows that the measure $\tilde{\mu}$ annihilates all polynomials. Since every function analytic in the range of f can be approximated uniformly on compact subsets by polynomials, it follows that $\tilde{\mu}$ annihilates all analytic functions. It then follows by a change of variable that

$$M(h) = \iint_E h(w)\, d\mu(w) = 0$$

for every $h \in \mathscr{A}$. Thus L has the form

$$L(h) = L(1)h(0) + L(f)h'(0)$$

and is constant on S, contrary to hypothesis. This proves the theorem.

Corollary. *Each support point of S has dense range.*

§9.5. Extreme Points of S

We now turn to the result of Brickman [1] that each extreme point of S has the *monotonic modulus property*: it maps the disk onto the complement of an arc which extends to ∞ with increasing modulus. We begin with an elementary theorem whose elegant proof is essentially due to Brickman and Wilken [1]. This will give as corollaries both Brickman's result and the fact that each support point of S has the monotonic modulus property.

Theorem 9.5. *If a function $f \in S$ omits two values of equal modulus, then f has the form*

$$f = tf_1 + (1 - t)f_2, \qquad 0 < t < 1,$$

where f_1 and f_2 are distinct functions in S which omit open sets.

Proof. Let D be the range of f. If f omits α and β, $\alpha \neq \beta$, then some branch of the function

$$\psi(w) = \{(w - \alpha)(w - \beta)\}^{1/2}$$

is analytic and single-valued in D. We claim that the two functions $w \pm \psi(w)$ are univalent and have disjoint ranges. To prove the univalence, suppose

$$w_1 \pm \psi(w_1) = w_2 \pm \psi(w_2),$$

or

$$\psi(w_1) - \psi(w_2) = \pm(w_2 - w_1).$$

Squaring both sides, we have

$$2\psi(w_1)\psi(w_2) = (w_1 - \alpha)(w_1 - \beta) + (w_2 - \alpha)(w_2 - \beta) - (w_2 - w_1)^2.$$

Squaring again, we obtain after some labor

$$(\alpha - \beta)^2(w_1 - w_2)^2 = 0,$$

which is impossible unless $w_1 = w_2$. Thus both of the functions $w \pm \psi(w)$ are univalent in D. A similar argument shows they have disjoint ranges. Indeed, if

$$w_1 + \psi(w_1) = w_2 - \psi(w_2),$$

essentially the same calculation shows that $w_1 = w_2$, which implies $\psi(w_1) = \psi(w_2) = 0$. This is clearly impossible.

In particular, we have shown that the functions $w \pm \psi(w)$ are univalent and omit open sets. Both properties are preserved under the normalizations

$$\psi_1(w) = \frac{w + \psi(w) - \psi(0)}{1 + \psi'(0)}, \qquad \psi_2(w) = \frac{w - \psi(w) + \psi(0)}{1 - \psi'(0)}.$$

These functions ψ_1 and ψ_2 are analytic and univalent in D, omit open sets, and satisfy $\psi_j(0) = 0$ and $\psi_j'(0) = 1, j = 1, 2$. Therefore, the compositions $f_1 = \psi_1 \circ f$ and $f_2 = \psi_2 \circ f$ are distinct functions in S which omit open sets. Furthermore, since

$$[1 + \psi'(0)]\psi_1(w) + [1 - \psi'(0)]\psi_2(w) = 2w,$$

the function f can be expressed by

$$f(z) = tf_1(z) + (1 - t)f_2(z),$$

where $t = \frac{1}{2}[1 + \psi'(0)]$.

It remains to show that $0 < t < 1$ under the additional assumption that $|\alpha| = |\beta|$. Equivalently, it is to be shown that $-1 < \psi'(0) < 1$ if $\alpha = re^{i\theta}$ and $\beta = re^{i\varphi}$, where $0 < \theta - \varphi < 2\pi$. But an easy calculation gives

$$\psi'(0) = -(\alpha + \beta)/2\psi(0) = \pm \cos \tfrac{1}{2}(\theta - \varphi),$$

which proves $-1 < \psi'(0) < 1$. This completes the proof.

Corollary 1. *Each extreme point of S has the monotonic modulus property.*

Proof. If a function $f \in S$ does not have the monotonic modulus property, it must omit two values of equal modulus. According to the theorem, f is then a proper convex combination of two different functions in S. This shows that f is not an extreme point of S.

Corollary 2. *Each support point of S has the monotonic modulus property.*

Proof. Let f be a support point of S, and let L be the corresponding linear functional. Thus f maximizes $\text{Re}\{L\}$ over S. If f does not have the monotonic modulus property, it must have the form

$$f = tf_1 + (1 - t)f_2, \qquad 0 < t < 1,$$

where f_1 and f_2 are functions in S which omit open sets. In particular, neither f_1 nor f_2 is a support point of S, since the support points have dense range (see §9.4). Thus

$$\operatorname{Re}\{L(f_j)\} < \operatorname{Re}\{L(f)\}, \qquad j = 1, 2.$$

By the linearity of L, this implies

$$\operatorname{Re}\{(L(f)\} = t \operatorname{Re}\{L(f_1)\} + (1 - t) \operatorname{Re}\{L(f_2)\} < \operatorname{Re}\{L(f)\}.$$

This contradiction completes the proof.

It should be observed that this last argument extends easily to convex functionals which have a Fréchet derivative.

§9.6. Extreme Points of Σ

Strictly speaking, the class Σ has no extreme points at all. The constant term b_0 in the expansion

$$g(z) = z + b_0 + b_1 z^{-1} + b_2 z^{-2} + \dots, \qquad |z| > 1,$$

may be adjusted to represent an arbitrary function $g \in \Sigma$ as a proper convex combination of two functions in Σ which differ only in their constant terms. In order to avoid this trivial complication, it is natural to consider the subclass Σ_0 consisting of all functions $g \in \Sigma$ for which $b_0 = 0$. The set Σ_0 has an abundance of extreme points.

Recall that each $g \in \Sigma$ maps Δ onto the complement of a compact set E, and that g is called a *full mapping* if E has measure zero. The subclass of full mappings is denoted by $\tilde{\Sigma}$. Springer [2] proved in 1955 that the set of extreme points of Σ_0 contains all of $\tilde{\Sigma}_0 = \tilde{\Sigma} \cap \Sigma_0$.

Theorem 9.6. *Every full mapping is an extreme point of* Σ_0.

Proof. Suppose that a function $f \in \tilde{\Sigma}_0$ has a representation

$$f = tg + (1 - t)h, \qquad 0 < t < 1,$$

as a proper convex combination of two functions g and h in Σ_0 with coefficients b_n and c_n, respectively. Then the case of equality in the area theorem (§2.2) gives

$$\begin{aligned} 1 &= \sum_{n=1}^{\infty} n|tb_n + (1-t)c_n|^2 \\ &= \sum_{n=1}^{\infty} n\{t|b_n|^2 + (1-t)|c_n|^2 - t(1-t)|b_n - c_n|^2\} \\ &\leq t + (1-t) - t(1-t)\sum_{n=1}^{\infty} n|b_n - c_n|^2. \end{aligned}$$

But since $0 < t < 1$, this implies

$$\sum_{n=1}^{\infty} n|b_n - c_n|^2 \leq 0,$$

so that $b_n = c_n$, $n = 1, 2, \ldots$. In other words, $g = h$, which proves that f is an extreme point of Σ_0 if it is a full mapping.

In the converse direction, it is easy to see that the range of every extreme point is dense in the complex plane.

Theorem 9.7. *Every extreme point of* Σ_0 *has dense range.*

Proof. If $f \in \Sigma_0$ and f omits a disk $|w - w_0| < \rho$, then each of the functions $G(w) = w + \rho^2/(w - w_0)$ and $H(w) = w - \rho^2/(w - w_0)$ is univalent on the range of f. Thus the functions $g = G \circ f$ and $h = H \circ f$ both belong to Σ_0, and $f = \frac{1}{2}(g + h)$. This proves that f is not an extreme point if it omits an open set, an equivalent statement of the theorem.

The complete converse lies deeper.

Theorem 9.8. *Every extreme point of* Σ_0 *is a full mapping.*

Proof. According to a theorem of Nguyen [1], every set $E \subset \mathbb{C}$ of positive measure supports a finite nonzero measure μ whose Cauchy transform

$$F(w) = \iint_E \frac{d\mu(\zeta)}{\zeta - w} = \alpha_1 w^{-1} + \alpha_2 w^{-2} + \ldots$$

satisfies a Lipschitz condition in the complement of E:

$$|F(w_1) - F(w_2)| \leq A|w_1 - w_2|, \qquad w_1 \notin E, \quad w_2 \notin E.$$

It follows immediately that for $0 < \varepsilon < 1/A$, the functions $w \pm \varepsilon F(w)$ are univalent on the complement of E. Now suppose that a function $f \in \Sigma_0$ is not a full mapping, so that its omitted set E has positive measure. Then the functions $w \pm \varepsilon F(w)$ just constructed are univalent on the range of f, so that the composed functions

$$g(z) = f(z) + \varepsilon F(f(z)) = z + b_1 z^{-1} + \cdots$$

and

$$h(z) = f(z) - \varepsilon F(f(z)) = z + c_1 z^{-1} + \cdots$$

both belong to Σ_0. Therefore, $f = \frac{1}{2}(g + h)$ is a proper convex combination of distinct functions in Σ_0, and so is not an extreme point. This proves the theorem.

It is evident that the proof leans heavily on the deep result of Nguyen, which will not be proved here. This argument was found by Hamilton [2] in response to a question raised by G. Schober.

Exercises

1. Use the Herglotz representation of starlike functions (Chapter 2, Exercise 11) to show that $\{\log[f(z)/z] : f \in S^*\}$ is a convex set of functions whose extreme points correspond to the rotations of the Koebe function.

2. As in §2.8, let P_R be the class of functions $\varphi \in P$ with real coefficients.

   (a) Show that the extreme points of P_R are the functions

   $$\varphi(z) = \frac{1 - z^2}{1 - 2z\cos t + z^2}, \qquad 0 \le t \le \pi.$$

   (*Suggestion*: Refer to the integral representation in Chapter 2, Exercise 26.)

   (b) Deduce from Rogosinski's theorem (§2.8) that the extreme points of the class T of typically real functions have the form

   $$f(z) = z[(z - e^{it})(z - e^{-it})]^{-1}, \qquad 0 \le t \le \pi.$$

   Show that these functions are starlike, hence univalent in $\mathbb{D}$.

   (c) Conclude (either by direct reasoning or from the Krein–Milman theorem) that T is the closure of the convex hull of S_R. (Brickman, MacGregor, and Wilken [1].)

3. Show that the most general continuous linear functional L which is constant on S has the form

$$L(h) = \alpha h(0) + \beta h'(0), \qquad \alpha, \beta \in \mathbb{C}.$$

4. Let $\mathcal{S}_0$ denote the class of functions f analytic and univalent in $\mathbb{D}$, with $f(z) \neq 0$ there and $f(0) = 1$. Show that every support point of $\mathcal{S}_0$ has dense range.

5. Show that if a nonvanishing univalent function $f \in \mathcal{S}_0$ omits two values on the same (possibly degenerate) ellipse with foci at 0 and 1, then f is a proper convex combination of two distinct functions in $\mathcal{S}_0$ which omit open sets. Conclude that each extreme point or support of $\mathcal{S}_0$ must map $\mathbb{D}$ onto the complement of a continuous arc Γ extending from 0 to ∞ monotonically with respect to the family of ellipses with common foci at 0 and 1. In other words, Γ intersects each such ellipse exactly once, and 0 is an endpoint of Γ. (Duren and Schober [2].)

Chapter 10

Boundary Variation

The method of boundary variation, introduced by Schiffer in 1938, was the first variational method devised for application to univalent functions. In many respects it is still the most effective. It applies readily to a wide variety of extremal problems, where it typically provides the information that the range of an extremal function is the complement of a system of analytic arcs satisfying a certain differential equation.

This chapter contains a proof of Schiffer's basic theorem. The method is then applied to some specific extremal problems. Among other things, it leads to further information about support points of the class S.

§10.1. Preliminary Remarks

The calculus of variations is a classical subject whose origins may be traced to Euler and Lagrange. It was developed primarily to solve certain extremal problems which arose in mathematical physics and geometry. Broadly speaking, an *extremal problem* asks for the maximum or minimum of a specified functional over some class of functions. The main idea of a variational method is to obtain information about an extremal function by comparing it with its "neighbors" in the given class. This information is often sufficient to describe the extremal functions and thus to solve the extremal problem.

Any variational approach to an extremal problem must consist of four main steps: (1) a proof of the existence of an extremal function; (2) the construction of a "large" neighborhood of comparison functions; (3) the use of these comparison functions to deduce properties of the extremal function; and (4) the use of this information to identify the extremal functions or to find the extremal value of the functional.

In the classical calculus of variations the first step may present a major obstacle. It is easy to formulate extremal problems which have no solution at all; a finite supremum or infimum fails to be attained within the given class of functions. Historically, this was the objection Weierstrass raised to Riemann's treatment of the Dirichlet principle, and Hilbert filled the gap only with great labor. (Courant [1] gives a full account of this famous episode.) On the other hand, in the classical theory the second step is usually a triviality, because these problems typically involve linear families of functions.

In extremal problems for univalent functions the situation is reversed. The typical extremal problem has a solution simply because a continuous functional is to be maximized over a compact normal family of analytic functions. However, the construction of suitable comparison functions may require great ingenuity. Linear perturbations are not allowed because most families of univalent functions are highly nonlinear. For example, the sum of two functions in S need not be univalent and may even have infinite valence (see §9.1). Elementary variations are of little use because they produce very restricted families of comparison functions. The Marty variation (§2.9), for instance, leads to the relatively weak result that a function in S which maximizes $\mathrm{Re}\{a_n\}$ must satisfy the Marty relation

$$(n - 1)\overline{a_{n-1}} + 2a_2 a_n - (n + 1)a_{n+1} = 0.$$

More sophisticated variational methods show that this extremal function actually satisfies a differential equation which implies the Marty relation and gives much more information.

Such a differential equation arises in a variety of extremal problems involving univalent functions. It is analogous to the Euler equation in the classical calculus of variations. Like the Euler equation, it expresses the fact that the extremal function is a critical point of the given functional. Additional ingenuity is usually required to deduce the desired information about the extremal function. The same difficulty arises in the classical setting, where the Euler equation may be difficult to integrate.

The basic idea of the method of boundary variation is to perturb the extremal function by composing it with functions univalent on its range which are near the identity. This gives a rich collection of nearby functions whose functional values may be compared with that of the extremal function. A geometric theorem due to Schiffer [4] then converts the resulting collection of inequalities into the conclusion that the set Γ omitted by the extremal function is a system of analytic arcs which are trajectories of a certain quadratic differential. From this it is generally possible to derive a differential equation for the extremal function itself.

Before considering Schiffer's theorem we must discuss the concept of *conformal radius* (or *transfinite diameter*) of a set of points in the plane.

§10.2. Conformal Radius

Let E be a compact connected set in the complex plane, and let D be the component of its complement which contains the point at infinity. Let $\zeta = \varphi(z)$ map D conformally onto $|\zeta| > \rho$ and have the form

$$\varphi(z) = z + c_0 + c_1 z^{-1} + c_2 z^{-2} + \cdots$$

near ∞. The radius $\rho = \rho(E)$ of the omitted disk is uniquely determined and is called the *conformal radius* of E.

There are several other approaches to this same measure of the "spread" of E. For example, the concept of *diameter* can be generalized by considering

$$\Delta_n(E) = \sup_{z_1, \ldots, z_n \in E} \prod_{\substack{j,k=1 \\ j<k}}^{n} |z_k - z_j|$$

and forming

$$\delta_n(E) = [\Delta_n(E)]^{2/n(n-1)}, \qquad n = 2, 3, \ldots.$$

The diameter of E is $d(E) = \delta_2(E) = \Delta_2(E)$. It can be shown that as $n \to \infty$, the numbers $\delta_n(E)$ decrease to a limit $\delta(E)$, called the *transfinite diameter* of E, and $\delta(E) = \rho(E)$. (See Goluzin [28] or Tsuji [1].) Other approaches involve polynomial approximation (*Chebyshev's constant*) or potential theory (*logarithmic capacity, Robin's constant*).

Several properties of the conformal radius are readily apparent. If E^* denotes the complement of D, or "E with its holes filled," then obviously $\rho(E^*) = \rho(E)$. If E is a circular disk of radius r, then $\rho(E) = r$. If E is a line segment of length ℓ, then $\rho(E) = \ell/4$. (See Exercise 1.) The conformal radius is *conformally invariant* in the following sense. If the unbounded component $\tilde{D}$ of the complement of another compact connected set $\tilde{E}$ is the image of D under a conformal mapping of the form

$$w = z + b_0 + b_1 z^{-1} + b_2 z^{-2} + \cdots$$

near ∞, then $\rho(\tilde{E}) = \rho(E)$. We now proceed to show that ρ is a monotonic function of E.

Lemma 1. *Let E and $\tilde{E}$ be compact connected sets with $\tilde{E} \subset E$. Then $\rho(\tilde{E}) \le \rho(E)$.*

Proof. Let $\rho = \rho(E)$ and $\tilde{\rho} = \rho(\tilde{E})$. Let D and $\tilde{D}$ be the unbounded components of E and $\tilde{E}$, respectively. Then $D \subset \tilde{D}$, since $\tilde{E} \subset E$. By conformal invariance and the fact that $\rho(E^*) = \rho(E)$, we may assume that E is the disk $|z| \le \rho$. Let

$$\zeta = \tilde{\varphi}(z) = z + c_0 + c_1 z^{-1} + \cdots$$

map $\tilde{D}$ onto $|\zeta| > \tilde{\rho}$. Then the function $f(\zeta) = \tilde{\varphi}^{-1}(\zeta)/\zeta$ is analytic in $|\zeta| > \tilde{\rho}$, and $f(\infty) = 1$. Since

$$\tilde{D} \supset D = \{z : |z| > \rho\},$$

it follows from the maximum modulus principle that $|f(\zeta)| \le \rho/\tilde{\rho}$ throughout the region $|\zeta| > \tilde{\rho}$. In particular, $1 = |f(\infty)| \le \rho/\tilde{\rho}$, or $\tilde{\rho} \le \rho$.

Lemma 2. *The set E omitted by a function $g \in \Sigma$ has diameter $d(E) \le 4$.*

Proof. Suppose first that a function $f \in S$ omits two finite values α and β. Then

$$F(z) = f(z)\{1 - f(z)/\alpha\}^{-1}$$

belongs to S and omits $\beta\{1 - \beta/\alpha\}^{-1}$. It follows from the Koebe one-quarter theorem that $|\beta\{1 - \beta/\alpha\}^{-1}| \ge \frac{1}{4}$, or $|\alpha - \beta| \le 4|\alpha\beta|$. Now suppose $g \in \Sigma$ omits a, b, and 0. Then its inversion $f(z) = 1/g(1/z)$ belongs to S and omits $\alpha = 1/a$ and $\beta = 1/b$. Thus by what we have just proved, $|a - b| \le 4$.

Lemma 3. *If a compact connected set E of conformal radius ρ is contained in a disk of radius r and has diameter $d \ge r$, then $\rho \le r \le 4\rho$.*

Proof. By Lemma 1, $\rho \le r$. Let D be the unbounded component of the complement of E, and let

$$z = \psi(\zeta) = \zeta + b_0 + b_1\zeta^{-1} + b_2\zeta^{-2} + \cdots$$

map $|\zeta| > \rho$ conformally onto D. Let $g(\zeta) = \psi(\rho\zeta)/\rho$. Then $g \in \Sigma$ and it omits the set E^*/ρ, so Lemma 2 gives $d = d(E^*) \le 4\rho$. Since $d \ge r$, this proves $r \le 4\rho$.

§10.3. Schiffer's Theorem

The leading idea of Schiffer's method of boundary variation is to compare an extremal function with nearby functions obtained by composing it with functions univalent on its range and near the identity. We shall begin with the explicit construction of two such families of functions which will play an important part in the proof of the main theorem.

Let Γ be a closed simply connected set in the extended complex plane, consisting of more than a single point. Thus the complement of Γ is a simply connected domain D. (In most of the applications, D will be the range of an extremal function.)

Example 1. Let α and β be distinct finite points on Γ. Then some branch of the function $\log\{(w - \alpha)/(w - \beta)\}$ is single-valued, analytic, and univalent in D. Near infinity this function is analytic and has an expansion

$$\log\frac{w - \alpha}{w - \beta} = -\log\left\{1 - \frac{\beta - \alpha}{w - \alpha}\right\} = \frac{\beta - \alpha}{w - \alpha} + \frac{1}{2}\left(\frac{\beta - \alpha}{w - \alpha}\right)^2 + \cdots$$

in powers of $(w - \alpha)^{-1}$. Fix $\alpha = w_0$ and choose $\beta = \beta_r$ on the circle $|w - w_0| = r$. Then the functions

$$\begin{aligned} G_r(w) &= (\beta_r - w_0)\left\{\log\frac{w - w_0}{w - \beta_r}\right\}^{-1} + \tfrac{1}{2}(\beta_r + w_0) \\ &= w - \frac{(\beta_r - w_0)^2}{12(w - w_0)} + O(r^3), \qquad r \to 0, \end{aligned} \tag{1}$$

are analytic and univalent in D. Here "$O(r^3)$" denotes an error term which is bounded by a constant multiple of r^3 as r tends to zero, uniformly in each set $|w - w_0| \geq \varepsilon > 0$.

Example 2. Fix a finite point $w_0 \in \Gamma$ and let $r > 0$ be small enough to make the circle $|w - w_0| = r$ meet Γ. Let Γ_r be the largest connected subset of Γ which contains w_0 and lies in the disk $|w - w_0| \leq r$. Let D_r be the complement of Γ_r, and let

$$\zeta = \varphi(\omega) = \omega + c_0 + c_1\omega^{-1} + c_2\omega^{-2} + \cdots, \qquad \omega = w - w_0,$$

map D_r conformally onto a domain $|\zeta| > \rho$. Let

$$\omega = \psi(\zeta) = \zeta + b_0 + b_1\zeta^{-1} + b_2\zeta^{-2} + \cdots$$

be the inverse map. By Lemma 3 of §10.2 we have $\rho \leq r \leq 4\rho$. Since $\psi(\rho\zeta)/\rho$ belongs to the class Σ, the area theorem (§2.2) gives

$$\sum_{n=1}^{\infty} n|b_n|^2\rho^{-2(n+1)} \leq 1. \tag{2}$$

In particular,

$$|b_n| \leq n^{-1/2}\rho^{n+1} \leq \rho^{n+1}, \qquad n = 1, 2, \ldots.$$

The same argument applies to φ, which is analytic and univalent in $|\omega| > r$, and yields $|c_n| \leq r^{n+1}$, $n = 1, 2, \ldots$. A simple application of the mean-value theorem gives $|b_0| \leq r$. A direct calculation, based on the fact that φ and ψ are inverse functions, shows that $c_0 = -b_0$ and $c_1 = -b_1$. Thus $|c_0| \leq r$ and $|c_1| \leq \rho^2$. Now compose φ with the function

$$h(\zeta) = \zeta + e^{i\gamma}\rho^2/\zeta,$$

which maps $|\zeta| > \rho$ onto the complement of a line segment. This produces the functions

$$H_r(w) = h(\varphi(w - w_0)) + (w_0 - c_0)$$
$$= w + \frac{c_1 + e^{i\gamma}\rho^2}{w - w_0} + O(r^3). \tag{3}$$

The one-parameter families of functions G_r and H_r given by (1) and (3) have the common structure

$$F_r(w) = w + \lambda_r(w - w_0)^{-1} + O(r^3),$$

where F_r is analytic and univalent in D for each $r > 0$ sufficiently small, $\lambda_r = O(r^2)$, and the error term "$O(r^3)$" is uniform in each set $|w - w_0| \geq \varepsilon > 0$. In the case of G_r the coefficient λ_r actually has modulus $|\lambda_r| = r^2/12$; in particular,

$$\liminf_{r \to 0} r^{-2}|\lambda_r| > 0.$$

The same is effectively true for H_r because of the flexibility in the choice of $e^{i\gamma}$.

We are now prepared to state the main theorem.

Theorem 10.1 (Schiffer's Theorem). *Let Γ be a closed simply connected set consisting of more than one point, and let D be its complement. Let s be a function analytic in a neighborhood of Γ and not identically zero. Suppose that for each finite point $w_0 \in \Gamma$ and for each family of functions*

$$F_r(w) = w + \lambda_r(w - w_0)^{-1} + O(r^3)$$

analytic and univalent in D, the inequality

$$\operatorname{Re}\{\lambda_r s(w_0) + O(r^3)\} \leq 0 \tag{4}$$

holds for all $r > 0$ sufficiently small. Then Γ is the union of analytic arcs $w = w(t)$ satisfying the differential equation

$$s(w(t))\left(\frac{dw}{dt}\right)^2 > 0. \tag{5}$$

It should be observed that at each point where $s(w) \neq 0$, the condition (5) determines the tangent direction $\pm s(w)^{-1/2}$ for the arc Γ. The parametrization of Γ can be chosen so that (for instance) $s(w)(dw/dt)^2 = 1$, so we are justified in calling (5) a differential equation.

It will become apparent that the hypotheses of the theorem can be considerably weakened. The proof will apply the inequality (4) not for the most general family of functions, but only for the special families (1) and (3).

The proof will use a special concept which we now define. Let E be a closed connected set in the complex plane. A number $e^{i\theta}$ is called a *limit direction* of E at $z_0 \in E$ if there is a sequence of points $z_n \in E$ with $z_n \neq z_0$ and $z_n \to z_0$, such that $\operatorname{sgn}\{z_n - z_0\} \to e^{i\theta}$ as $n \to \infty$. Here "sgn" denotes the signum, defined by $\operatorname{sgn} z = z/|z|, z \neq 0$. By the Bolzano–Weierstrass theorem, E has at least one limit direction at z_0.

A topological lemma due to Haslam-Jones [1] will play a crucial role in the proof of the theorem. It is intuitively obvious but surprisingly difficult to verify. A proof may be found in Schober [1] or in Campbell and Lamoreaux [1].

Haslam-Jones' Lemma. *Let E be a compact connected subset of the complex plane whose only limit directions are ± 1. Then E is a horizontal line segment.*

Proof of Schiffer's Theorem. Fix an arbitrary point $w_0 \in \Gamma$ where $s(w_0) \neq 0$ and set

$$s(w_0) = |s(w_0)|e^{-2i\sigma}.$$

We shall prove that Γ has at most the two limit directions $\pm e^{i\sigma}$ at w_0.

Let $e^{i\theta}$ be a limit direction of Γ at w_0, and choose points $w_n \in \Gamma$ with $w_n \neq w_0$, $w_n \to w_0$, and $\operatorname{sgn}\{w_n - w_0\} \to e^{i\theta}$. Let $r_n = |w_n - w_0|$. Referring to Example 1, let $w_n = \beta_{r_n}$ and consider the functions G_r defined by (1), with $\lambda_r = -(\beta_r - w_0)^2/12$. With this choice the condition (4) gives

$$\operatorname{Re}\{(w_n - w_0)^2 e^{-2i\sigma} + O(r_n^3)\} \geq 0.$$

Divide by r_n^2 and pass to the limit to obtain

$$\operatorname{Re}\{e^{2i(\theta - \sigma)}\} \geq 0.$$

This inequality says that every limit direction $e^{i\theta}$ of Γ at w_0 lies in one of the sectors

$$|\theta - \sigma| \leq \frac{\pi}{4}, \qquad |\theta - (\sigma + \pi)| \leq \frac{\pi}{4}.$$

Thus the set Γ is locally confined to these sectors. More precisely, to each $\varepsilon > 0$ there corresponds a radius $r > 0$ so small that the intersection of Γ with the disk $|w - w_0| \leq r$ is contained in the union of the two sectors

$$|\theta - \sigma| < \frac{\pi}{4} + \varepsilon, \qquad |\theta - (\sigma + \pi)| < \frac{\pi}{4} + \varepsilon.$$

This shows already that Γ has no interior points.

The next step is to show that if Γ extends from w_0 into the sector $|\theta - \sigma| < \pi/4 + \varepsilon$, then $e^{i\sigma}$ is itself a limit direction of Γ at w_0. (A similar argument proves that $-e^{i\sigma}$ is a limit direction if Γ extends into the opposite sector.) For this we invoke the functions of Example 2. Applying the condition (4) to the functions H_r given by (3), where $\lambda_r = c_1 + e^{i\gamma}\rho^2$, we find

$$\text{Re}\{(c_1 + e^{i\gamma}\rho^2)e^{-2i\sigma} + O(r^3)\} \leq 0.$$

Now divide by ρ^2 and let r tend to zero, recalling that $r \leq 4\rho$ and $|c_1| \leq \rho^2$. If r tends to zero through a sequence $\{r_n\}$ for which $c_1(r_n)[\rho(r_n)]^{-2} \to \alpha$, then $|\alpha| \leq 1$ and

$$\text{Re}\{(\alpha + e^{i\gamma})e^{-2i\sigma}\} \leq 0.$$

Since this inequality must hold for each γ, the conclusion is that $\alpha e^{-2i\sigma} = -1$, or $\alpha = -e^{2i\sigma}$. Because the limit is independent of the choice of sequence $\{r_n\}$, this proves

$$\lim_{r \to 1} b_1(r)[\rho(r)]^{-2} = e^{2i\sigma}, \tag{6}$$

since $b_1 = -c_1$.

Referring again to Example 2, let the mapping ψ of $|\zeta| > \rho$ onto D_r be denoted by ψ_r, to emphasize its dependence on r. In view of (6), ψ_r has the form

$$\psi_r(\zeta) = \chi_r(\zeta) + \Delta_r(\zeta),$$

where

$$\chi_r(\zeta) = \zeta + b_0(r) + e^{2i\sigma}[\rho(r)]^2\zeta^{-1}$$

and

$$\Delta_r(\zeta) = \sum_{n=2}^{\infty} b_n(r)\zeta^{-n} + o(r^2)\zeta^{-1}.$$

This formula represents ψ_r as a slight perturbation of the function χ_r, which maps the region $|\zeta| > \rho$ onto the complement of a line segment inclined in the direction $e^{i\sigma}$. For fixed $\delta > 0$, let C_r be the circle $|\zeta| = (1 + \delta)\rho$. Then χ_r maps C_r onto an ellipse E_r. The error term $\Delta_r(\zeta)$ may be estimated on C_r by the Cauchy–Schwarz inequality and the area theorem (2):

$$
\begin{aligned}
|\Delta_r(\zeta)| &\le \rho \sum_{n=2}^{\infty} |b_n| \rho^{-(n+1)}(1+\delta)^{-n} + o(r) \\
&\le \rho\delta^{-1}\left\{\sum_{n=2}^{\infty} |b_n|^2 \rho^{-2(n+1)}\right\}^{1/2} + o(r) \\
&\le \rho\delta^{-1}\{1 - |b_1|^2\rho^{-4}\}^{1/2} + o(r).
\end{aligned}
$$

In view of (6), we conclude that $r^{-1}\Delta_r(\zeta) \to 0$ uniformly on C_r as $r \to 0$.

The image under ψ_r of the circle C_r is a Jordan curve J_r which may be viewed as a slight perturbation of the ellipse E_r. Clearly, Γ_r lies inside J_r. Under the assumption that Γ_r extends into the sector $|\theta - \sigma| < \pi/4 + \varepsilon$, choose a point $w_r \in \Gamma_r$ in this sector and on the circle $|w - w_0| = r$. (Such a point exists for every r sufficiently small.) Now extend the line segment joining w_0 and w_r in both directions until it intersects J_r in points w_r' and w_r'', respectively. (See Figure 10.1.) Let e_r' and e_r'' be the points on E_r which are

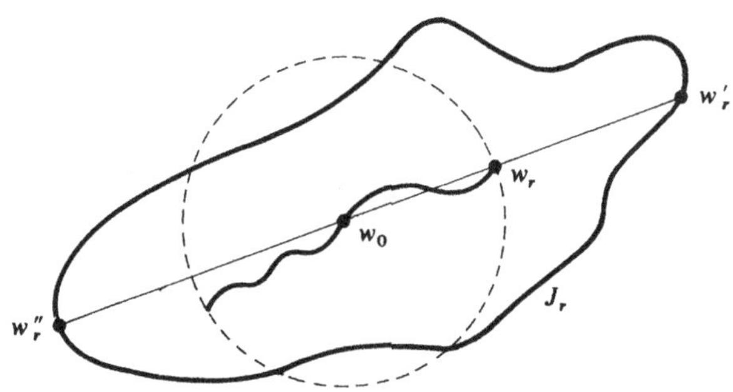

Figure 10.1. The perturbed ellipse.

the images under $\chi_r \circ \psi_r^{-1}$ of w_r' and w_r'', respectively. Note that w_r' and e_r' are images under ψ_r and χ_r of the same point on C_r, and similarly for w_r'' and e_r''. Since $\psi_r = \chi_r + \Delta_r$, the uniform estimate $\Delta_r(\zeta) = o(r)$ on C_r gives

$$|w_r' - e_r'| = o(r), \qquad |w_r'' - e_r''| = o(r).$$

On the other hand, it is clear from the construction of w_r' and w_r'' that

$$|w_r' - w_r''| \ge |w_r - w_0| = r.$$

Combining these estimates, we conclude that $|e_r' - e_r''| > r/2$ for all r sufficiently small. Since the major axis of the ellipse E_r has length greater than $4\rho \geq r$ and the minor axis has length $o(r)$, it follows that the segment joining e_r' and e_r'' has limiting direction $e^{i\sigma}$, the inclination of the major axis of E_r. Consequently, the segment joining w_r' and w_r'' also has limiting direction $e^{i\sigma}$. More precisely, $\operatorname{sgn}\{w_r' - w_r''\} \to e^{i\sigma}$. This obviously implies that $\operatorname{sgn}\{w_r - w_0\} \to e^{i\sigma}$. Thus $e^{i\sigma}$ is a limiting direction of Γ at w_0 if Γ extends from w_0 into the sector $|\theta - \sigma| < \pi/4 + \varepsilon$. A similar argument shows that $-e^{i\sigma}$ is a limit direction if Γ extends into the opposite sector.

The next step is to show that if Γ extends from w_0 into the sector $|\theta - \sigma| < \pi/4 + \varepsilon$, then $e^{i\sigma}$ is the *only* limit direction in this sector. (A similar argument applies to the opposite sector.) Suppose, on the contrary, that Γ has some other limit direction $e^{i\theta}$ in the same sector. Then there is a sequence of points $\alpha_n \in \Gamma$ with $\alpha_n \neq w_0$, $\alpha_n \to w_0$, and $\operatorname{sgn}\{\alpha_n - w_0\} \to e^{i\theta}$. Let $r_n = |\alpha_n - w_0|$ and choose points $\beta_n \in \Gamma$ with $|\beta_n - w_0| = r_n$ and $\operatorname{sgn}\{\beta_n - w_0\} \to e^{i\sigma}$. Since $e^{i\theta} \neq e^{i\sigma}$, we may assume that $\alpha_n \neq \beta_n$ for all n. Consider the functions

$$
\begin{aligned}
F_{r_n}(w) &= (\beta_n - \alpha_n)\left\{\log\frac{w - \alpha_n}{w - \beta_n}\right\}^{-1} + \tfrac{1}{2}(\alpha_n + \beta_n) \\
&= w - \frac{(\beta_n - \alpha_n)^2}{12(w - w_0)} + O(r_n^3),
\end{aligned}
$$

which are analytic and univalent in D. For these functions the condition (4) gives

$$
\operatorname{Re}\{(\beta_n - \alpha_n)^2 e^{-2i\sigma} + O(r_n^3)\} \geq 0.
$$

Now divide by r_n^2 and let $n \to \infty$ to obtain

$$
\operatorname{Re}\{(1 - e^{i(\theta - \sigma)})^2\} \geq 0,
$$

which implies $e^{i(\theta - \sigma)} = 1$, since $|\theta - \sigma| \leq \pi/4$. Hence $e^{i\theta} = e^{i\sigma}$, a contradiction. This proves that $e^{i\sigma}$ is the only possible limit direction in the sector $|\theta - \sigma| \leq \pi/4$. A similar argument shows that $-e^{i\sigma}$ is the only possible limit direction in the opposite sector. We have therefore proved that $\pm e^{i\sigma}$ are the only possible limit directions of Γ at w_0. In other words, Γ has a tangent line at each point where $s(w_0) \neq 0$.

The final step is to show that Γ is locally an analytic arc satisfying the differential equation (5). Let $\alpha \in \Gamma$ be a point where $s(\alpha) \neq 0$, and consider the function

$$
W = \Phi(w) = \int_{\alpha}^{w} \sqrt{s(\omega)}\, d\omega.
$$

Each branch of Φ is single-valued, analytic, and univalent in some neighborhood of α. Choose points $w_0 \in \Gamma$ and $w_n \in \Gamma$ in this neighborhood with $w_n \neq w_0$, $w_n \to w_0$, and $\operatorname{sgn}\{w_n - w_0\} \to \pm e^{i\sigma}$, where $\operatorname{sgn}\{s(w_0)\} = e^{-2i\sigma}$. Let $W_0 = \Phi(w_0)$ and $W_n = \Phi(w_n)$, and write

$$\operatorname{sgn}\{W_n - W_0\} = \frac{W_n - W_0}{w_n - w_0} \cdot \frac{|w_n - w_0|}{|W_n - W_0|} \cdot \frac{w_n - w_0}{|w_n - w_0|}.$$

This shows that

$$\lim_{n \to \infty} \operatorname{sgn}\{W_n - W_0\} = \pm e^{i\sigma} \operatorname{sgn} \Phi'(w_0) = \pm 1.$$

In other words, Φ maps Γ locally onto a connected set whose only limit directions are ± 1. It follows from Haslam-Jones' lemma that the image of Γ under Φ is locally a horizontal segment. Since $\Phi(\alpha) = 0$, it is in fact a real interval $a < t < b$. Thus Γ is locally the image of an interval under an analytic locally univalent function $\Psi = \Phi^{-1}$. In other words, Γ is locally an analytic arc with parametrization $w = \Psi(t)$, $a < t < b$.

To derive the differential equation (5), differentiate the identity $t = \Phi(\Psi(t))$ to obtain

$$1 = \Phi'(\Psi(t))\Psi'(t) = s(\Psi(t))^{1/2}\Psi'(t),$$

or

$$s(\Psi(t))[\Psi'(t)]^2 = 1.$$

This completes the proof of Schiffer's theorem.

§10.4. Local Structure of Trajectories

In the applications of Schiffer's theorem, the set Γ will be the complement of the range of an extremal function. The conclusion is that Γ consists of a collection of analytic arcs satisfying a differential equation of the form $Q(w)\,dw^2 > 0$, where Q is analytic on Γ. In general, such an expression $Q(w)\,dw^2$ is called a *quadratic differential* and the arcs for which $Q(w)\,dw^2 > 0$ are called its *trajectories*. The arcs defined by $Q(w)\,dw^2 < 0$ are called the *orthogonal trajectories*. In many important cases the function Q is actually a rational function.

With a view to later applications, we wish now to investigate the local structure of the trajectories of a quadratic differential near a given point w_0. We assume that Q is meromorphic in a neighborhood of w_0. Suppose first that w_0 is a finite point.

The point w_0 is called a *regular point* of the quadratic differential $Q(w)\,dw^2$ if Q has neither a zero nor a pole there; that is, if Q is analytic at w_0 and $Q(w_0) \neq 0$. At a regular point w_0 the condition $Q(w)\,dw^2 > 0$ determines a tangent direction, and there is a unique locally analytic trajectory passing through w_0.

If Q has a simple pole at w_0, there is a unique analytic trajectory which terminates at w_0. To see this, assume for convenience that $w_0 = 0$ and make the substitution $w = v^2$. Then the condition $Q(w)\,dw^2 > 0$ takes the form $q(v)\,dv^2 > 0$, where $q(v) = v^2Q(v^2)$. Since Q has a simple pole at the origin, the quadratic differential $q(v)\,dv^2$ has a regular point at $v = 0$. Thus $q(v)\,dv^2$ has a unique trajectory passing through the origin. Furthermore, this trajectory is symmetric with respect to the origin, since $q(-v) = q(v)$. The squaring operation $w = v^2$ therefore transforms it to a single trajectory of $Q(w)\,dw^2$, terminating at the origin and having a tangent direction there.

At a point w_0 where Q has a zero of order m, there are $m + 2$ analytic trajectories which join at equal angles. To see this, we assume $w_0 = 0$ and make the formal substitution $w = u^{2/(m+2)}$. More precisely, we first make the substitution $w = v^2$ and find as above that the trajectories of the induced quadratic differential $q(v)\,dv^2$ are symmetric with respect to the origin. Next we let $u = v^{m+2}$ and find that these trajectories are mapped into a set in the u-plane with a uniquely determined tangent line at the origin, because Q has a zero of order m at $w = 0$. This determines a set of $m + 2$ tangent lines at $v = 0$, intersecting at equal angles. Thus the trajectories of $q(v)\,dv^2$ consist locally of $m + 2$ analytic arcs intersecting at the origin at equal angles. Since each arc is symmetric with respect to the origin, the transformation $w = v^2$ carries them into $m + 2$ analytic arcs joining at $w = 0$ with equal angles.

These three kinds of local trajectory structure are illustrated in Figure 10.2. The structure near a pole of higher order is more complicated and will not be discussed here. (See Jenkins [7] or Jensen [1].)

The structure of the trajectories near ∞ may be deduced through the transformation $w = 1/v$. This carries the trajectories of $Q(w)\,dw^2$ near ∞ to the trajectories of $v^{-4}Q(1/v)\,dv^2$ near 0. Thus we may say that the quadratic differential $Q(w)\,dw^2$ has a regular point at ∞ if Q has a zero of order 4 there.

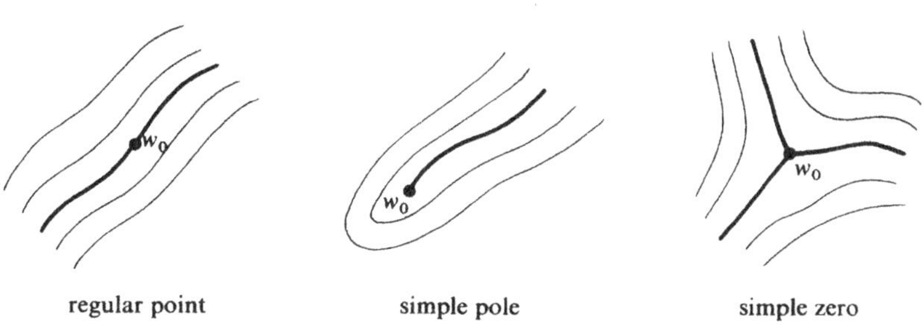

Figure 10.2. Local trajectory structures.

The quadratic differential has a zero of order m at ∞ if Q has a zero of order $m + 4$ there, and a pole of order m at ∞ if Q has a pole of order $m - 4$ there. Thus $Q(w)\,dw^2$ has a simple pole at ∞ if Q has a zero of order 3 there, a double pole if Q has a double zero, etc.

The case where the quadratic differential has a simple pole at ∞ is especially important for the applications to extremal problems over the class S of univalent functions. The variational method will provide the information that the extremal function maps the disk onto the complement of a system Γ of trajectories of some quadratic differential $Q(w)\,dw^2$. It is clear that $\infty \in \Gamma$, so at least one arc of Γ must extend to ∞. If the quadratic differential $Q(w)\,dw^2$ has a simple pole at ∞, we can conclude that this arc terminates at ∞. In other words, only one arc of Γ extends to ∞ in this case.

The zeros and poles of a quadratic differential are called its *singular points.*

§10.5. Application to Extremal Problems

The method of boundary variation will now be applied to a general extremal problem in the class S. Let ϕ be a continuous complex-valued functional, linear or not, on the space $\mathscr{A}$ of all functions analytic in the unit disk, endowed with the usual topology of local uniform convergence. Consider the problem of finding the maximum of $\mathrm{Re}\{\phi(f)\}$ for all $f \in S$. Because S is a compact subset of $\mathscr{A}$, there is a function $f \in S$ where the maximum is attained. We have previously shown (Corollary to Theorem 9.4) that under a mild differentiability assumption on ϕ, this extremal function has dense range. The variational method gives much more precise information.

Theorem 10.2. *Let ϕ be a continuous functional on $\mathscr{A}$, and let $f \in S$ be a point where $\mathrm{Re}\{\phi\}$ attains its maximum value on S. Suppose that ϕ has a Fréchet differential $\ell(\cdot\,;f)$ at f which is not constant on S. Then f maps the unit disk onto the complement of a system of finitely many analytic arcs $w = w(t)$ satisfying the differential equation*

$$\frac{1}{w^2}\,\ell\left(\frac{f^2}{f-w};f\right)\left(\frac{dw}{dt}\right)^2 > 0. \tag{7}$$

Proof. Let D be the range of the extremal function f, and let Γ be the complement of D. Choose a finite point $w_0 \in \Gamma$ and let

$$F_r(w) = w + \lambda_r(w - w_0)^{-1} + o(r^2)$$

be analytic and univalent in D. Normalize F_r by forming

$$G_r(w) = [F_r(w) - F_r(0)]/F_r'(0).$$

Then G_r is analytic and univalent in D, with $G_r(0) = 0$ and $G_r'(0) = 1$. Hence $f^* = G_r \circ f \in S$. Simple calculations give

$$F_r'(0) = 1 - \lambda_r w_0^{-2} + o(r^2)$$

and

$$G_r(w) = w + \lambda_r \left[\frac{w^2}{w_0^2(w - w_0)} \right] + o(r^2).$$

Thus by a general property of the Fréchet differential (see §9.1, Example 4),

$$\phi(f^*) = \phi(f) + \lambda_r \frac{1}{w_0^2} \ell\left(\frac{f^2}{f - w_0}; f \right) + o(r^2).$$

But $\text{Re}\{\phi(f^*)\} \leq \text{Re}\{\phi(f)\}$, since f is extremal. This gives the inequality

$$\text{Re}\{\lambda_r s(w_0) + o(r^2)\} \leq 0$$

for all $w_0 \in \Gamma$, where

$$s(w) = \frac{1}{w^2} \ell\left(\frac{f^2}{f - w}; f \right).$$

This function s is analytic on Γ, since the linear functional $\ell(\cdot; f)$ has an integral representation with respect to a measure supported on a compact subset E of the open unit disk (see §9.2). In fact, s is analytic on the complement of the compact set $f(E)$. To show that s is not the zero function, we have only to apply the argument previously used to show that an extremal function has dense range (Theorem 9.4). Thus we may conclude from Schiffer's theorem that Γ is composed of analytic arcs satisfying the differential equation (7).

It remains to show that Γ consists of only a finite number of analytic arcs. We have just shown that these arcs are trajectories of the quadratic differential $s(w)\,dw^2$. The function s is analytic on Γ and has a zero of order 3 or higher at ∞. This means that the quadratic differential $s(w)\,dw^2$ has at worst a simple pole at ∞. (In fact, it has a simple pole at ∞ if and only if $\ell(f^2; f) \neq 0$.) Since s is analytic at ∞ and does not vanish identically on Γ, it can have at most a finite number of zeros on Γ. The general theory of the local trajectory structure (§10.4) shows that finitely many of the analytic arcs of Γ may join at each of these zeros, but elsewhere there can be no such junctions. In particular, only finitely many arcs can extend to ∞. This proves that Γ consists of a finite number of analytic arcs.

§10.6. Support Points of S

When the preceding theorem is specialized to a linear functional, we obtain new information about the support points of S. Recall that we have already proved (§9.5) that each support point of S maps the disk onto the complement of an arc which extends to infinity with increasing modulus. We can now say much more about the omitted arc.

Theorem 10.3. *Let L be a continuous linear functional on $\mathscr{A}$ which is not constant on S, and let f maximize* $\mathrm{Re}\{L\}$ *on S. Then f maps the unit disk onto the complement of a single analytic arc Γ which satisfies the differential equation*

$$\frac{1}{w^2} L\left(\frac{f^2}{f-w}\right) dw^2 > 0. \tag{8}$$

At each point $w \in \Gamma$ except perhaps the finite tip, the tangent line makes an angle of less than $\pi/4$ with the radial line from 0 to w.

Proof. Since a continuous linear functional is its own Fréchet differential, Theorem 10.2 shows that Γ consists of finitely many analytic arcs satisfying the differential equation (8). In fact, Γ is a single unbranched arc which extends to infinity with monotonic modulus (Theorem 9.5, Corollary 2). These properties alone do not imply, however, that Γ is an analytic arc, for it may conceivably have corners where analytic subarcs join.

Choose a point $w \in \Gamma$, not an endpoint, and consider the function $g = wf/(w - f)$. Observe that g belongs to S and maps the disk onto the complement of two disjoint arcs extending to infinity. Thus g is not a support point, and so $\mathrm{Re}\{L(g)\} < \mathrm{Re}\{L(f)\}$. Since L is linear, this inequality is equivalent to

$$\mathrm{Re}\left\{L\left(\frac{f^2}{f-w}\right)\right\} > 0, \qquad w \in \Gamma, \tag{9}$$

where w is not an endpoint of Γ.

The inequality (9) has two consequences. First, the fact that $L(f^2/(f - w)) \neq 0$ assures that the quadratic differential has no singularities on Γ, except perhaps at the endpoints, so that Γ has no corners. In other words, Γ is a single analytic arc. Second, the inequality (9) may be combined with the differential equation (8) to show that $\mathrm{Re}\{(dw/w)^2\} > 0$ on Γ, which is equivalent to the $\pi/4$-property. This completes the proof.

The angle $\arg\{dw/w\}$ between the tangent line and the radial line is called the *radial angle* of the curve. We have shown that the radial angle is less than $\pi/4$ in magnitude everywhere on Γ except perhaps at the tip. The question arises whether the bound $\pi/4$ is best possible. After Brown [1] gave an affirmative answer based on numerical calculation, Pearce [1] supplied a proof by producing a simple example where a radial angle of $\pi/4$ is attained at the tip of the slit. In this example the linear functional has the form $L(f) = -f'(\sin \pi/8)$ and the omitted arc is a half-line. (See §10.7 for further discussion.) A recent result of Duren, Leung, and Schiffer [1] shows that under very general conditions the omitted arc is a half-line whenever it has a radial angle of $\pm\pi/4$ at its tip.

Theorem 10.3 has been known for some time, at least in the case of a coefficient functional. Schaeffer and Spencer [6, p. 149] proved the single-arc assertion, and the $\pi/4$-property may be found in the book of Goluzin [28, Ch. IV, §4]. Schiffer [15] gave another proof of the single-arc property by showing that $L(f^2) \neq 0$, so that the associated quadratic differential has a simple pole at infinity. Although Schiffer discussed only the nth coefficient, he remarked that the proof extends without change to the case of a general linear functional. Pfluger [3] explicitly discussed general linear extremal problems. Then Brickman and Wilken [1] found the amazingly simple proof, which we have followed here, that Γ consists of a single arc. Using the stronger result that $L(f^2) \neq 0$, they showed further that Γ is asymptotic to a line at infinity. However, their approach does not lead to a proof that $L(f^2) \neq 0$. To establish this important property we shall now follow Schiffer's argument, which is based on a new variation at infinity.

Theorem 10.4. *Let L be a continuous linear functional on $\mathscr{A}$ which is not constant on S, and let f maximize* $\text{Re}\{L\}$ *on S. Then $L(f^2) \neq 0$.*

Proof. By Theorem 10.3, the function f maps $\mathbb{D}$ onto the complement of an analytic arc Γ satisfying $\Phi(w)(dw/w)^2 > 0$, where

$$\Phi(w) = L\left(\frac{f^2}{f - w}\right) = -\sum_{n=1}^{\infty} L(f^{n+1})w^{-n}$$

near infinity and $\Phi(w) \neq 0$ on Γ. This quadratic differential has a simple pole at infinity if and only if $L(f^2) \neq 0$. If $L(f^2) = 0$, then infinity is either a regular point (if $L(f^3) \neq 0$) or a zero (if $L(f^3) = 0$). In any of these cases, the general theory of the local trajectory structure of a quadratic differential (§10.4) allows us to conclude that Γ has an asymptotic direction at infinity. In other words, sgn $w \to e^{i\sigma}$ as $w \to \infty$ along Γ.

Now choose a subarc $\gamma \subset \Gamma$ which extends from some finite point $\hat{w} \in \Gamma$ to infinity. Let

$$w = F(z) = z + B_2 z^2 + B_3 z^3 + \cdots$$

map the disk $|z| < \rho$ conformally onto the complementary domain $\Delta = \mathbb{C} - \gamma$. Then $1 \leq \rho < \infty$, and $\rho \to \infty$ as $\hat{w} \to \infty$ along Γ. It is convenient to introduce the more explicit notation $\hat{w} = w_\rho$, $\gamma = \Gamma_\rho$, and $F = F_\rho$. Note that Γ is parametrized by $w = w_\rho$, $1 \leq \rho < \infty$. Furthermore, because Γ has the asymptotic direction $e^{i\sigma}$ at infinity, it is clear that for each $\varepsilon > 0$, the arc Γ_ρ lies entirely in the sector $|\arg w - \sigma| < \varepsilon$ if ρ is sufficiently large.

Consider now the family of functions $G_\rho \in S$ defined by

$$G_\rho(z) = F_\rho(\rho z)/\rho = z + A_2 z^2 + A_3 z^3 + \cdots, \qquad |z| < 1.$$

The function G_ρ maps $\mathbb{D}$ onto the complement of a radial magnification of Γ_ρ, which eventually lies in each sector about $e^{i\sigma}$. Thus it follows from the Carathéodory convergence theorem (§3.1) that as $\rho \to \infty$, the functions G_ρ converge uniformly on each compact subset of $\mathbb{D}$ to the corresponding rotation of the Koebe function,

$$\omega = k_\beta(z) = e^{-i\beta}k(e^{i\beta}z) = z(1 - e^{i\beta}z)^{-2}.$$

where $\beta = \pi - \sigma$. Furthermore, the inverse functions $G_\rho^{-1}(\omega) \to k_\beta^{-1}(\omega)$ uniformly on compact subsets of the range of k_β.

Now choose an arbitrary direction $e^{i\alpha}$ and consider the function

$$\begin{aligned} \omega^* &= v(\omega) = v(\omega; \rho, \alpha) = k_\alpha(G_\rho^{-1}(\omega)) \\ &= \omega + \sum_{n=2}^{\infty} c_n(\rho, \alpha)\omega^n. \end{aligned}$$

Observe that as $\rho \to \infty$,

$$v(\omega; \rho, \alpha) \to k_\alpha(k_\beta^{-1}(\omega)) = \omega + \sum_{n=2}^{\infty} c_n(\alpha)\omega^n,$$

uniformly on compact subsets of $|\omega| < \frac{1}{4}$. Thus

$$\lim_{\rho \to \infty} c_n(\rho, \alpha) = c_n(\alpha), \qquad n = 2, 3, \ldots.$$

For α near β and for large ρ, the function $\omega^* = v(\omega)$ is near the identity mapping. The same is true of the function

$$\begin{aligned} w^* &= V(w) = V(w; \rho, \alpha) = \rho v(w/\rho; \rho, \alpha) \\ &= w + \sum_{n=2}^{\infty} c_n(\rho, \alpha)\rho^{1-n}w^n, \end{aligned}$$

which is analytic and univalent in Δ. Consequently, the function $f^* \in S$ defined by

$$f^*(z) = V(f(z)) = f(z) + \sum_{n=2}^{\infty} c_n(\rho, \alpha)\rho^{1-n}[f(z)]^n$$

is a variation of the given function f.

We are to prove that $L(f^2) \neq 0$. If we assume, on the contrary, that $L(f^n) = 0$ for $2 \leq n < m$ but $L(f^m) \neq 0$, the inequality $\operatorname{Re}\{L(f^*)\} \leq \operatorname{Re}\{L(f)\}$ gives

$$\operatorname{Re}\left\{\sum_{n=m}^{\infty} c_n(\rho, \alpha)\rho^{1-n}L(f^n)\right\} \leq 0.$$

Multiplying by ρ^{m-1} and letting $\rho \to \infty$, we deduce that

$$\operatorname{Re}\{c_m(\alpha)L(f^m)\} \leq 0$$

for every value of α. In order to calculate $c_m(\alpha)$ for $m > 2$, we write

$$k_\alpha(k_\beta^{-1}(\omega)) = e^{-i\beta}\frac{1}{a}k(ak^{-1}(e^{i\beta}\omega)), \qquad a = e^{i(\alpha-\beta)}.$$

But

$$k^{-1}(\zeta) = \frac{(1 + 4\zeta)^{1/2} - 1}{(1 + 4\zeta)^{1/2} + 1},$$

and so for a near 1,

$$\frac{1}{a}k(ak^{-1}(\zeta)) = 4\zeta[(1 + a) + (1 - a)(1 + 4\zeta)^{1/2}]^{-2}$$

$$= \frac{4\zeta}{(1 + a)^2}\left\{1 - 2\frac{1 - a}{1 + a}(1 + 4\zeta)^{1/2} + 3\left(\frac{1 - a}{1 + a}\right)^2(1 + 4\zeta) + O((1 - a)^3)\right\}$$

$$= \zeta - 8\frac{1 - a}{(1 + a)^3}[(1 + 4\zeta)^{1/2} - 1]\zeta + 3(1 - a)^2\zeta^2 + O((1 - a)^3).$$

The binomial expansion

$$(1 + 4\zeta)^{1/2} = 1 + \sum_{n=1}^{\infty}\binom{\frac{1}{2}}{n}(4\zeta)^n$$

therefore gives the asymptotic formula

$$c_m(\alpha) = -2\frac{1-a}{(1+a)^3}\binom{\frac{1}{2}}{m-1}4^m e^{i(m-1)\beta} + O((1-a)^3),$$

where $a = e^{i(\alpha-\beta)}$ and $m > 2$. For $a = -1$, we obtain the simple expansion

$$-k(-k^{-1}(\zeta)) = \zeta(1+4\zeta)^{-1} = \sum_{n=1}^{\infty}(-4)^{n-1}\zeta^n,$$

so

$$c_m(\beta+\pi) = (-4)^{m-1}e^{i(m-1)\beta}.$$

Now choose α near β and write $a = e^{i\varepsilon}$. Because

$$\operatorname{sgn}\binom{\frac{1}{2}}{m-1} = (-1)^m,$$

the inequality $\operatorname{Re}\{c_m(\alpha)L(f^m)\} \le 0$ then gives

$$\operatorname{Re}\left\{L(f^m)(-1)^m e^{i(m-1)\beta}\frac{1-e^{i\varepsilon}}{(1+e^{i\varepsilon})^3}\right\} + O(\varepsilon^3) \ge 0.$$

But

$$\frac{1-e^{i\varepsilon}}{(1+e^{i\varepsilon})^3} = -\tfrac{1}{8}(i\varepsilon + \varepsilon^2) + O(\varepsilon^3),$$

so we find by sending ε to 0 first through positive values, then through negative values, that

$$\lambda = L(f^m)e^{i(m-1)\beta}$$

is a real number. A consideration of the ε^2 term now shows that $(-1)^m\lambda \le 0$. On the other hand, the choice $a = -1$ gives

$$\operatorname{Re}\{c_m(\beta+\pi)L(f^m)\} \le 0,$$

which implies that $(-1)^m\lambda \ge 0$. Thus it follows that $\lambda = 0$, or $L(f^m) = 0$. This contradiction completes the proof of the theorem.

We can now show that the omitted arc is asymptotic to a line at infinity.

Theorem 10.5. *Let L be a continuous linear functional on $\mathscr{A}$ which is not constant on S, and let f maximize* $\operatorname{Re}\{L\}$ *on S. Then the arc Γ omitted by f is asymptotic to the half-line*

$$w = \frac{1}{3}\frac{L(f^3)}{L(f^2)} - L(f^2)t, \qquad t \geq 0, \tag{10}$$

at infinity. Furthermore, the radial angle of Γ tends to zero at infinity.

Proof. Let Γ be parametrized by $w = w(t)$, $0 < t < \infty$, in such a way that $w(t) \to \infty$ as $t \to 0$ and the differential equation (8) takes the form

$$\frac{1}{w^2} L\left(\frac{f^2}{f - w}\right)\left(\frac{dw}{dt}\right)^2 = 1.$$

Because $L(f^2) \neq 0$, the substitution $w = v^{-2}$ transforms Γ to an analytic curve

$$v = b_1 t + b_3 t^3 + \cdots$$

through the origin which satisfies

$$-4L\left(\frac{f^2}{1 - fv^2}\right)\left(\frac{dv}{dt}\right)^2 = 1,$$

or

$$(c_0 + c_1 b_1^2 t^2 + \cdots)(b_1^2 + 6b_1 b_3 t^2 + \cdots) = -\tfrac{1}{4},$$

where $c_n = L(f^{n+2})$, $n = 0, 1, 2, \ldots$. Equating coefficients, we obtain

$$c_0 b_1^2 = -\tfrac{1}{4}, \qquad c_1 b_1^4 + 6c_0 b_1 b_3 = 0. \tag{11}$$

On the other hand,

$$w = v^{-2} = b_1^{-2}t^{-2} - 2b_1^{-3}b_3 + O(t^2), \qquad t \to 0.$$

Thus Γ is asymptotic to the line

$$w = \alpha + \beta t, \qquad t \to \infty,$$

where $\alpha = -2b_1^{-3}b_3$ and $\beta = b_1^{-2}$. But the equations (11) give

$$b_1^{-2} = -4c_0, \qquad b_1^{-3}b_3 = -c_1/6c_0.$$

This proves that Γ approaches the half-line (10) near infinity.

In particular, $\arg w \to \arg\{-L(f^2)\}$ as $w \to \infty$ along Γ. Because

$$\Phi(w) = L\left(\frac{f^2}{f - w}\right) = -\frac{L(f^2)}{w} + O\left(\frac{1}{w^2}\right),$$

it follows that $\arg\{\Phi(w)\} \to 0$. Thus the differential equation $\Phi(w)(dw/w)^2 > 0$ shows that the radial angle $\arg\{dw/w\} \to 0$ as $w \to \infty$ along Γ.

The half-line (10) is called the *asymptotic half-line* of Γ. It distinguishes Γ from all other trajectories of the quadratic differential.

In the proof of Theorem 10.3 we showed that $\mathrm{Re}\{\Phi(w)\} > 0$ everywhere on Γ except perhaps at the finite tip w_0. In view of the differential equation (8), this is equivalent to the assertion that the radial angle of Γ is less than $\pi/4$ in magnitude except perhaps at w_0. In particular, $\Phi(w) \neq 0$ for $w \in \Gamma$, $w \neq w_0$. By a more subtle argument (due to Schiffer) we shall now show that $\Phi(w_0) \neq 0$ unless f is a rotation of the Koebe function. In the process we shall also establish a recent result of Duren, Leung, and Schiffer [1] that if $\mathrm{Re}\{\Phi(w_0)\} = 0$, or equivalently if Γ has a radial angle of $\pm\pi/4$ at its tip, then Γ satisfies a second differential equation.

To prove these assertions, consider the function $\tilde{f} = f(1 - f/w_0)^{-1} \in S$, and observe that $f - \tilde{f} = f^2/(f - w_0)$. Thus if $\mathrm{Re}\{\Phi(w_0)\} = 0$, it follows that $\mathrm{Re}\{L(\tilde{f})\} = \mathrm{Re}\{L(f)\}$, so that $\tilde{f}$ also maximizes $\mathrm{Re}\{L\}$. As a consequence, $\tilde{f}$ has an omitted arc $\tilde{\Gamma}$ which satisfies $\tilde{\Phi}(\tilde{w})(d\tilde{w}/\tilde{w})^2 > 0$, where $\tilde{\Phi}(\tilde{w}) = L(\tilde{f}^2/(\tilde{f} - \tilde{w}))$. The arcs Γ and $\tilde{\Gamma}$ are related by the mapping $\tilde{w} = w(1 - w/w_0)^{-1}$, and the identity

$$\frac{\tilde{w}\tilde{f}}{\tilde{w} - \tilde{f}} = \frac{wf}{w - f}$$

is easily verified. In view of the relation

$$f - \frac{wf}{w - f} = \frac{f^2}{f - w}$$

and the definition of $\tilde{f}$, we deduce that

$$\Phi(w_0) = L(f) - L(\tilde{f}) = \Phi(w) - \tilde{\Phi}(\tilde{w}).$$

On the other hand, a simple calculation gives

$$\frac{d\tilde{w}^2}{\tilde{w}^2} = \left(1 - \frac{w}{w_0}\right)^{-2} \frac{dw^2}{w^2},$$

so the differential equation for $\tilde{\Gamma}$ is transformed to

$$[\Phi(w) - \Phi(w_0)]\left(1 - \frac{w}{w_0}\right)^{-2}\frac{dw^2}{w^2} > 0, \qquad w \in \Gamma.$$

This is a second differential equation for Γ. If $\Phi(w_0) = 0$, we may divide it by the first differential equation $\Phi(w)(dw/w)^2 > 0$ to obtain

$$\left(1 - \frac{w}{w_0}\right)^2 > 0, \qquad w \in \Gamma, \quad w \neq w_0.$$

Thus w/w_0 is real, and Γ is a radial arc.

The class $\sigma(S)$ of support points of S is preserved under several important transformations. Obviously, every rotation of a support point is again a support point. It is also clear that $\sigma(S)$ is preserved under conjugation: if $f \in \sigma(S)$, then $\bar{f} \in \sigma(S)$, where $\bar{f}(z) = \overline{f(\bar{z})}$. This is perhaps most easily seen from the Toeplitz representation of a linear functional (Theorem 9.3).

It is less obvious that $\sigma(S)$ is preserved under a truncation operation defined as follows. Let $f \in \sigma(S)$ and let its omitted arc Γ have the nonsingular parametrization $w = w(t)$, $0 \leq t < \infty$, where $w \to \infty$ as $t \to \infty$. For fixed $t > 0$, let Γ_t be the subarc of Γ starting at $w(t)$, and let F be the conformal mapping of the unit disk onto the complement of Γ_t, normalized by $F(0) = 0$ and $F'(0) > 0$. The function $g = F/F'(0)$ then belongs to S. We claim that $g \in \sigma(S)$. To see this, observe that f is subordinate to F, and so (*cf.* §6.1) $f(z) = F(\omega(z))$ for some analytic univalent function ω with $|\omega(z)| \leq |z|$. Given that f maximizes $\text{Re}\{L\}$ over S, define the continuous linear functional M by

$$M(h) = L(h \circ \omega/\omega'(0)), \qquad h \in \mathscr{A}.$$

If $h \in S$, then $h \circ \omega/\omega'(0) \in S$, and it follows that

$$\text{Re}\{M(h)\} \leq \text{Re}\{L(f)\} = \text{Re}\{M(g)\},$$

since $F'(0)\omega'(0) = 1$. Thus g maximizes $\text{Re}\{M\}$ over S, and $g \in \sigma(S)$.

This result was discovered by Schaeffer and Spencer [6], Pfluger [3], Pell [1], and others. Kirwan and Schober [3] have systematically applied it to solve a number of extremal problems. The main idea is very simple: if a function maximizes a given linear functional, its truncations will maximize related linear functionals which can be calculated. The results are surprisingly deep. The method applies also to certain nonlinear extremal problems.

A few other general properties of support points are known. Hengartner and Schober [1] used the monotonic modulus property to show that for every support point or extreme point $f \in S$, both $f(z)/z$ and $\log[f(z)/z]$ are univalent. In [4] they used the $\pi/4$-property to show that $|a_2| > 1$ and

$|a_3| > \frac{3}{8}$ for every support point. Kirwan and Pell [1] improved these estimates to $|a_2| > \sqrt{2}$ and $|a_3| > 1$, and they produced an example for which $|a_2| < 1.774$. The sharp lower bounds are unknown.

§10.7. Point-Evaluation Functionals

Every rotation of the Koebe function is both a support point and an extreme point of S. It can be shown indirectly that other examples exist, but they are not easy to construct. For instance, a result of Brickman, MacGregor, and Wilken [1] asserts that the closed convex hull of the rotations of the Koebe function contains the starlike functions, but not all of S. Thus by the Krein–Milman theorem, S must have other extreme points and other support points.

The point-evaluation functionals $L(f) = f(\zeta)$ are a source of nonelementary support points (and extreme points) of S. Despite the invariance of S under rotations, the problem

$$\max_{f \in S} \operatorname{Re}\{f(\zeta)\}, \qquad 0 < |\zeta| < 1, \tag{12}$$

is not equivalent to the maximum modulus problem, and if ζ is nonreal it admits no rotation of the Koebe function as a solution. An equivalent form of the problem (12), more suggestive geometrically, is

$$\max_{f \in S} \operatorname{Re}\{e^{-i\alpha} f(\zeta)\}, \qquad 0 < \zeta < 1.$$

According to Theorem 10.3, a function f which solves the extremal problem (12) must map the disk onto the complement of an analytic arc Γ which satisfies

$$\frac{B^2}{B - w}\frac{dw^2}{w^2} > 0, \qquad B = f(\zeta). \tag{13}$$

The asymptotic half-line (10) reduces to

$$w = \tfrac{1}{3}B - B^2 t, \qquad t \geq 0, \tag{14}$$

and the $\pi/4$-property (9) gives

$$\operatorname{Re}\left\{\frac{B^2}{B - w}\right\} > 0, \qquad w \in \Gamma, \tag{15}$$

except at the endpoints.

If f were a rotation of the Koebe function, Γ would be a half-line $w = e^{i\theta}t$, $t \geq \frac{1}{4}$. Such a curve satisfies (13) only if $e^{i\theta} = -1$ and either $B > 0$ or $-\frac{1}{4} < B < 0$. In both cases, f is the Koebe function itself and ζ is real. This shows that if ζ is not real, no rotation of the Koebe function can maximize $\mathrm{Re}\{f(\zeta)\}$. Furthermore, the case $B \leq -\frac{1}{4}$ cannot occur because the asymptotic half-line would then be the unique solution to (13) near infinity, and Γ would lie on the negative real axis, forcing f to be the Koebe function.

We now propose to show that for every support point f arising from (12), the omitted arc Γ has monotonic argument. This is obviously true if f is the Koebe function, so we may assume that B is not real. In fact, we may suppose without loss of generality that $\mathrm{Im}\{B\} > 0$, because S is preserved under conjugation. The condition (15) says that Γ lies in a half-plane bounded by the line through B perpendicular to the asymptotic half-line (see Figure 10.3).

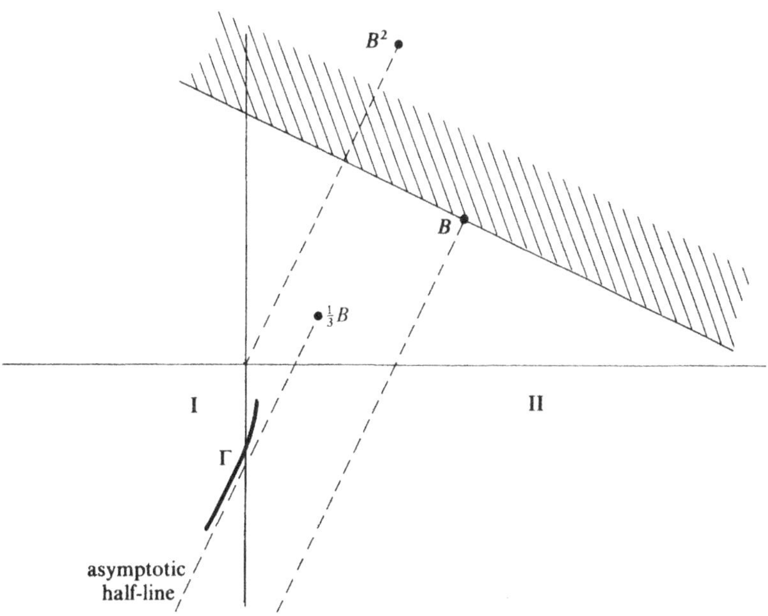

Figure 10.3. The problem max $\mathrm{Re}\{f(\zeta)\}$.

The half-line

$$w = B - B^2 t, \qquad t \geq 0 \tag{16}$$

divides the half-plane (15) into a quarter-plane I containing the asymptotic half-line (14) and a quarter-plane II. Whatever the value of B, the asymptotic half-line lies between the half-line (16) and the origin. Furthermore, the direction field determined by the differential equation (13) is radial on the half-line (16), because $B^2/(B - w) > 0$ there. Thus if Γ ever enters the quarter-plane II, it must eventually cross the half-line (16) radially, with decreasing modulus, in order to approach its asymptotic half-line near infinity. This violation of the monotonic modulus property shows that Γ lies entirely in the quarter-plane I. In this region, however, it is clear that

$$\operatorname{Im}\left\{\frac{B^2}{B - w}\right\} > 0,$$

so it follows from the differential equation (13) that $\operatorname{Im}\{dw^2/w^2\} < 0$ on Γ. Now let Γ be parametrized by $w = w(t), 0 \le t < \infty$, with $w(t) \to \infty$ as $t \to \infty$. Since the radial angle $\alpha(t) = \arg\{w'(t)/w(t)\}$ lies in the interval $-\pi/4 < \alpha(t) < \pi/4$, we conclude that

$$\frac{d}{dt} \arg w(t) = \frac{d}{dt} \operatorname{Im}\{\log w(t)\} = \operatorname{Im}\left\{\frac{w'(t)}{w(t)}\right\} < 0$$

for $t > 0$, showing that $\arg w(t)$ is decreasing. This completes the proof that Γ has a monotonic argument.

The preceding result is due to Brown [1], who also showed that the radial angle α is monotonic on Γ for every support point arising from the problem (12). His proof was based on the observation that the differential equation (13) can actually be integrated. Indeed, with an appropriate parametrization, the equation (13) takes the form

$$B^2[w'(t)]^2 = [w(t)]^2[B - w(t)].$$

Under the transformation $w = -4Bk(s)$, where k is the Koebe function, it then reduces to

$$B[s'(t)/s(t)]^2 = 1,$$

whose solutions are logarithmic spirals. Thus Γ is essentially the image under the Koebe function of a logarithmic spiral.

It can be shown that the Koebe function is the unique solution to (12) if and only if $0 < \zeta < 1$ or

$$0 > \zeta > \frac{1 - e}{1 + e} = -0.462\ldots.$$

(This was observed by Schober [1].) For other values of ζ the problem has exactly two solutions, related by conjugation. It follows that both of these functions are extreme points of S. With the help of a computer, Brown [1] found numerically that for ζ near -0.56, the radial angle α assumes a value within 10^{-5} of $\pi/4$ in magnitude at the tip of Γ. This was evidence that the bound $\pi/4$ in the general description of support points (Theorem 10.3) is best possible. Pearce [1] later established this analytically by appeal to a different linear extremal problem.

Pearce's observation may be described as follows. Since $f'(0) = 1$, the region of values

$$R_\zeta = \{f'(\zeta) : f \in S\}, \qquad 0 < \zeta < 1,$$

will lie in the right half-plane if ζ is sufficiently small. In fact, the rotation theorem (Theorem 3.7) says that this is true precisely for $\zeta \leq \sin \pi/8$. But if R_ζ is in the right half-plane, it is evident geometrically that the extremal problem

$$\max_{f \in S} \arg f'(\zeta), \qquad 0 < \zeta < 1, \tag{17}$$

is equivalent to some *linear* problem of the form

$$\max_{f \in S} \operatorname{Re}\{e^{-i\sigma} f'(\zeta)\}, \qquad 0 < \zeta < 1. \tag{18}$$

On the other hand, for $\zeta \leq 1/\sqrt{2}$, the function

$$f(z) = \frac{z - \zeta z^2}{(1 - e^{i\varphi}z)^2}, \qquad \varphi = \cos^{-1}\zeta,$$

is known to solve the rotation problem (17). (See §3.7.) Thus f is a support point of S for each $\zeta \leq \sin \pi/8$. A calculation gives

$$f'(z) = \frac{1 - e^{-i\varphi}z}{(1 - e^{i\varphi}z)^3}.$$

Let Γ be parametrized by $w = f(e^{i\theta})$, and express its tangent vector in the form

$$izf'(z) = \frac{ie^{-2i\varphi}z}{(1 - e^{i\varphi}z)^2} \cdot \frac{z - e^{i\varphi}}{z - e^{-i\varphi}}, \qquad z = e^{i\theta}.$$

This is essentially the product of a half-plane mapping and a rotation of the Koebe function. Simple geometric considerations reveal that

$$ie^{i\theta}f'(e^{i\theta}) = ie^{-2i\varphi}\rho(\theta),$$

where $\rho(\theta) > 0$ for $-\varphi < \theta < \varphi$ and $\rho(\theta) < 0$ elsewhere. This shows that Γ is a half-line with inclination $-ie^{-2i\varphi}$. The finite tip of Γ is the image of $z = e^{i\varphi}$, since $f'(e^{i\varphi}) = 0$. Another calculation gives $f(e^{i\varphi}) = (i/4)\csc\varphi$. Now choose $\zeta = \sin\pi/8$, so that $\varphi = 3\pi/8$. Then the radial angle at the tip of Γ is

$$\alpha = \arg\{-ie^{-2i\varphi}/f(e^{i\varphi})\} = \pi/4.$$

For each of the support points just discussed, Γ is a half-line and so has monotonic argument and monotonic radial angle. Duren [15] has used a variational method to show that for every support point arising from a problem of the form (18), the omitted arc Γ has monotonic argument. It turns out that Γ is a nonradial half-line only if the problem (18) is equivalent to a rotation problem (see Exercise 6).

Duren, Leung, and Schiffer [1] have shown that no point-evaluation functional can generate a support point whose omitted arc Γ has a radial angle of $\pm\pi/4$ at its tip, an unexpected discovery in view of Brown's numerical evidence described above. The same is true for coefficient functionals.

§10.8. The Coefficient Problem

The general results on support points of S apply to the coefficient problem and shed new light on the Bieberbach conjecture. The coefficient functional Λ_n defined by $\Lambda_n(h) = c_n$, where

$$h(z) = c_0 + c_1 z + c_2 z^2 + \cdots, \qquad |z| < 1,$$

is linear and continuous on the space $\mathscr{A}$. The nth coefficient problem is

$$\max_{f \in S} \operatorname{Re}\{\Lambda_n(f)\}, \qquad n = 2, 3, \ldots.$$

In more familiar notation, the problem is to find the functions

$$f(z) = z + a_2 z^2 + a_3 z^3 + \cdots$$

of class S which maximize $\operatorname{Re}\{a_n\}$ for a given integer $n \geq 2$.

According to Theorem 10.3, an extremal function f must map the disk onto the complement of an analytic arc Γ which satisfies

$$\Lambda_n\left(\frac{f^2}{f-w}\right)\frac{dw^2}{w^2} > 0.$$

The expansion

$$\frac{[f(z)]^2}{f(z) - w} = -\sum_{k=1}^{\infty} [f(z)]^{k+1} w^{-k} = -\sum_{k=2}^{\infty} P_k\left(\frac{1}{w}\right) z^k \tag{19}$$

shows that

$$\Lambda_n\left(\frac{f^2}{f - w}\right) = -\sum_{k=1}^{n-1} \Lambda_n(f^{k+1}) w^{-k} = -P_n\left(\frac{1}{w}\right),$$

where

$$P_n(\zeta) = \zeta^{n-1} + \cdots + C_n \zeta^2 + B_n \zeta$$

is a monic polynomial of degree $n - 1$ without constant term, and

$$B_n = \Lambda_n(f^2), \qquad C_n = \Lambda_n(f^3).$$

Thus Γ satisfies the differential equation

$$P_n\left(\frac{1}{w}\right) \frac{dw^2}{w^2} < 0. \tag{20}$$

The asymptotic half-line (11) takes the form

$$w = \tfrac{1}{3} C_n / B_n - B_n t, \qquad t \geq 0,$$

and the $\pi/4$-property (9) gives

$$\operatorname{Re}\left\{P_n\left(\frac{1}{w}\right)\right\} < 0, \qquad w \in \Gamma, \tag{21}$$

except at the endpoints.

The first few polynomials P_n are easily calculated:

$$P_2(\zeta) = \zeta, \qquad P_3(\zeta) = \zeta^2 + 2a_2 \zeta,$$

$$P_4(\zeta) = \zeta^3 + 3a_3 \zeta^2 + (a_2^2 + 2a_3)\zeta.$$

The general polynomial P_n involves the coefficients $a_2, \ldots, a_{n-1}$ of the extremal function as parameters.

The differential equation (20) for the omitted arc leads naturally to a differential equation for the extremal function. Let $e^{i\alpha}$ and $e^{i\beta}$ be the points on the unit circle which correspond to the endpoints of Γ. Thus $f(e^{i\alpha}) = w_0$

and $f(e^{i\beta}) = \infty$, where w_0 is the finite tip of Γ. Since Γ is an analytic arc, f has an analytic continuation across each of the two arcs of the unit circle bounded by $e^{i\alpha}$ and $e^{i\beta}$. In particular, Γ has the parametrization $w = f(e^{it})$, and

$$\frac{dw}{dt} = ie^{it}f'(e^{it}).$$

The differential equation (20) therefore asserts that the meromorphic function

$$R_n(z) = \left[\frac{zf'(z)}{f(z)}\right]^2 P_n\left(\frac{1}{f(z)}\right) \tag{22}$$

is real and positive on the unit circle, except at $e^{i\alpha}$. A square-root transformation shows that f is analytic at $e^{i\alpha}$, and that $(f - w_0)$ has a double zero there. Hence R_n is analytic at $e^{i\alpha}$, and it has a zero of order two or higher there. (In fact, R_n has a double zero at $e^{i\alpha}$ if f is not a rotation of the Koebe function, since it then follows from the general result of §10.6 that $P_n(1/w_0) \neq 0$.) A similar argument shows that f has a double pole at $e^{i\beta}$. Because the polynomial P_n has no constant term, it follows that R_n is analytic at $e^{i\beta}$.

Consequently, the function R_n is analytic in the closed unit disk except for a pole of order $n - 1$ at the origin, and it is real on the unit circle. This allows R_n to be continued analytically according to the formula $R_n(1/\bar{z}) = \overline{R_n(z)}$. The resulting function R_n is analytic in the extended complex plane except for poles of order $n - 1$ at zero and infinity. In other words, R_n is a rational function of the form

$$R_n(z) = \sum_{k=1-n}^{n-1} c_k z^k.$$

It is clear from the symmetry of R_n that c_0 is real and that $c_{-k} = \bar{c}_k$ for $k = 1, 2, \dots, n - 1$.

In order to compute the coefficients c_k, we now establish a connection between the polynomials P_k and the Faber polynomials F_k associated with the inverted function $g \in \Sigma$ defined by $g(\zeta) = 1/f(1/\zeta)$. The Faber polynomials of g are generated by the formula (*cf.* §4.1)

$$\frac{\zeta g'(\zeta)}{g(\zeta) - w} = \sum_{k=0}^{\infty} F_k(w)\zeta^{-k}.$$

Integration with respect to ζ and differentiation with respect to w leads to the formula (*cf.* §5.2)

$$\frac{1}{g(\zeta) - w} = \sum_{k=1}^{\infty} \frac{1}{k} F_k'(w)\zeta^{-k}. \tag{23}$$

On the other hand, the expansion (19) gives

$$\frac{1}{g(\zeta) - w} - f\left(\frac{1}{\zeta}\right) = \frac{[f(z)]^2}{1/w - f(z)} = \sum_{k=2}^{\infty} P_k(w) z^k, \qquad z = \frac{1}{\zeta}. \tag{24}$$

Comparison of (23) and (24) yields the relation

$$P_k(w) = \frac{1}{k} F_k'(w) - a_k, \qquad k = 2, 3, \ldots. \tag{25}$$

Recall now that the Faber polynomials have the property

$$F_n(g(\zeta)) = \zeta^n + \sum_{k=1}^{\infty} \beta_{nk} \zeta^{-k}, \qquad n = 1, 2, \ldots.$$

Hence

$$F_n'\left(\frac{1}{f(z)}\right) \frac{f'(z)}{[f(z)]^2} = nz^{-n-1} - \sum_{k=1}^{\infty} k\beta_{nk} z^{k-1}.$$

This combined with (22) and (25) gives

$$R_n(z) = \sum_{k=1}^{n-1} k a_k z^{k-n} + (n-1)a_n + \sum_{k=1}^{\infty} d_k z^k \tag{26}$$

for some coefficients d_k, where $a_1 = 1$. In view of the symmetry of the coefficients of R_n, it follows that

$$R_n(z) = (n-1)a_n + \sum_{k=1}^{n-1} (k a_k z^{k-n} + k\overline{a_k} z^{n-k}). \tag{27}$$

The structural formula (27) may be substituted into (22) to produce an explicit differential equation for the extremal function f. This result is due to Schiffer [4, 5] and is known as the *Schiffer differential equation*. Unfortunately, the equation involves the initial coefficients of the unknown extremal function.

It is instructive to go a step further and make a direct calculation of the coefficient $d_1 = (n-1)\overline{a_{n-1}}$ in the expansion (26) of R_n. One computes

$$d_1 = (n+1)a_{n+1} - 2a_2 a_n.$$

Thus the coefficients of f must satisfy the equation

$$(n+1)a_{n+1} - 2a_2 a_n = (n-1)\overline{a_{n-1}},$$

which is the Marty relation (§2.9). The direct calculation of $d_2, d_3, \dots$ would lead to similar relations among the coefficients of the extremal function. This suggests that the Schiffer differential equation may be viewed as an infinite succession of "higher Marty relations."

A function $f \in S$ which satisfies the Schiffer differential equation is called a *critical point* of the nth coefficient functional Λ_n. The functions which maximize or minimize $\operatorname{Re}\{a_n\}$ are among the critical points, but in general there are others (see Exercise 11).

It is not entirely obvious that the Koebe function is a critical point. This can be verified as follows. The generating function (19) for the polynomials P_ν gives

$$\sum_{\nu=1}^{\infty} \nu z^\nu + \sum_{\nu=2}^{\infty} P_\nu\left(\frac{1}{k(\zeta)}\right) z^\nu = k(z) - \frac{[k(z)]^2}{k(z) - k(\zeta)}$$

$$= \frac{z\zeta}{(\zeta - z)(1 - z\zeta)} = \frac{\zeta}{1 - \zeta^2}\left\{\frac{\zeta}{\zeta - z} - \frac{1}{1 - \zeta z}\right\},$$

where $k(\zeta) = \zeta(1 - \zeta)^{-2}$ is the Koebe function. Comparing coefficients of z^n, one obtains the elegant formula

$$P_n\left(\frac{1}{k(\zeta)}\right) = \frac{\zeta^n - \zeta^{-n}}{\zeta - \zeta^{-1}} - n. \tag{28}$$

On the boundary this reduces to

$$P_n\left(\frac{1}{w}\right) = \frac{\sin n\theta}{\sin \theta} - n, \qquad w = k(e^{i\theta}),$$

which shows that $P_n(1/w) < 0$ for $-\infty < w < -\frac{1}{4}$. This implies that omitted arc Γ of the Koebe function satisfies the corresponding differential equation (20), and Schiffer's equation (22) follows with

$$R_n(z) = (n - 1)n + \sum_{\nu=1}^{n-1} \nu^2(z^{\nu-n} + z^{n-\nu}). \tag{29}$$

Thus the Koebe function is a critical point of Λ_n.

A more direct approach is to carry out the summation in (29) and to compare the result with (28). This is left as an exercise.

It should be observed that the definition of a critical point is purely algebraic. A function $f \in S$ may satisfy the Schiffer differential equation (or an appropriate generalization of it) although its omitted arc Γ does not satisfy the corresponding differential equation of the form (20). This will happen trivially for a function which *minimizes* the functional in question, so that the sign of (20) is reversed. However, a critical point may have an omitted arc which in a more essential way fails to satisfy the appropriate differential equation. See, for instance, Exercise 10.

Further necessary conditions for a local maximum can be obtained with the aid of the second variation as developed by Duren and Schiffer [2], Babenko [1], Tsao [1], and Chang, Schiffer, and Schober [1]. The second variation can be effective in disproving certain conjectures by showing that the conjectured extremal function is not even a local maximum, but is a saddle point. For the nth coefficient problem, however, it is known that the Koebe function is a local maximum. References are given in the *Notes* at the end of Chapter 4.

§10.9. Region of Values of $\log f(\zeta)/\zeta$

For each fixed point ζ in the unit disk, the set of values of $\log[f(\zeta)/\zeta]$ as f ranges over S is precisely a closed disk. This remarkable result was discovered by Grunsky [1] in 1932. We shall now give a proof by the method of boundary variation. As usual, the logarithm is the branch for which $\log 1 = 0$.

Theorem 10.6 (Grunsky's Theorem). *For each point ζ with $|\zeta| = \rho < 1$, the region*

$$W_\zeta = \{\log[f(\zeta)/\zeta] : f \in S\}$$

is the disk

$$D_\rho = \left\{ w : \left| w - \log \frac{1}{1-\rho^2} \right| \le \log \frac{1+\rho}{1-\rho} \right\}.$$

Observe that the theorem contains the sharp inequalities

$$\frac{\rho}{(1+\rho)^2} \le |f(\zeta)| \le \frac{\rho}{(1-\rho)^2}$$

and

$$\left| \arg \frac{f(\zeta)}{\zeta} \right| \le \log \frac{1+\rho}{1-\rho},$$

which we have already established by other methods (see §2.3 and §3.6).

Proof of Theorem. We begin with a few direct observations about the region W_ζ. First of all, W_ζ is a compact set. It is bounded because both $\log|f(\zeta)|$ and $\arg[f(\zeta)/\zeta]$ are bounded on S for fixed ζ. It is closed because S is a compact normal family. Note also that W_ζ depends only on $|\zeta|$, because S is preserved under rotations. Hence we may assume that $0 < \zeta < 1$. Finally, because S is preserved under dilations, it is easy to see that W_ζ expands as ζ increases: $W_{\zeta_1} \subset W_{\zeta_2}$ if $\zeta_1 < \zeta_2$.

Fixing a direction $e^{i\sigma}$, we now pose the extremal problem

$$\max_{f \in S} \operatorname{Re}\left\{e^{-i\sigma} \log \frac{f(\zeta)}{\zeta}\right\}, \qquad 0 < \zeta < 1. \tag{30}$$

Geometrically, this problem asks for the point in W_ζ whose projection onto the ray with direction $e^{i\sigma}$ has maximum distance from the origin. A complete solution for every choice of $e^{i\sigma}$ will determine at least the closed convex hull of W_ζ.

Let $f \in S$ be an extremal function for the problem (30), and let Γ be its omitted set. Let

$$f^*(z) = f(z) + \frac{\lambda_r [f(z)]^2}{w_0^2[f(z) - w_0]} + O(r^3)$$

be a variation of f with respect to a point $w_0 \in \Gamma$ (*cf.* §10.6). A simple calculation gives

$$\log \frac{f^*(\zeta)}{\zeta} = \log \frac{f(\zeta)}{\zeta} + \frac{\lambda_r f(\zeta)}{w_0^2[f(\zeta) - w_0]} + O(r^3).$$

Because f is extremal, we conclude that

$$\operatorname{Re}\left\{\lambda_r \frac{e^{-i\sigma}B}{w_0^2(B - w_0)} + O(r^3)\right\} \leq 0, \qquad B = f(\zeta).$$

Thus by Schiffer's theorem (§10.3), Γ is a system of analytic arcs satisfying

$$\frac{e^{-i\sigma}B}{w^2(B - w)} dw^2 > 0. \tag{31}$$

This quadratic differential has a simple pole at infinity and no other zeros or poles on Γ, since 0 and B are in the range of f. It follows that Γ is a single analytic arc.

Representing Γ by $w = f(e^{it})$, we conclude from (31) that the function

$$F(z) = \frac{e^{-i\sigma}B}{f(z) - B}\left[\frac{zf'(z)}{f(z)}\right]^2 \tag{32}$$

is real and nonnegative on the circle $|z| = 1$. Since F is analytic in $|z| < 1$ aside from a simple pole at ζ, it can be continued analytically by Schwarz reflection: $F(1/\bar{z}) = \overline{F(z)}$. The extended function F has simple poles at ζ and $1/\zeta$, and a double zero at the point $e^{i\alpha}$ which corresponds to the finite tip of Γ. Thus F is a rational function of the form

$$F(z) = \frac{A(z - e^{i\alpha})^2}{(z - \zeta)(1 - \zeta z)} \tag{33}$$

for some constant A.

Equating the two expressions (32) and (33) for F and letting z tend to zero, we obtain

$$\zeta e^{-i\sigma} = Ae^{2i\alpha}. \tag{34}$$

On the other hand, recalling that F is real and positive on the unit circle except at $e^{i\alpha}$, we find

$$0 < F(e^{i\theta}) = -4Ae^{i\alpha}|e^{i\theta} - \zeta|^{-2} \sin^2\left(\frac{\theta - \alpha}{2}\right),$$

so that $Ae^{i\alpha} < 0$. Combining this with (34), we have

$$\zeta e^{-i\sigma} = -|A|e^{i\alpha},$$

so that $e^{i\alpha} = -e^{-i\sigma}$ and $A = \zeta e^{i\sigma}$.

Having determined the parameters $e^{i\alpha}$ and A, we can equate (32) and (33) and express the resulting differential equation for $w = f(z)$ in the form

$$\frac{B\, dw^2}{w^2(w - B)} = \frac{\zeta(1 + e^{i\sigma}z)^2\, dz^2}{z^2(z - \zeta)(1 - \zeta z)},$$

or

$$\int_B^w \frac{dw}{w\sqrt{1 - w/B}} = \int_\zeta^z \frac{(1 + e^{i\sigma}z)\, dz}{z\sqrt{(1 - z/\zeta)(1 - \zeta z)}},$$

where the branches of the square-root functions are determined by requiring that the limits of integration z and w be sufficiently near zero.

This last equation leads to an implicit formula for the extremal function f and allows the explicit evaluation of $B = f(\zeta)$. Integrating the left-hand side and letting z tend to 0, we obtain

$$\lim_{z\to 0} \log\left\{\frac{1 - \sqrt{1 - f(z)/B}}{1 + \sqrt{1 - f(z)/B}} \cdot \frac{\zeta}{z}\right\} = \int_0^\zeta \left\{1 - \frac{1}{\sqrt{P(z)}}\right\}\frac{dz}{z} - e^{i\sigma}\int_0^\zeta \frac{dz}{\sqrt{P(z)}},$$

where $P(z) = (1 - z/\zeta)(1 - \zeta z)$. The limit is easily found to be

$$-\log \frac{f(\zeta)}{\zeta} - \log 4,$$

while the integrals can be calculated as

$$\log(1 - \zeta^2) - \log 4 \quad \text{and} \quad \log \frac{1 + \zeta}{1 - \zeta},$$

respectively. These calculations give

$$\log \frac{f(\zeta)}{\zeta} = \log \frac{1}{1 - \zeta^2} + e^{i\sigma} \log \frac{1 + \zeta}{1 - \zeta}. \tag{35}$$

The formula (35) has an obvious geometric interpretation. As $e^{i\sigma}$ runs around the unit circle, the extremal functions of the problem (30) give values of $\log[f(\zeta)/\zeta]$ which traverse the circle C_ζ bounding the disk D_ζ. This proves that D_ζ is the closed convex hull of W_ζ, and that its full boundary C_ζ lies in W_ζ. To complete the proof, we have only to observe that the disks D_r expand as r increases, so that each interior point of D_ζ lies on a circle C_r for some r, $0 \leq r < \zeta$. Since $C_r \subset W_r \subset W_\zeta$, this shows that W_ζ fills the disk D_ζ, and the proof is complete.

§10.10. Multiply Connected Domains

One advantage of the method of boundary variation is that it applies readily to extremal problems for families of univalent functions over multiply connected domains. As an illustration we shall use it to prove the existence of certain canonical mappings. A more thorough discussion of conformal mappings of multiply connected domains may be found in the book of Nehari [2].

Let D be a finitely connected domain in the extended complex plane, none of whose boundary components is a single point. Assume for convenience that both 0 and ∞ lie in D. Let $\mathscr{F}$ be the family of functions g analytic and univalent in D with an expansion near infinity of the form

$$g(z) = z + b_0 + b_1 z^{-1} + b_2 z^{-2} + \cdots, \qquad |z| > R.$$

This family $\mathscr{F}$ may be viewed as a generalization of the family Σ. The area theorem provides the crude bounds

$$|b_n| \leq n^{-1/2} R^{n+1}, \qquad n = 1, 2, \ldots.$$

Let $\mathscr{F}_0$ be the subfamily of all $g \in \mathscr{F}$ for which $b_0 = 0$. It can be shown that $\mathscr{F}_0$ is a compact normal family.

Now let $g \in \mathscr{F}$ be a function which maximizes $\operatorname{Re}\{b_1\}$, and let Γ be the set omitted by g. Choose $w_0 \in \Gamma$ and let

$$F_r(w) = w + \lambda_r(w - w_0)^{-1} + O(r^3)$$

be analytic and univalent on the complement of the component of Γ which contains w_0. Then the function

$$g^*(z) = F_r(g(z)) = z + b_0^* + b_1^* z^{-1} + b_2^* z^{-2} + \cdots$$

belongs to $\mathscr{F}$, and $b_1^* = b_1 + \lambda_r$. Thus $\operatorname{Re}\{b_1 + \lambda_r\} \le \operatorname{Re}\{b_1\}$, and Schiffer's theorem gives the simple differential equation $dw^2 > 0$ for Γ. This implies that Γ is a union of horizontal line segments. In other words, g is a horizontal slit mapping. Similarly, the minimum of $\operatorname{Re}\{b_1\}$ is provided by a vertical slit mapping.

Next let g be a function in $\mathscr{F}$ for which $|g'(0)|$ is maximal. Let g^* be as above and compute

$$g^{*\prime}(0) = g'(0)[1 - \lambda_r w_0^{-2} + O(r^3)].$$

The inequality $|g^{*\prime}(0)| \le |g'(0)|$ therefore gives

$$\operatorname{Re}\{\log[1 - \lambda_r w_0^{-2} + O(r^3)]\} \le 0,$$

or

$$\operatorname{Re}\{\lambda_r w_0^{-2} + O(r^3)\} \ge 0.$$

It follows from Schiffer's theorem that Γ consists of arcs satisfying $(dw/w)^2 < 0$, whose solutions are circular arcs centered at the origin. Thus g is a circular slit mapping. The problem of minimizing $|g'(0)|$ leads analogously to the differential equation $(dw/w)^2 > 0$, whose solutions are radial arcs. This proves the existence of a radial slit mapping.

As a general rule the maximum and minimum problems always lead to trajectories and orthogonal trajectories, respectively, of the same quadratic differential. This is an immediate consequence of Schiffer's theorem.

The method of boundary variation is quite versatile and may be applied to obtain qualitative solutions of other problems in multiply connected domains. Some further applications may be found in Schiffer [11].

For certain classes of functions univalent in an annulus, the method of boundary rotation leads to specific solutions of extremal problems, with actual bounds expressed in terms of elliptic functions. See, for instance, Duren and Schiffer [1], Duren [2, 3,], and McLaughlin [2, 3].

§10.11. Other Variational Methods

In addition to the method of boundary variation, several other variational methods have been developed and applied with success to extremal problems for univalent functions. The various methods typically lead to the same differential equation for an extremal function, or to the same quadratic differential for its omitted arcs. However, each method has special advantages for certain purposes.

The method of *interior variation* as developed by Schiffer [6, 7] is based on potential theory. We shall now give an outline of the main ideas. Detailed expositions may be found in Schiffer [11, 12], Duren and Schiffer [2], and Hummel [6].

Let a function $f \in S$ map the disk onto a domain D bounded by a rectifiable Jordan curve C. Let v be a function analytic outside a compact subset of D. Then for small values of the real or complex parameter ε, the mapping $w^* = w + \varepsilon v(w)$ sends C onto a nearby rectifiable Jordan curve C^* which bounds a domain D^*. (The univalence of this mapping is equivalent to the fact that v satisfies a Lipschitz condition on C.) One can express Green's function of D^* in the form

$$g^*(w, \omega) = g(w, \omega) + \varepsilon h(w, \omega) + O(\varepsilon^2),$$

where $g(w, \omega)$ is Green's function of D. (See §1.8 for the definition and properties of Green's function.) Potential-theoretic formulas permit the explicit calculation of the first variation h in terms of g and v. Now the close connection between Green's function of D and the inverse of f allows this expression to be translated to formulas for the first variation of f^{-1} and ultimately for the first variation of any differentiable functional ϕ of f. The initial requirement that the range of f be bounded by a rectifiable Jordan curve is removed by a continuity argument. Now if f maximizes ϕ, the first variation must vanish. This is a strong necessary condition for an extremal function which involves an arbitrary variation function v. The special choice $v(w) = (w - w_0)^{-1}$, where w_0 is an arbitrary point in the range of f, again leads to the Schiffer differential equation for the extremal function f.

More general choices such as

$$v(w) = \sum_{j=1}^{m} \alpha_j(w - w_j)^{-1}, \qquad w_j \in D, \quad \alpha_j \in \mathbb{C},$$

permit the variation of f within certain subclasses of S, preserving given constraints. (See, for instance, Schaeffer, Schiffer, and Spencer [1].) In fact, the flexibility of interior variation is its main advantage over boundary variation. Duren and Schiffer [2] used the method of interior variation to develop workable formulas for the second variation. This could not have been done with boundary variation. On the other hand, boundary variation leads directly to a quadratic differential for the omitted arcs and is generally more convenient to use than interior variation. An additional disadvantage of interior variation is its confinement to families of functions in simply connected domains, because the exponential relation between Green's function and the Riemann mapping function does not extend to multiply connected domains.

In a general finitely connected domain, the method of interior variation provides a generalization of the classical formula of Hadamard for the first variation of Green's function. Hadamard's formula is based on normal displacement and must assume *a priori* that the boundary is smooth. Schiffer and others have systematically applied the method of interior variation to various physical and geometric problems which can be stated as extremal problems for classes of functions in multiply connected domains.

Goluzin [13, 14, 15, 27] developed another powerful variational method which may be regarded as a new version of interior variation. Although it is perhaps less natural in conception, it is often more convenient to apply than Schiffer's version. It is now known as the *Goluzin variation*. An outline of the method is given below. Details may be found in the book of Goluzin [28]. An alternate proof of the main theorem, based on the Carathéodory convergence theorem, was given by Pommerenke [11]. The specific connection with the interior variation was explored by Poole [1], who derived the Goluzin variation from Schiffer's interior variation.

Goluzin's result is as follows. Let $f \in S$ and suppose there is a function

$$F(z, \lambda) = f(z) + \lambda q(z) + O(\lambda^2)$$

analytic in z and λ for $|\lambda| < \rho$ and for z in an annulus $r < |z| < 1$. Suppose also that for each fixed λ, the function F is univalent in the annulus. Let D_λ be the simply connected domain consisting of the range of F with its "hole" adjoined. Let f^* map the unit disk onto D_λ, with $f^*(0) = 0$ and $f^{*\prime}(0) > 0$. Then

$$f^*(z) = f(z) + \lambda q(z) - \lambda z f'(z) h(z) + \lambda z f'(z) \overline{h(1/\bar{z})} + O(\lambda^2),$$

where h is the principal part of the Laurent expansion of $q(z)/zf'(z)$ in the annulus $r < |z| < 1$.

Viewing λ as a small parameter, we may regard f^* as a small perturbation of f. Functions of the form

$$F(z, \lambda) = f(z) + \lambda \sum_{j=1}^{m} \frac{\alpha_j [f(z)]^2}{f(z) - w_j},$$

for points w_j in the range of f, are adequate for most applications. Specializing the sum to a single term, one easily derives the Schiffer differential equation for f.

Tsao [1] obtained an expression for the second variation in Goluzin's method and applied it to some coefficient problems.

A new variational method has recently been developed by Chang, Schiffer, and Schober [1]. This is a comparatively elementary method which gives a strikingly simple derivation of Schiffer's differential equation and a more attractive approach to the second variation.

The deeper variational methods all reveal that a function $f \in S$ which solves a reasonable extremal problem must map the disk onto the complement of a system of analytic arcs which are trajectories of a certain quadratic differential. Further information can be obtained from certain auxiliary variations. One is the *spire variation*, which is obtained by composing f with a mapping of the unit disk onto the unit disk minus a short radial segment, or spire, emanating from some point on the unit circle (*cf.* §6.5). Another is the *Julia variation*, which deforms the analytic boundary arcs by inner normal displacements. This idea was introduced in 1922 by Julia [1], whose discussion was based on the Hadamard variational formula. For a derivation of Julia's formula via Schiffer's method of boundary variation, see Hummel [6].

Because an extremal function is a slit mapping, it can be represented by Loewner's differential equation in terms of a function $\kappa(t)$ of unit modulus (see Chapter 3). It is natural to try to use the differential equation for the extremal function to obtain more specific information about κ, or perhaps to determine κ completely. This idea first occurred to Schiffer [8], who obtained a differential equation for κ. Similar results were later given by Lebedev [1, 2] and Kufarev [4, 5, 6]. Kufarev successfully applied the method to solve several extremal problems. Garabedian and Schiffer [1] used it in their original proof of the Bieberbach conjecture for the fourth coefficient. (See also Schiffer [13].) More recently, Friedland and Schiffer [1] have applied the method in conjunction with Pontryagin's theory of optimal control. The book of Aleksandrov [3] develops the method and presents other applications.

Exercises

1. Prove that a line segment of length ℓ has conformal radius $\ell/4$. (*Hint*: Consider the mapping $z = \zeta + 1/\zeta$.)

2. Show that every support point of S has positive Hayman index.

3. For the point-evaluation problem (12), show that the asymptotic half-line (14) is not a trajectory (near infinity) of the quadratic differential (13) unless B is real. Thus for point-evaluation functionals a support point cannot omit a half-line unless it is the Koebe function.

4. Show that the arc omitted by a solution to the point-evaluation problem (12) cannot have a radial angle of $\pm\pi/4$ at its tip. (Duren, Leung, and Schiffer [1].)

5. Show that for fixed $\zeta \in \mathbb{D}$ no rotation of the Koebe function maximizes $\operatorname{Re}\{f(\zeta)\}$ in S unless $0 < \zeta < 1$ or $(1-e)/(1+e) \leq \zeta < 0$. Conclude that S has extreme points other than the Koebe function and its rotations. (Schober [1].)

6. For the derivative problem (18), compute the differential equation for the arc Γ omitted by the extremal function f, and find its asymptotic half-line. Show that the asymptotic half-line is itself a trajectory (near infinity) if and only if $[e^{-i\sigma}C]^2$ is real, where $C = f'(\zeta)$. Deduce that Γ is a half-line if and only if either the problem (18) is equivalent to the problem of maximizing or minimizing $\arg g'(\zeta)$, or f is a rotation of the Koebe function. (Duren [15].)

7. Let Γ be the arc omitted by a support point $f \in S$ corresponding to a coefficient functional $L(f) = \sum_{j=2}^{n} \lambda_j a_j$. Show that the normalized truncation of Γ is the arc omitted by a support point which again corresponds to a coefficient functional. Obtain the analogous result for point-evaluation functionals.

8. Show that for each fixed $z \in \mathbb{D}$ the region of values of $\log[zf'(z)/f(z)]$ for $f \in S$ is the disk $|w| \leq \log[(1+|z|)/(1-|z|)]$. (*Hint*: Apply the transformation used in the proof of Theorem 3.6.)

9. Show directly that the Koebe function is a critical point of the nth coefficient functional Λ_n by evaluating the sum in the expression (29) for R_n and comparing the result with (28).

10. For fixed complex constants $\lambda_2, \ldots, \lambda_n$, consider the coefficient functional $L(f) = \lambda_2 a_2 + \cdots + \lambda_n a_n$.

(a) Show that each function $f \in S$ which maximizes $\operatorname{Re}\{L\}$ must satisfy a generalized Schiffer differential equation

$$[zf'(z)/f(z)]^2 \mathscr{P}_n(1/f(z)) = \mathscr{R}_n(z),$$

where $\mathscr{P}_n = \lambda_2 P_2 + \cdots + \lambda_n P_n$ and $\mathscr{R}_n = \lambda_2 R_2 + \cdots + \lambda_n R_n$.

(b) Show that the Koebe function always satisfies this differential equation. Show, however, that the omitted arc of the Koebe function need not satisfy $\mathscr{P}_n(1/w)(dw^2/w^2) < 0$.

(c) Show that no rotation of the Koebe function maximizes $\operatorname{Re}\{a_3 + ia_2\}$ in S.

11. Show that $\sqrt{k(z^2)} = z/(1-z^2)$ is a critical point of the third coefficient functional Λ_3 but does not maximize or minimize $\operatorname{Re}\{a_3\}$ in S. Show further

that the omitted arc Γ satisfies the differential equation (20) with $n = 3$. Generalize to $n > 3$.

12. Show that the function $P_3(1/w) = -\Lambda_3(k^2/(k - w))$ actually vanishes at the tip of the arc omitted by the Koebe function k. (See the discussion at the end of §10.6.) Generalize to $n > 3$.

13. Apply the method of boundary variation to prove the Grunsky inequalities (see §4.3)

$$\left| \sum_{n=1}^{N} \sum_{k=1}^{N} \gamma_{nk} \lambda_n \lambda_k \right| \leq \sum_{n=1}^{N} \frac{1}{n} |\lambda_n|^2.$$

(*Hint*: Show that equality occurs when the real part of the double sum is maximized.)

14. Try to prove the rotation theorem (Theorem 3.7) by a variational method.

(a) Show that each function $f \in S$ which maximizes $\arg f'(\zeta)$ for fixed $\zeta \in \mathbb{D}$ must map $\mathbb{D}$ onto the complement of a system Γ of analytic arcs satisfying

$$\frac{iA(2w - A)}{(w - A)^2 w^2} dw^2 > 0, \qquad A = f(\zeta).$$

(b) Assuming without loss of generality that $\zeta = r > 0$, infer that $w = f(z)$ satisfies a differential equation of the form

$$\frac{iA(A - 2w)}{(w - A)^2} \left(\frac{zw'}{w}\right)^2 = Q(z),$$

where

$$Q(z) = \frac{a}{(z - r)^2} + \frac{b}{z - r} + c + \frac{\bar{b}z}{1 - rz} + \frac{\bar{a}z^2}{(1 - rz)^2},$$

a and b are complex constants, c is real, and $Q(z) \geq 0$ for $|z| = 1$.

(c) Obtain the expressions $a = -ir^2$ and $b = rc - 2ir$. Show that $c = 4r/\sqrt{1 - r^2}$ if $r \leq 1/\sqrt{2}$, while $c = 2/(1 - r^2)$ if $r \geq 1/\sqrt{2}$.

(d) Show that Q is a perfect square if $r \geq 1/\sqrt{2}$:

$$Q(z) = -i\left[\frac{r(z - e^{i\theta})(z - e^{i\varphi})}{(z - r)(1 - rz)}\right]^2.$$

(e) Integrate the differential equation and determine $A = f(r)$.

15. For α, $\beta \in \mathbb{C}$ with $\alpha \neq \beta$, show that the trajectories and orthogonal trajectories of the quadratic differential $dw^2/[(w - \alpha)(w - \beta)]$ are the hyperbolas and ellipses, respectively, with foci at α and β.

16. Let $\mathscr{S}_0$ be the class of functions f analytic and univalent in $\mathbb{D}$, with $f(z) \neq 0$ and $f(0) = 1$. Let L be a continuous linear functional which is not constant on $\mathscr{S}_0$.

(a) Show that a function $f \in \mathscr{S}_0$ which maximizes $\text{Re}\{L\}$ must map $\mathbb{D}$ onto the complement of an analytic arc Γ which satisfies

$$L\left(\frac{f(f-1)}{f-w}\right)\frac{dw^2}{w(w-1)} > 0.$$

(*Suggestion*: Apply a boundary variation which fixes the points 0 and 1. Refer to Chapter 9, Exercise 5.)

(b) Show that Γ is monotonic with respect to the family of ellipses with foci at 0 and 1, and that it makes an angle of less than $\pi/4$ with each of the confocal hyperbolas it meets. (Duren and Schober [2].)

17. Use Goluzin's variational formula (see §10.11) to derive the Schiffer differential equation for a function $f \in S$ which maximizes $\text{Re}\{a_n\}$.

Chapter 11

Coefficient Regions

The coefficient region $\mathbb{V}_n$ is the subset of $\mathbb{C}^{n-1}$ consisting of all points $(a_2, \ldots, a_n)$ corresponding to the initial coefficients of some function $f \in S$. This chapter is a detailed study of the regions $\mathbb{V}_n$ and the associated functions. Much of the material is adapted from the book of Schaeffer and Spencer [6]. The most striking phenomenon is the sharp dichotomy between the properties of interior points and boundary points of $\mathbb{V}_n$.

§11.1. Elementary Properties

For an arbitrary function

$$f(z) = z + \sum_{k=2}^{\infty} a_k z^k$$

analytic in $\mathbb{D}$, let

$$A_n = A_n(f) = (a_2, \ldots, a_n), \qquad n = 2, 3, \ldots,$$

be the corresponding point in $\mathbb{C}^{n-1}$ determined by the initial coefficients of f. Such a function f is said to *belong to* the point A_n. For each integer $n \geq 2$, the nth *coefficient region* $\mathbb{V}_n$ is defined as the set of points $A_n(f)$ which correspond to functions $f \in S$. Thus $\mathbb{V}_n$ consists of all vectors $(a_2, \ldots, a_n)$ of initial coefficients of univalent functions of class S. The set of all functions $f \in S$ which belong to a given point $A_n \in \mathbb{V}_n$ will be denoted by $S(A_n)$.

The complex m-dimensional space $\mathbb{C}^m$ will be given the usual Euclidean topology induced by the norm

$$\|z\| = \left\{ \sum_{k=1}^{m} |z_k|^2 \right\}^{1/2}, \qquad z = (z_1, \ldots, z_m).$$

For any set $E \subset \mathbb{C}^m$, the notation $\mathring{E}$ will indicate the interior, ∂E the boundary, and $\bar{E}$ the closure of E.

The first theorem characterizes the interior points of $\mathbb{V}_n$.

Theorem 11.1. *For any point $A_n \in \mathbb{V}_n$, the following are equivalent:*

(i) $A_n \in \overset{\circ}{\mathbb{V}}_n$;
(ii) *there exists a bounded function $f \in S(A_n)$;*
(iii) *there exists a function $f \in S(A_n)$ whose range is not dense in $\mathbb{C}$.*

Proof. (i) ⇒ (ii). If $A_n = (a_2, \ldots, a_n) \in \overset{\circ}{\mathbb{V}}_n$, then for some $\rho > 1$ the point $(\rho a_2, \rho^2 a_3, \ldots, \rho^{n-1} a_n)$ is in $\mathbb{V}_n$. In other words, there is a function $g(z) = \Sigma\, b_k z^k$ in S with $b_k = \rho^{k-1} a_k$ for $k = 2, \ldots, n$. Thus the dilation $f(z) = \rho g(z/\rho) = \Sigma\, c_k z^k$ is a bounded function in S with $c_k = a_k$ for $k = 2, \ldots, n$; that is, $f \in S(A_n)$.

(ii) ⇒ (iii). This is obvious.

(iii) ⇒ (i). If the range of $f \in S(A_n)$ is not dense, it must omit some disk $|w - w_0| \leq \rho$ of positive radius. Choose distinct points $\beta_1, \ldots, \beta_n$ in the interior of this disk, and consider the function

$$G(w) = w + \sum_{j=1}^{n} \frac{\alpha_j w}{w - \beta_j}, \qquad \alpha_j \in \mathbb{C}.$$

Since each function $w(w - \beta_j)^{-1}$ satisfies a Lipschitz condition, G is univalent in $|w - w_0| > \rho$ if the coefficients α_j are sufficiently small. Therefore, the composed function

$$g(z) = G(f(z)) = \sum_{k=1}^{\infty} b_k z^k$$

is analytic and univalent in $\mathbb{D}$, and $g(0) = 0$.

The coefficients b_k may be computed in terms of the parameters α_j and β_j, and the coefficients a_k of f. For $v = 1, 2, \ldots$, let

$$[f(z)]^v = \sum_{k=v}^{\infty} a_k^{(v)} z^k.$$

Thus $a_k^{(1)} = a_k$, $a_v^{(v)} = 1$, and $a_k^{(v)} = 0$ for $k < v$. Since

$$\frac{w}{w - \beta_j} = -\sum_{v=1}^{\infty} \beta_j^{-v} w^v,$$

one finds after slight manipulation

$$b_k = a_k + \sum_{v=1}^{k} a_k^{(v)} \sum_{j=1}^{n} \alpha_j \beta_j^{-v}, \qquad k = 1, 2, \ldots. \tag{1}$$

Because the points β_j are distinct, the Vandermonde matrix $(\beta_j^{-\nu})$, $j, \nu = 1, \ldots, n$, is nonsingular. The matrix $(a_k^{(\nu)})$, $k, \nu = 1, \ldots, n$, is also nonsingular, since it is triangular and has 1's along the main diagonal. Therefore, as the point $\alpha = (\alpha_1, \ldots, \alpha_n) \in \mathbb{C}^n$ varies within a ball $\|\alpha\| < r$ which ensures the univalence of G, the corresponding points $\gamma = (\gamma_1, \ldots, \gamma_n) \in \mathbb{C}^n$ defined by

$$\gamma_\nu = \sum_{j=1}^{n} \alpha_j \beta_j^{-\nu}, \qquad \nu = 1, \ldots, n, \tag{2}$$

fill a region which contains some ball $\|\gamma\| < s$. Furthermore, by suitable choice of γ in this ball, every point $\delta = (\delta_1, \ldots, \delta_n) \in \mathbb{C}^n$ in some ball $\|\delta\| < t$ is attained in the form

$$\delta_k = \sum_{\nu=1}^{k} \gamma_\nu a_k^{(\nu)}, \qquad k = 1, \ldots, n. \tag{3}$$

The equations (1), (2), and (3) may be combined to give

$$b_k = a_k + \delta_k, \qquad k = 1, 2, \ldots, n. \tag{4}$$

The above considerations show that for each prescribed point $\delta = (\delta_1, \ldots, \delta_n)$ in the ball $\|\delta\| < t$, the parameters α_j may be chosen so that $g = G \circ f$ is univalent and has coefficients b_k satisfying (4). In particular, it is possible to require that $\delta_1 = 0$ and to prescribe $\delta = (0, \delta_2, \ldots, \delta_n)$ otherwise arbitrarily in the ball $\|\delta\| < t$. In other words, each point $B_n = (b_2, \ldots, b_n) \in \mathbb{C}^{n-1}$ which satisfies $\|B_n - A_n\| < t$ must correspond to a function $g \in S$. Thus $A_n \in \mathring{\mathbb{V}}_n$, which was to be shown.

It may be remarked that the proof of the first assertion ((i) $\Rightarrow$ (ii)) actually reveals that each interior point of $\mathbb{V}_n$ corresponds to a function which maps the disk onto a region bounded by an analytic Jordan curve.

We shall see later that exactly one function $f \in S$ belongs to each boundary point of $\mathbb{V}_n$. This is a deep theorem. It is fairly easy to prove, however, that infinitely many functions belong to each interior point.

Theorem 11.2. *For each point $A_n \in \mathring{\mathbb{V}}_n$, there are infinitely many functions in $S(A_n)$.*

Proof. If $A_n \in \mathring{\mathbb{V}}_n$, then by Theorem 11.1, some bounded function f is in $S(A_n)$. Suppose $|f(z)| < M$ for $z \in \mathbb{D}$, and consider the function

$$G(w) = w + \varepsilon M^{-n} w^{n+1},$$

where ε is a complex parameter. We claim that G is univalent in $|w| < M$ if $|\varepsilon| < 1/(n+1)$. Indeed,

$$|G(w_1) - G(w_2)| = |w_1 - w_2|\left|1 + \varepsilon M^{-n}\sum_{k=0}^{n} w_1^k w_2^{n-k}\right|$$

$$\geq |w_1 - w_2|\left\{1 - |\varepsilon| M^{-n}\sum_{k=0}^{n} |w_1|^k |w_2|^{n-k}\right\} > 0$$

if $|\varepsilon| < 1/(n+1)$, $|w_1| < M$, and $|w_2| < M$, unless $w_1 = w_2$. Thus the function

$$g(z) = G(f(z)) = f(z) + \varepsilon M^{-n}[f(z)]^{n+1}$$

is univalent for each ε with $|\varepsilon| < 1/(n+1)$, and it is clear that $g \in S(A_n)$.

The next theorem gives a topological description of the set $\mathbb{V}_n$.

Theorem 11.3. *For each $n \geq 2$, the coefficient region $\mathbb{V}_n$ is simply connected and compact, and it coincides with the closure of its interior.*

Proof. It is clear that $\mathbb{V}_n$ is a bounded set. Indeed, Littlewood's theorem (Theorem 2.8) says that every point $A_n = (a_2, \ldots, a_n) \in \mathbb{V}_n$ satisfies $|a_k| < ek$ for $k = 2, \ldots, n$. That $\mathbb{V}_n$ is closed follows from the fact that S is a compact normal family. Because the identity mapping $f(z) = z$ is a bounded function in S, we may conclude from Theorem 11.1 that the origin $0 = (0, \ldots, 0) \in \mathring{\mathbb{V}}_n$. For an arbitrary function $f(z) = \sum a_k z^k$ in S, each dilation

$$\rho^{-1} f(\rho z) = \sum_{k=1}^{\infty} \rho^{k-1} a_k z^k, \qquad 0 < \rho < 1,$$

is a bounded function in S, and so

$$(\rho a_2, \rho^2 a_3, \ldots, \rho^{n-1} a_n) \in \mathring{\mathbb{V}}_n, \qquad 0 \leq \rho < 1.$$

This simple device defines a path from each point $A_n = (a_2, \ldots, a_n) \in \mathbb{V}_n$ to the origin, and it shows that each point in $\mathbb{V}_n$ can be approximated by interior points. Thus $\mathbb{V}_n$ is pathwise connected, and $\mathbb{V}_n = \overline{\mathring{\mathbb{V}}_n}$.

We shall now exhibit a homeomorphism between $\mathbb{V}_n$ and a certain starlike subset of $\mathbb{C}^{n-1}$. Given $A_n \in \mathbb{V}_n$, let $a_k = r_k e^{i\theta_k}$ and define $B_n = T(A_n)$ by

$$b_k = r_k^{1/(k-1)} e^{i\theta_k}, \qquad k = 2, \ldots, n.$$

Note that T is a homeomorphism of $\mathbb{V}_n$ onto a set $\mathbb{W}_n = T(\mathbb{V}_n)$, and that

$$T: (\rho a_2, \rho^2 a_3, \ldots, \rho^{n-1} a_n) \to \rho(b_2, \ldots, b_n), \qquad 0 \leq \rho \leq 1.$$

Thus $\mathbb{W}_n$ is starlike with respect to the origin, and so is simply connected. Because simple connectivity is a topological invariant, this shows that $\mathbb{V}_n$ is simply connected. The proof is complete.

The most general form of the nth coefficient problem for the class S is to describe the region $\mathbb{V}_n$ precisely. This is obviously more ambitious than trying to find the sharp upper bound for $|a_n|$. In view of Bieberbach's theorem (Theorem 2.2), the region $\mathbb{V}_2$ is simply the disk $|z| \leq 2$. (Rotate and dilate the Koebe function to see that the full disk is covered.) The region $\mathbb{V}_3$ is much more complicated. Schaeffer and Spencer [6] computed it and portrayed two of its three-dimensional projections with plaster models whose photographs are shown in their book. Although $\mathbb{V}_2$ is convex, $\mathbb{V}_3$ is not. More recently, Haario [1] used the Loewner formulas to obtain an independent determination of $\mathbb{V}_3$.

§11.2. Boundary Points

Using a variational method, we shall now prove that to each point A_n on the boundary of $\mathbb{V}_n$ there belongs a function $f \in S(A_n)$ which satisfies a differential equation similar to the Schiffer differential equation (§10.8) for functions maximizing $\text{Re}\{a_n\}$. As a matter of fact, this function f is unique. In other words, $S(A_n)$ consists of a single function when $A_n \in \partial\mathbb{V}_n$. The question of uniqueness will be deferred to the next section.

Here is the main result.

Theorem 11.4. *To each point $A_n \in \partial\mathbb{V}_n$ there belongs a function $f \in S(A_n)$ which maps $\mathbb{D}$ onto $\mathbb{C}$ minus a system of finitely many analytic arcs $w = w(t)$ satisfying*

$$\frac{1}{w^2}\mathscr{P}_n\left(\frac{1}{w}\right)\left(\frac{dw}{dt}\right)^2 < 0, \tag{5}$$

where $\mathscr{P}_n$ is a polynomial of the form

$$\mathscr{P}_n(\zeta) = \sum_{k=1}^{n-1} \alpha_k \zeta^k. \tag{6}$$

Furthermore, f satisfies a differential equation

$$\left[\frac{zf'(z)}{f(z)}\right]^2 \mathscr{P}_n\left(\frac{1}{f(z)}\right) = \mathscr{R}_n(z), \qquad |z| < 1, \tag{7}$$

where $\mathscr{R}_n$ *is a rational function of the form*

$$\mathscr{R}_n(z) = \beta_0 + \sum_{k=1}^{n-1} (\beta_k z^{-k} + \overline{\beta_k} z^k) \tag{8}$$

with the properties $\beta_0 > 0$, $\mathscr{R}_n(z) \geq 0$ *for all* $z \in \mathbb{T}$, *and* $\mathscr{R}_n(z) = 0$ *for some* $z \in \mathbb{T}$.

In preparation for the proof, we consider an arbitrary real-valued function $F(z) = F(z_2, \ldots, z_n)$ which has continuous first partial derivatives

$$F_k = \frac{\partial F}{\partial z_k} = \frac{1}{2}\left(\frac{\partial F}{\partial x_k} - i\frac{\partial F}{\partial y_k}\right), \qquad z_k = x_k + iy_k, \qquad k = 2, \ldots, n,$$

on some open set $\Omega \supset \mathbb{V}_n$, and which has no critical points in Ω:

$$\sum_{k=2}^{n} |F_k(z)|^2 > 0 \qquad \text{for all } z \in \Omega.$$

Since $\mathbb{V}_n$ is compact, this continuous function F will attain a maximum value on $\mathbb{V}_n$. Because F has no critical points in $\mathbb{V}_n$, it must attain this maximum on the boundary $\partial\mathbb{V}_n$. We begin with a lemma which specializes the theorem to boundary points of this type.

Lemma. *Let* F *be as above and let* $f \in S(A_n)$ *belong to a point* $A_n \in \partial\mathbb{V}_n$ *where* F *attains its maximum on* $\mathbb{V}_n$. *Then* f *satisfies the conclusions of Theorem* 11.4 *with*

$$\alpha_k = \sum_{j=k+1}^{n} F_j(A_n) a_j^{(k+1)}, \qquad k = 1, \ldots, n-1, \tag{9}$$

$$\beta_0 = \sum_{j=2}^{n} (j-1) a_j F_j(A_n), \tag{10}$$

and

$$\beta_k = \sum_{j=1}^{n-k} j a_j F_{k+j}(A_n), \qquad k = 1, \ldots, n-1. \tag{11}$$

Proof. Let $f(z) = \sum a_k z^k$ belong to a point $A_n \in \partial\mathbb{V}_n$ where F has its maximum value. Apply the method of boundary variation (Chapter 10) to obtain a family of nearby functions $f^* \in S$ of the form

$$f^*(z) = G_r(f(z)) = z + \sum_{k=2}^{\infty} a_k^* z^k,$$

where

$$G_r(w) = w + \frac{\lambda_r w^2}{w_0^2(w - w_0)} + o(r^2)$$

and w_0 is a point outside the range of f, as in §10.5. In terms of the expansion of $[f(z)]^2/[f(z) - w]$ in powers of z (see equation (19) in §10.8), we obtain

$$a_k^* = a_k - \lambda_r w_0^{-2} P_k(1/w_0) + o(r^2),$$

where P_k is a monic polynomial of degree $k - 1$ without constant term. Thus by Taylor's formula,

$$F(A_n^*) = F(A_n) - 2\,\mathrm{Re}\{\lambda_r w_0^{-2} \mathscr{P}_n(1/w_0)\} + o(r^2),$$

where

$$\mathscr{P}_n(\zeta) = \sum_{k=2}^{n} F_k(A_n) P_k(\zeta). \tag{12}$$

Since $F(A_n^*) \leq F(A_n)$, it follows that

$$\mathrm{Re}\{\lambda_r w_0^{-2} \mathscr{P}_n(1/w_0) + o(r^2)\} \geq 0.$$

Because of the hypothesis that F has no critical point in $\mathbb{V}_n$, the polynomial $\mathscr{P}_n$ does not reduce to the zero function. The analytic function $\mathscr{P}_n(1/w)$ therefore has only isolated zeros. This allows us to apply Schiffer's theorem (§10.3) to conclude that f maps $\mathbb{D}$ onto the complement of a set Γ consisting of analytic arcs $w = w(t)$ satisfying (5). Thus f is analytic except for poles on the unit circle $\mathbb{T}$, and $f(z) \neq 0$ on $\mathbb{T}$. This permits us to introduce the parametrization $w = f(e^{it})$ for the curve Γ, and we deduce from (5) that

$$\mathscr{R}_n(z) = \left[\frac{zf'(z)}{f(z)}\right]^2 \mathscr{P}_n\left(\frac{1}{f(z)}\right)$$

is real and nonnegative on $\mathbb{T}$. Actually, this function $\mathscr{R}_n$ is analytic in $\mathbb{D}$ except for a pole of order at most $n - 1$ at the origin, and it is analytic on $\mathbb{T}$ because $\mathscr{P}_n(1/f(z))$ has at least a double zero wherever f has a pole. Consequently, $\mathscr{R}_n$ can be continued analytically to Δ according to the formula $\mathscr{R}_n(z) = \overline{\mathscr{R}_n(1/\bar{z})}$. This produces a function $\mathscr{R}_n$ analytic in the extended complex plane except for poles of order $m \leq n - 1$ at the origin and at infinity. Thus $\mathscr{R}_n$ is a rational function which has, in view of its symmetry, the form (8) with β_0 real. This proves that f satisfies the differential equation (7), where $\mathscr{P}_n$ is a polynomial of the form (6) and $\mathscr{R}_n$ is a rational function of the form (8).

The coefficients α_k and β_k in $\mathscr{P}_n$ and $\mathscr{R}_n$ are easily computed in terms of F and A_n. The generating formula for the polynomials P_k gives (*cf.* equation (19) in §10.8)

$$\sum_{k=2}^{\infty} P_k(\zeta)z^k = \sum_{v=1}^{\infty} [f(z)]^{v+1}\zeta^v$$
$$= \sum_{v=1}^{\infty} \sum_{k=v+1}^{\infty} a_k^{(v+1)} z^k \zeta^v = \sum_{k=2}^{\infty} \sum_{v=1}^{k-1} a_k^{(v+1)} \zeta^v z^k.$$

Hence

$$P_k(\zeta) = \sum_{v=1}^{k-1} a_k^{(v+1)} \zeta^v, \qquad k = 2, \dots, n,$$

and a simple calculation based on the definition (12) of $\mathscr{P}_n$ leads to the formula (9) for α_k. On the other hand, a comparison with the calculations in §10.8 gives

$$\mathscr{R}_n(z) = \sum_{k=2}^{n} F_k(A_n) R_k(z),$$

where R_k is defined by equation (27) in §10.8. From this expression it is easy to derive the formulas (10) and (11) for β_k.

Because $\mathscr{R}_n(z) \geq 0$ on $\mathbb{T}$, it is clear that

$$\beta_0 = \frac{1}{2\pi} \int_0^{2\pi} \mathscr{R}_n(e^{i\theta})\, d\theta \geq 0.$$

If $\beta_0 = 0$, then $\mathscr{R}_n(z) \equiv 0$ on $\mathbb{T}$, which implies that all $\beta_k = 0, k = 1, \dots, n-1$. It would then follow from the equations (11) that $F_k(A_n) = 0$ for $k = 2, \dots, n$, contradicting the hypothesis that F has no critical points on $\mathbb{V}_n$. Thus $\beta_0 > 0$.

Finally, it is clear from the differential equation (7) that $\mathscr{R}_n(z) = 0$ at each point $z \in \mathbb{T}$ which corresponds to a finite tip of Γ, because $f'(z) = 0$ there. This completes the proof of the lemma.

Proof of Theorem. The strategy is to show that the boundary points of the type described in the lemma are dense in $\partial\mathbb{V}_n$, and to deduce the general result by continuity.

Consider the set $E \subset \partial\mathbb{V}_n$ of points where some ball disjoint from $\mathring{\mathbb{V}}_n$ meets $\mathbb{V}_n$. In other words, a point $\zeta \in \partial\mathbb{V}_n$ is said to belong to E if there exists a point $\gamma \in \mathbb{C}^{n-1}$ such that $\|z - \gamma\| \geq \|\zeta - \gamma\| > 0$ for all $z \in \mathbb{V}_n$. We claim that E is dense in $\partial\mathbb{V}_n$. To prove this, choose an arbitrary point $a \in \partial\mathbb{V}_n$. Then to each $\varepsilon > 0$ there corresponds a point $c \notin \mathbb{V}_n$ with $\|a - c\| < \varepsilon$. Let δ be the distance from c to $\mathbb{V}_n$. Obviously, $\delta \leq \varepsilon$. Because $\mathbb{V}_n$ is closed, $\delta > 0$ and

$\|b - c\| = \delta$ for some point $b \in \partial\mathbb{V}_n$. It is clear from the definition of δ that $\|z - c\| > \delta$ for all $z \in \mathring{\mathbb{V}}_n$. Thus $b \in E$, and

$$\|b - a\| \leq \|b - c\| + \|a - c\| < \delta + \varepsilon \leq 2\varepsilon.$$

This proves that E is dense in $\partial\mathbb{V}_n$.

Now observe that each $A_n \in E$ is a point where some function

$$F(z) = \|z - \gamma\|^{-2}, \qquad \gamma \notin \mathbb{V}_n,$$

attains its maximum over $\mathbb{V}_n$. Thus the lemma may be applied, for it is evident that F has continuous partial derivatives and no critical points on some neighborhood of $\mathbb{V}_n$. The conclusion is that each $f \in S(A_n)$ satisfies a differential equation of the form (7) with $\mathscr{P}_n$ and $\mathscr{R}_n$ having the respective structures (6) and (8), where the coefficients α_k and β_k are given by (9), (10), and (11). Furthermore, $\beta_0 > 0$, $\mathscr{R}_n(z) \geq 0$ on $\mathbb{T}$, and $\mathscr{R}_n(z) = 0$ for some $z \in \mathbb{T}$. After multiplying F by a suitable positive constant, we may assume that $\sum_{k=1}^{n-1} |\beta_k|^2 = 1$. This amounts to a normalization of the differential equation (7).

Now let A_n be an arbitrary point of $\partial\mathbb{V}_n$ and choose a sequence of points $\zeta_j \in E$ converging to A_n. Each function $f_j \in S(\zeta_j)$ satisfies a differential equation (7) with coefficients α_{jk} and β_{jk}, normalized so that $\sum_{k=1}^{n-1} |\beta_{jk}|^2 = 1$, $j = 1, 2, \ldots$. The equations (11) ensure that the derivatives $F_k^{(j)}(\zeta_j)$ of the corresponding functions

$$F^{(j)}(z) = c_j\|z - \gamma_j\|^{-2}$$

remain bounded as $j \to \infty$. The equations (9) then imply that the coefficients α_{jk} also remain bounded. After passing to a suitable subsequence, we may therefore assume that f_j converges uniformly on each compact subset of $\mathbb{D}$ to a function $f \in S(A_n)$, and that $\alpha_{jk} \to \alpha_k$ $(k = 1, \ldots, n)$ and $\beta_{jk} \to \beta_k$ $(k = 0, 1, \ldots, n-1)$. This function f must satisfy the differential equation (7) with $\mathscr{P}_n$ and $\mathscr{R}_n$ defined by (6) and (8). It is clear that $\mathscr{R}_n(z) \geq 0$ on $\mathbb{T}$ and that $\beta_0 > 0$, since $\sum_{k=1}^{n-1} |\beta_k|^2 = 1$. Because the unit circle $\mathbb{T}$ is compact, we are assured that $\mathscr{R}_n(z) = 0$ for some $z \in \mathbb{T}$.

Finally, an integration of the differential equation (7) shows that f has an analytic continuation across arcs of $\mathbb{T}$, which f must map onto arcs satisfying (5). Since $A_n \in \partial\mathbb{V}_n$, it follows from Theorem 11.1 that f maps $\mathbb{D}$ onto a region dense in $\mathbb{C}$. Thus the range of f is the entire plane $\mathbb{C}$ minus the system of analytic arcs which constitute the image of $\mathbb{T}$. The proof of Theorem 11.4 is now complete.

Theorems 11.1 and 11.4 have a rather surprising corollary. Suppose a function $f \in S$ maps the disk onto the complement of a nowhere analytic Jordan arc. Then each of its initial coefficient vectors A_n must lie in the interior of $\mathbb{V}_n$, and so for each n there is a bounded function in S (in fact, a mapping onto the inside of an analytic Jordan curve) with the same initial coefficients A_n.

§11.3. Canonical Differential Equation

We have shown that for each point $A_n \in \partial\mathbb{V}_n$, there exists a function $f \in S(A_n)$ which satisfies a differential equation of the form

$$\left[\frac{zf'(z)}{f(z)}\right]^2 P\left(\frac{1}{f(z)}\right) = Q(z), \qquad |z| < 1, \tag{13}$$

where

$$P(\zeta) = \sum_{k=1}^{n-1} \alpha_k \zeta^k, \qquad Q(z) = \sum_{k=1-n}^{n-1} \beta_k z^{-k}, \tag{14}$$

$\beta_0 > 0, \beta_{-k} = \overline{\beta_k}, Q(z) \geq 0$ everywhere on $\mathbb{T}$, and $Q(z) = 0$ somewhere on $\mathbb{T}$. Because $f(0) = 0$ and $f'(0) = 1$, it is clear from the structure of the equation (13) that if $m - 1$ is the largest index for which $\beta_k \neq 0$ (*i.e.*, if $\beta_{m-1} \neq 0$ but $\beta_k = 0$ for $k = m, \ldots, n-1$), then $\alpha_{m-1} = \beta_{m-1}$ and $\alpha_k = 0$ for $k = m, \ldots, n-1$. A differential equation of the form (13), where P and Q have all of the properties just indicated, is called a *$\mathcal{D}_n$-equation* of *degree m*. Because of the requirement that $\beta_0 > 0$ and $Q(z) = 0$ somewhere on $\mathbb{T}$, the degree m must lie in the range $2 \leq m \leq n$. Note that a $\mathcal{D}_n$-equation of degree m is also a $\mathcal{D}_\nu$-equation of degree m for every integer $\nu \geq m$.

If a function f is analytic near the origin and satisfies a $\mathcal{D}_n$-equation of degree m, it must have the properties $f(0) = 0$ and $[f'(0)]^{m-1} = 1$. Any such function f is called a *$\mathcal{D}_n$-function* if it is analytic in $\mathbb{D}$ and has the property $f'(0) = 1$. It can be shown with some difficulty (*cf.* Schaeffer and Spencer [6]) that every $\mathcal{D}_n$-function is univalent.

Two $\mathcal{D}_n$-functions cannot belong to the same point $A_n \in \mathbb{V}_n$ and satisfy the same $\mathcal{D}_n$-equation unless they are identical. Indeed, the $\mathcal{D}_n$-equation (13) allows the iterative calculation of all higher coefficients of f in terms of $a_2, \ldots, a_n$. In particular, to each point A_n in the set $E \subset \partial\mathbb{V}_n$ defined above (see the proof of Theorem 11.4) there belongs exactly one function $f \in S$. This is true because any two members of $S(A_n)$ will maximize the same function F and so will satisfy the same $\mathcal{D}_n$-equation, as is shown by the lemma in §11.2.

This uniqueness result can be generalized to the entire boundary of $\mathbb{V}_n$. However, it cannot be deduced as in the proof of Theorem 11.4 from the fact that E is dense in $\partial\mathbb{V}_n$. The proof rests instead on a fundamental theorem of Teichmüller [1] which is a special case of Jenkins' general coefficient theorem (see Jenkins [7, 8]).

Theorem 11.5 (Teichmüller's Theorem). *Let a function $f \in S$ map $\mathbb{D}$ onto the complement of a set Γ which consists of analytic arcs satisfying $w^{-2}P(1/w)\,dw^2 < 0$, where*

$$P(\zeta) = \alpha_1\zeta + \cdots + \alpha_{n-1}\zeta^{n-1}, \qquad \alpha_{n-1} \neq 0.$$

Let

$$\psi(w) = w + c_n w^n + c_{n+1}w^{n+1} + \cdots$$

be analytic and univalent in the complement of Γ. Then $\operatorname{Re}\{c_n\alpha_{n-1}\} \leq 0$, with equality if and only if $\psi(w) \equiv w$.

The proof of Teichmüller's theorem is beyond the scope of this book. See, for example, Schaeffer and Spencer [6], Jenkins [7], Jensen [1], or Ahlfors [1].

The following lemma is useful in the applications of Teichmüller's theorem.

Lemma. *If $f(z) = \sum a_k z^k$ and $g(z) = \sum b_k z^k$ are in the same set $S(A_{n-1})$ for some point $A_{n-1} \in \mathbb{V}_{n-1}$, then $g \circ f^{-1}$ has the form*

$$g(f^{-1}(w)) = w + (b_n - a_n)w^n + \cdots.$$

Proof. Let

$$z = f^{-1}(w) = \sum_{k=1}^{\infty} c_k w^k, \qquad c_1 = 1,$$

and

$$z^\nu = [f^{-1}(w)]^\nu = \sum_{k=\nu}^{\infty} c_k^{(\nu)} w^k, \qquad c_\nu^{(\nu)} = 1.$$

Then

$$\begin{aligned} w = f(f^{-1}(w)) &= \sum_{\nu=1}^{\infty} a_\nu \sum_{k=\nu}^{\infty} c_k^{(\nu)} w^k \\ &= \sum_{k=1}^{\infty} \sum_{\nu=1}^{k} a_\nu c_k^{(\nu)} w^k, \end{aligned}$$

and so

$$\sum_{\nu=1}^{k} a_\nu c_k^{(\nu)} = 0, \qquad k = 2, 3, \ldots. \tag{15}$$

Similarly,

$$g(f^{-1}(w)) = \sum_{k=1}^{\infty} \sum_{\nu=1}^{k} b_\nu c_k^{(\nu)} w^k.$$

Thus if $g \in S(A_{n-1})$; that is, if $b_\nu = a_\nu$ for $\nu = 2, \ldots, n-1$, then

$$g(f^{-1}(w)) = w + \sum_{k=n}^{\infty} d_k w^k,$$

where

$$\begin{aligned} d_n &= \sum_{\nu=1}^{n} b_\nu c_n^{(\nu)} = b_n c_n^{(n)} + \sum_{\nu=1}^{n} a_\nu c_n^{(\nu)} - a_n c_n^{(n)} \\ &= b_n - a_n, \end{aligned}$$

in view of (15) and the fact that $c_n^{(n)} = 1$.

We are now prepared for the uniqueness theorem.

Theorem 11.6. *To each point $A_n \in \partial\mathbb{V}_n$ there belongs only one function $f \in S(A_n)$.*

Proof. Let $A_n \in \partial\mathbb{V}_n$ and let $f \in S(A_n)$. By Theorem 11.4, this function f maps $\mathbb{D}$ onto the complement of a set Γ consisting of analytic arcs satisfying $w^{-2}P(1/w)\,dw^2 < 0$, where

$$P(\zeta) = \alpha_1\zeta + \cdots + \alpha_{m-1}\zeta^{m-1}, \qquad \alpha_{m-1} \neq 0, \quad 2 \leq m \leq n.$$

For each $g \in S$, the function $\psi = g \circ f^{-1}$ is analytic and univalent in the complement of Γ. If $g(z) = \sum b_k z^k$ and $g \in S(A_n)$, the lemma shows that ψ has the form

$$\psi(w) = g(f^{-1}(w)) = w + \sum_{k=1}^{\infty} c_k w^k,$$

where $c_k = 0$ for $k = 1, 2, \ldots, n$. In particular, $c_m = 0$ and so $\mathrm{Re}\{c_m\alpha_{m-1}\} = 0$. Thus it follows from Teichmüller's theorem that $\psi(w) \equiv w$, or $g = f$. This proves the uniqueness.

Essentially the same proof gives another result also known as Teichmüller's theorem. Recall that each function $f \in S$ which maximizes $\mathrm{Re}\{a_n\}$ must satisfy the Schiffer differential equation (see §10.8)

$$\left[\frac{zf'(z)}{f(z)}\right]^2 P_n\left(\frac{1}{f(z)}\right) = R_n(z), \tag{16}$$

where $R_n(z) \geq 0$ on $\mathbb{T}$ and $P_n(\zeta) = \zeta^{n-1} + \cdots$ is a *monic* polynomial of degree $n-1$. This equation is known to have extraneous solutions with the property that $R_n(z) \geq 0$ on $\mathbb{T}$, which nevertheless fail to maximize $\operatorname{Re}\{a_n\}$ in S. The following theorem asserts, however, that each such function does maximize $\operatorname{Re}\{a_n\}$ within the subclass of all functions in S which have the same initial coefficients.

Theorem 11.7. *Let $f \in S$ satisfy the Schiffer differential equation* (16) *for some $n \geq 3$, with $R_n(z) \geq 0$ on $\mathbb{T}$. Let $A_{n-1} = (a_2, \ldots, a_{n-1})$ be the point determined by the initial coefficients of f. Then $\operatorname{Re}\{b_n\} \leq \operatorname{Re}\{a_n\}$ for each function $g \in S(A_{n-1})$ with $g(z) = \sum b_k z^k$. Equality occurs only for $g = f$.*

Proof. The hypothesis implies that f maps the disk onto the complement of a set Γ consisting of arcs satisfying $P_n(1/w)(dw/w)^2 < 0$. For each $g \in S(A_{n-1})$, the function $\psi = g \circ f^{-1}$ is analytic and univalent on the range of f and (by the lemma) has the structure

$$\psi(w) = w + (b_n - a_n)w^n + \cdots.$$

Since $\alpha_{n-1} = 1$, it follows from Teichmüller's theorem that $\operatorname{Re}\{b_n\} \leq \operatorname{Re}\{a_n\}$, with equality only if $\psi(w) \equiv w$, or $g = f$.

The proof can be modified to yield the more general result that each section $\mathbb{V}_n(A_{n-1})$ is *strongly convex*: each supporting plane touches it at exactly one point. (Here $\mathbb{V}_n(A_k)$ denotes the set of points in $\mathbb{V}_n$ with initial coordinates fixed at $a_2, \ldots, a_k$, $k < n$.) The details are left as an exercise.

There is a generalized version of Teichmüller's theorem due to Jenkins [11, 15] which extends Theorem 11.7 to a wider class of sections. The conclusion is that for $n/2 \leq k < n$, a complex vector $(\lambda_{k+1}, \ldots, \lambda_n)$ exists such that

$$\operatorname{Re}\left\{\sum_{j=k+1}^{n} \lambda_j b_j\right\} \leq \operatorname{Re}\left\{\sum_{j=k+1}^{n} \lambda_j a_j\right\}$$

for all $g \in S(A_k)$, with equality only for $g = f$. (See also Pfluger [2]. A local version of this result was found by Duren and Schiffer [2].) As Pfluger [2] observed, this more general result implies that the sections $\mathbb{V}_n(A_k)$ are strongly convex for $n/2 \leq k < n$.

§11.4. Algebraic Functions

By definition, an *algebraic function* is one which satisfies an identity of the form $P(z, f(z)) \equiv 0$, where $P(z, w)$ is a polynomial in two variables. It is easy to see that any function f which satisfies two independent $\mathscr{D}_n$-equations must be an algebraic function. Indeed, we have only to divide one equation by the other to obtain a polynomial relation for f.

Under a small additional hypothesis, one can conclude that a univalent function which satisfies two independent $\mathscr{D}_n$-equations is not only an algebraic function, but is in fact a rational function of very special form. This surprisingly precise result is due to Schaeffer and Spencer [6].

Theorem 11.8. *Let $n-1$ be a prime number, and let $f \in S$ satisfy two independent $\mathscr{D}_n$-equations, one of degree n. Then f has the form*

$$f(z) = \frac{z}{(1-\alpha z)(1-\beta z)}, \qquad |\alpha| = |\beta| = 1. \tag{17}$$

It should be remarked that a function of the form (17) is either a rotation of the Koebe function (if $\alpha = \beta$) or a mapping of the disk onto the complement of two rays situated on a common line through the origin. In the absence of the hypothesis that $n-1$ is prime, Schaeffer and Spencer ([6], p. 153) constructed examples where f omits nonlinear arcs, and where f is not a rational function.

One part of the proof is independent of the assumption that $n-1$ is prime.

Lemma. *For $n = 2, 3, \ldots$, every rational $\mathscr{D}_n$-function has the form* (17).

Corollary. *For $n = 2, 3, \ldots$, let $f \in S$ be a rational function with $A_n(f) \in \partial \mathbb{V}_n$. Then f has the form* (17).

Proof of Lemma. A rational $\mathscr{D}_n$-function must have the form

$$f(z) = \frac{zG(z)}{H(z)},$$

where G and H are polynomials with no common factor and $G(0) = H(0) = 1$. Suppose f satisfies a $\mathscr{D}_n$-equation

$$(zw'/w)^2 P(1/w) = Q(z), \qquad w = f(z). \tag{18}$$

We may assume without loss of generality that this equation has degree n, so that P and Q are given by (14) with $\alpha_{n-1} = \beta_{n-1} \neq 0$. If $G(z_0) = 0$ for some $z_0 \in \mathbb{C}$, then $z_0 \neq 0$ and $f(z) \to 0$ as $z \to z_0$, so the left-hand side of (18) tends to infinity while the right-hand side approaches the finite limit $Q(z_0)$. This contradiction shows that G has no zeros, so $G(z) \equiv 1$. On the other hand, H is not constant because the identity function $w = z$ is obviously not a solution to (18). Similarly, no function of the form $w = z/(1-\alpha z)$ is a solution to (18), because the left-hand side would tend to zero as $z \to \infty$, while $Q(z) \to \infty$. Thus H has degree $\nu + 1 \geq 2$, and f has an expansion

$$f(z) = z/H(z) = cz^{-\nu} + dz^{-\nu-1} + \cdots, \qquad c \neq 0,$$

near infinity. When this is substituted into the $\mathscr{D}_n$-equation (18), a calculation shows that the left-hand side has

$$\nu^2 \alpha_{n-1}(z^\nu/c)^{n-1}$$

as its term of highest degree. Comparing this with Q, we conclude $\nu = 1$ and $\alpha_{n-1} = c^{n-1}\overline{\beta_{n-1}}$. Thus $|c| = 1$, since $\alpha_{n-1} = \beta_{n-1}$. This proves that H has the form

$$H(z) = (1 - \alpha z)(1 - \beta z), \qquad |\alpha\beta| = 1.$$

But then, because f is analytic in $\mathbb{D}$, the only possibility is that $|\alpha| = |\beta| = 1$. Thus f has the form (17).

Proof of Theorem. By hypothesis, the function $f \in S$ satisfies two linearly independent $\mathscr{D}_n$-equations. One is an equation (18), where P and Q are given by (14) with $\alpha_{n-1} \equiv \beta_{n-1} \neq 0$. This equation has degree n. The other is an equation

$$(zw'/w)^2 \tilde{P}(1/w) = \tilde{Q}(z), \qquad w = f(z), \tag{19}$$

where

$$\tilde{P}(\zeta) = \sum_{k=1}^{m-1} \tilde{\alpha}_k \zeta^k, \qquad \tilde{Q}(z) = \sum_{k=1-m}^{m-1} \tilde{\beta}_k z^{-k},$$

with $\tilde{\alpha}_{m-1} = \tilde{\beta}_{m-1} \neq 0$. This equation (19) has degree $m \leq n$. If $m = n$, some constant multiple of the equation (18) can be subtracted from (19) to produce

$$(zw'/w)^2 \hat{P}(1/w) = \hat{Q}(z), \qquad w = f(z), \tag{20}$$

where

$$\hat{P}(\zeta) = \sum_{k=1}^{s-1} \hat{\alpha}_k \zeta^k, \qquad \hat{Q}(z) = \sum_{k=-q}^{s-1} \hat{\beta}_k z^{-k}, \qquad 2 \leq s < n, \quad q < n,$$

and $\hat{\alpha}_{s-1} = \hat{\beta}_{s-1} \neq 0$. If $m < n$, the equation (19) already has this form and will be assumed to coincide with (20). Dividing (20) by (18), we obtain

$$\hat{P}(1/w)/P(1/w) = \hat{Q}(z)/Q(z). \tag{21}$$

After division by $\hat{\alpha}_{s-1}/\alpha_{n-1} = \hat{\beta}_{s-1}/\beta_{n-1}$, the equation (21) takes the form

$$w^{n-s}G(w) - z^{n-s}H(z), \tag{22}$$

where G and H are rational functions with $G(0) = H(0) = 1$. Extraction of $(n - s)$th roots now gives

$$\varphi(w) = e^{i\theta}\psi(z), \qquad e^{i\theta(n-s)} = 1,$$

where φ and ψ are analytic near the origin, with $\varphi(0) = \psi(0) = 0$ and $\varphi'(0) = \psi'(0) = 1$. Thus every solution to the algebraic equation (22) with the value $w = 0$ at $z = 0$ must have the form

$$w = \varphi^{-1}(e^{i\theta}\psi(z)) = b_1 z + b_2 z^2 + \cdots \tag{23}$$

in a neighborhood of the origin, with $b_1 = e^{i\theta}$ representing some $(n - s)$th root of unity.

We now take a global view and let the two $\mathscr{D}_n$-equations (18) and (19) generate a common analytic continuation $w = F(z)$ of the given function $f \in S$ beyond the unit disk. This continuation F satisfies the equation (22) and so is an algebraic function. It is analytic in the extended complex plane apart from a finite number of singular points which are either poles or algebraic branch-points.

We first show that $F(0) = 0$ on every sheet of the continuation. Indeed, if some continuation leads to a (possibly infinite) value $F(0) \neq 0$, then because F is an algebraic function it must have one of the local structures

$$F(z) = w_0 + \gamma z^{\lambda} + \cdots, \qquad w_0 \neq 0, \tag{24}$$

or

$$F(z) = \gamma z^{-\lambda} + \cdots \tag{25}$$

in a neighborhood of the origin, where $\lambda > 0$ and $\gamma \in \mathbb{C}$, $\gamma \neq 0$. Neither of these structures is compatible with the differential equation (18). For the form (24) we find

$$\left[\frac{zF'(z)}{F(z)}\right]^2 P\left(\frac{1}{F(z)}\right) = \left(\frac{\gamma\lambda}{w_0}\right)^2 P\left(\frac{1}{w_0}\right) z^{2\lambda} + \cdots = O(z^{2\lambda});$$

while the form (25) leads to

$$\left[\frac{zF'(z)}{F(z)}\right]^2 P\left(\frac{1}{F(z)}\right) = \lambda^2 \alpha_j \gamma^{-j} z^{j\lambda} + \cdots = O(z^{\lambda}),$$

where j is the smallest integer $(1 \le j \le n-1)$ for which $\alpha_j \ne 0$. Both of these conclusions contradict the fact that Q on the right-hand side of (18) has a pole at the origin. Thus $F(0) = 0$ for every continuation.

Next we show that every branch of F coincides with the original function f in some neighborhood of the origin. The function $w = F(z)$ satisfies both of the $\mathscr{D}_n$-equations (18) and (19), so it must also satisfy the algebraic equation (22). Because $F(0) = 0$, this implies that F has the form (23) near the origin, with $b_1^{n-s} = 1$. On the other hand, when the expansion (23) is inserted into the $\mathscr{D}_n$-equation (18) and the coefficients of z^{1-n} are compared, we find $b_1^{n-1} = 1$. (Recall that $\alpha_{n-1} = \beta_{n-1} \ne 0$.) Because $n-1$ is prime and $2 \le s < n$, this is consistent with the equation $b_1^{n-s} = 1$ if and only if $b_1 = 1$. The local solution (23) to the equation (22) is then uniquely determined as

$$w = F(z) = \varphi^{-1}(\psi(z)) = z + b_2 z^2 + \cdots.$$

This proves that all branches of F coincide in some neighborhood of the origin.

It now follows that F is single-valued in the entire complex plane. Suppose, on the contrary, that some pair of continuations from the origin to a point z_0 along paths C_1 and C_2 actually produce different values $F(z_0) = w_1$ and $F(z_0) = w_2$. We have just proved that the continuation along the closed path $C_1 - C_2$ must return F to its initial function element at the origin. A reversal of the part of this continuation which extends over $-C_2$ then shows that the continuation along C_2 from 0 to z_0 must produce the value $F(z_0) = w_1 \ne w_2$. This contradiction proves that F is single-valued.

Because F is single-valued, it can have no branch-points. Thus F is analytic apart from poles in the extended complex plane, and so is a rational function. According to the lemma, however, the only rational $\mathscr{D}_n$-functions are those of the form (17). This concludes the proof.

Corollary 1. *If $f \in S$ satisfies two $\mathscr{D}_n$-equations, one of degree n and the other of degree $m < n$, and if $n-1$ and $m-1$ are relatively prime, then f has the form* (17).

Proof. This is really a corollary of the proof of Theorem 11.8. The argument now shows that $b_1^{n-m} = 1$ and $b_1^{n-1} = 1$. But by hypothesis, $n-1$ and $n-m$ have no common divisor larger than 1, so $b_1 = 1$.

Corollary 2. *If $f \in S$ maximizes* $\operatorname{Re}\{a_n\}$ *and* $\operatorname{Re}\{a_m\}$, *and if $n-1$ and $m-1$ are relatively prime, then f is the Koebe function.*

Proof. Recall that every support point f maps the disk onto the complement of a single arc (see Theorem 9.5, Corollary 2). Thus the only admissible functions of the form (17) are rotations of the Koebe function. But because $n-1$ and $m-1$ are relatively prime, no nontrivial rotation of the Koebe function can give $a_n = n$ and $a_m = m$.

Corollary 2 can be strengthened dramatically. In the case of individual coefficient functionals it is not necessary to require that $n-1$ and $m-1$ be relatively prime. The following theorem is a recent discovery of Bahtin [1].

Theorem 11.9. *If $f \in S$ maximizes $\mathrm{Re}\{a_n\}$ and $\mathrm{Re}\{a_m\}$ for some pair of distinct integers $n > 2$ and $m > 2$, then f is a rotation of the Koebe function.*

Proof. According to the results of §10.8, a function $f \in S$ which maximizes $\mathrm{Re}\{a_n\}$ must satisfy the Schiffer differential equation

$$(zw'/w)^2 P_n(1/w) = R_n(z), \qquad w = f(z), \tag{26}$$

where

$$P_n(\zeta) = \sum_{j=1}^{n-1} \Lambda_n(f^{j+1})\zeta^j = \zeta^{n-1} + (n-1)a_2\zeta^{n-2} + \cdots$$

and

$$R_n(z) = (n-1)a_n + \sum_{j=1}^{n-1} (ja_j z^{j-n} + j\overline{a_j} z^{n-j}).$$

(Recall that Λ_n is the nth coefficient functional.) If f also maximizes $\mathrm{Re}\{a_m\}$, it must satisfy a second differential equation of the same form. As in the proof of Theorem 11.8, these two differential equations generate an analytic continuation of f to an algebraic function F. Each branch of F has the structure

$$w = F(z) = b_1 z + b_2 z^2 + \cdots, \qquad b_1 \neq 0, \tag{27}$$

in a neighborhood of the origin. As before, it will suffice to show that $b_1 = 1$ for each such branch. The rest of the proof is then identical to that of Theorem 11.8.

In order to show that $b_1 = 1$, we insert the expansion (27) into the equation (26). Simple calculations give

$$(zw'/w)^2 = 1 + 2b_1^{-1}b_2 z + \cdots$$

and

$$w^{-j} = b_1^{-j}z^{-j} - jb_1^{-j-1}b_2 z^{1-j} + \cdots, \qquad j = 1, 2, \ldots .$$

Thus

$$P_n(1/w) = b_1^{1-n}z^{1-n} + (n-1)b_1^{-n}(a_2 b_1^2 - b_2)z^{2-n} + \cdots,$$

and it follows from (26) that

$$b_1^{1-n}z^{1-n} + b_1^{-n}\{2b_2 + (n-1)(a_2 b_1^2 - b_2)\}z^{2-n} + \cdots \\ = z^{1-n} + 2a_2 z^{2-n} + \cdots.$$

Comparing coefficients, we obtain the relations $b_1^{n-1} = 1$ and

$$b_1^{-n}\{2b_2 + (n-1)(a_2 b_1^2 - b_2)\} = 2a_2,$$

or

$$2b_2 + (n-1)(a_2 b_1^2 - b_2) = 2a_2 b_1. \tag{28}$$

Similarly, $b_1^{m-1} = 1$ and

$$2b_2 + (m-1)(a_2 b_1^2 - b_2) = 2a_2 b_1. \tag{29}$$

Subtracting (29) from (28), we find

$$(n-m)(a_2 b_1^2 - b_2) = 0,$$

or $b_2 = a_2 b_1^2$, since $n \neq m$. Thus (28) reduces to $b_2 = a_2 b_1$, and these last two equations imply that either $a_2 = 0$ or $b_1 = 1$. However, it is already known (*cf.* Theorem 5.3) that $|a_n| < n$ for all n if $a_2 = 0$ (or more generally if $|a_2| < 0.867$). In particular, no function $f \in S$ with $a_2 = 0$ can maximize $\text{Re}\{a_n\}$ for any n. This allows us to conclude that $b_1 = 1$, which finishes the proof.

Exercises

1. Show that if $|a_2| + \cdots + |a_n| \leq 1/n$, then the polynomial

$$f(z) = z + a_2 z^2 + \cdots + a_n z^n$$

is univalent in $\mathbb{D}$. Conclude (without appeal to Theorem 11.1) that the origin is an interior point of $\mathbb{V}_n$.

2. Give an example of a $\mathcal{D}_n$-function of the form described in Theorem 11.8 with $\alpha \neq \beta$. (*Suggestion*: Show that $f(z) = z/(1-z^2)$ satisfies the Schiffer differential equation for $n = 3$.)

3. Prove that each section $\mathbb{V}_n(A_{n-1})$ is strongly convex. (See the end of §11.3.)

4. (a) Let k be the Koebe function and define

$$f(z) = \{k(z^m)\}^{1/m} = z + c_{m+1}z^{m+1} + c_{2m+1}z^{2m+1} + \cdots$$

for some integer $m \geq 2$. Let $g(z) = e^{-i\theta}f(e^{i\theta}z)$, where $e^{im\theta} = -1$. Show that although $f \in S$ and $g \in S$, their average $\frac{1}{2}(f + g) \notin S$. (*Hint*: Refer to Chapter 4, Exercise 3.)

(b) Let $A_m = (0, \ldots, 0)$ be the origin in $\mathbb{C}^{m-1}$. Show that the section $\mathbb{V}_{2m+1}(A_m)$ is not convex. (Pfluger [2].)

Appendix

Suggestions for Further Reading

There are several recent books which treat various aspects of univalent functions and related topics. The most comprehensive is Pommerenke [14], which gives an overview of the field and an excellent account of several topics virtually omitted from the present book. Milin [6] gives full details of his method and its applications. Schober [1] emphasizes extreme point theory and quasiconformal extensions. Babenko [1] gives a specialized treatment of variational methods and coefficient regions. Lebedev [7] discusses area methods and their applications. Aleksandrov [3] develops applications of the Loewner–Kufarev equation and of variational methods. Tammi [1] treats bounded univalent functions. The forthcoming book of Goodman [9] gives a leisurely introduction to the field, with emphasis on special classes of univalent and multivalent functions. The proceedings of the Instructional Conference at Durham, edited by Brannan and Clunie [1], contain some good expository articles on various subjects.

There are also several older books which are still quite valuable. Nehari [2] emphasizes the general theory of conformal mapping and has a good introductory section on univalent functions. Hayman [2] is concerned primarily with the generalization to multivalent functions but is also a source of information on univalent functions. Jenkins [7] gives an exposition of the general coefficient theorem and its many applications. Schmidt [1] and Jensen [1] give more restricted accounts of this method. The book of Schaeffer and Spencer [6] on coefficient regions is somewhat dated but is still a rich source of ideas. Goluzin [28] is an excellent reference for certain topics. In particular, it gives a coherent exposition of Goluzin's monumental contributions to the subject. The supplement for the second edition (1966), prepared by N. A. Lebedev, G. V. Kuzmina, and Yu. E. Alenicyn, brought the book up to date. An earlier monograph by Goluzin [5] contains some material not included in his later book. The books of Montel [1] and Biernacki [4] are now very much out of date but present some unusual material. Montel's book contains an interesting appendix by Cartan [1] on the possibility (and impossibility) of extending the theory of univalent functions to several complex variables.

In addition to the books, many shorter expository articles are available. Most of these articles discuss recent research and open problems. Coefficient problems are discussed in the survey articles of Hayman [6], Duren

[11], and Pfluger [7]. Schiffer [12, 13] gives a descriptive account of variational methods. Goodman [6, 8] discusses a variety of problems on univalent and multivalent functions. Duren [13] describes results on extreme points and support points. Campbell [4] surveys a variety of results on convex combinations of univalent functions. The article by Duren [12] on subordination is essentially a summary of Chapter 6 of this book. Aleksandrov [2] discusses covering problems and related geometric questions. Avhadiev and Aksentev [1] give a survey of sufficient conditions for univalence. The articles of Bernardi [1], MacGregor [6], and Goodman [8] are written at an introductory level and are accessible to beginners.

Several collections of research problems have been published. Of course, some of these problems have since been solved. The problem collections edited by Hayman [7, 8], Anderson, Barth, and Brannan [1], and Campbell, Clunie, and Hayman [1] are outgrowths of four research conferences in complex analysis held in Great Britain in 1964, 1973, 1976, and 1979. Each of the last three articles contains a report describing progress on problems posed earlier. The problem collections edited by Pommerenke [12, 13] arose from conferences at the Research Institute in Oberwolfach, Germany. The article by Goodman [6] is especially directed toward open problems.

Anyone who works in the field of univalent functions should be aware of the exhaustive bibliography prepared by Bernardi [2]. It contains an elaborate subject index.

Bibliography

Aharonov, D.

[1] A note on slit mappings. *Bull. Amer. Math. Soc.*, **75** (1969), 836–839.

[2] A generalization of a theorem of J. A. Jenkins. *Math. Z.*, **110** (1969), 218–222.

[3] On Bieberbach Eilenberg functions. *Bull. Amer. Math. Soc.*, **76** (1970), 101–104.

[4] Proof of the Bieberbach conjecture for a certain class of univalent functions. *Israel J. Math.*, **8** (1970), 103–104.

[5] *Special Topics in Univalent Functions.* Lecture Notes, Univ. of Maryland, 1971.

[6] On the Bieberbach conjecture for functions with a small second coefficient. *Israel J. Math.*, **15** (1973), 137–139.

[7] On pairs of functions and related classes. *Duke Math. J.*, **40** (1973), 669–676.

Aharonov, D. and Friedland, S.

[1] On an inequality connected with the coefficient conjecture for functions of bounded boundary rotation. *Ann. Acad. Sci. Fenn., Ser. AI*, no. 524 (1972), 14 pp.

Ahlfors, L. V.

[1] *Variational Methods in Function Theory.* Lecture notes by E. C. Schlesinger, Harvard Univ., 1953.

[2] An inequality between the coefficients a_2 and a_4 of a univalent function, in *Some Problems of Mathematics and Mechanics* (Izdat. "Nauka": Leningrad, 1970), 71–74 (in Russian) = *Amer. Math. Soc. Transl.* (2), **104** (1976), 57–60.

[3] *Conformal Invariants: Topics in Geometric Function Theory.* McGraw-Hill: New York, 1973.

[4] Sufficient conditions for quasi-conformal extension, in *Discontinuous Groups and Riemann Surfaces*, Annals of Math. Studies No. 79 (Princeton University Press, 1974), 23–29.

[5] *Complex Analysis*, 3rd ed. McGraw-Hill: New York, 1979.

Ahlfors, L. and Weill, G.

[1] A uniqueness theorem for Beltrami equations. *Proc. Amer. Math. Soc.*, **13** (1962), 975–978.

Aleksandrov, I. A.

[1] Boundary values of the functional $J = J(f, \bar{f}, f', \overline{f'})$ over the class of holomorphic univalent functions in the disk. *Sibirsk. Mat. Ž.*, **4** (1963), 17–31 (in Russian).

[2] Geometric properties of schlicht functions. *Trudy Tomsk. Gos. Univ. Ser. Meh.-Mat.*, **175** (1964), no. 2, 29–38 (in Russian).

[3] *Parametric Extensions in the Theory of Univalent Functions.* Izdat. "Nauka": Moscow, 1976 (in Russian).

[4] On a case of integration of the Loewner differential equation. *Sibirsk. Mat. Ž.*, **22** (1981), no. 2 (126), 207–209 (in Russian).

Aleksandrov, I. A. and Gutljanskii, V. Ja.

[1] On the coefficient problem in the theory of univalent functions. *Dokl. Akad. Nauk*, **188** (1969), 266–268 (in Russian) = *Soviet Math. Dokl.*, **10** (1969), 1091–1094.

Aleksandrov, I. A. and Kopanev, S. A.

[1] The region of values of the derivative in the class of holomorphic univalent functions. *Ukrain. Mat. Ž.*, **22** (1970), 660–664 (in Russian) = *Ukrainian Math. J.*, **22** (1970), 565–569.

Alenicyn, Yu. E.

[1] On the theory of univalent functions and Bieberbach–Eilenberg functions. *Dokl. Akad. Nauk SSSR*, **109** (1956), 247–249 (in Russian).

Alexander, J. W.
[1] Functions which map the interior of the unit circle upon simple regions. *Ann. of Math.*, **17** (1915–1916), 12–22.
Anderson, J. M., Barth, K. F., and Brannan, D. A.
[1] Research problems in complex analysis. *Bull. London Math. Soc.*, **9** (1977), 129–162.
Avhadiev, F. G. and Aksentev, L. A.
[1] Basic results on sufficient conditions for the univalence of analytic functions. *Uspehi Mat. Nauk*, **30** (1975), no. 4 (184), 3–60 (in Russian) = *Russian Math. Surveys*, **4** (1975), 1–64.
Babenko, K. I.
[1] *On the Theory of Extremal Problems for Univalent Functions of Class S*. Proc. Steklov Inst. Math. No. 101 (1972); English transl., Amer. Math. Soc., 1975.
Baernstein, A.
[1] Integral means, univalent functions and circular symmetrization. *Acta Math.*, **133** (1974), 139–169.
[2] Univalence and bounded mean oscillation. *Michigan Math. J.*, **23** (1976), 217–223.
Baernstein, A. and Brown, J. E.
[1] Integral means of derivatives of monotone slit mappings. *Comment. Math. Helv.*, **57** (1982), 331–348.
Baernstein, A. and Schober, G.
[1] Estimates for inverse coefficients of univalent functions from integral means. *Israel J. Math.*, **36** (1980), 75–82.
Bahtin, A. K.
[1] Some properties of functions of class S. *Ukrain. Mat. Ž.*, **33** (1981), 154–159 (in Russian) = *Ukrainian Math. J.*, **33** (1981), 122–126.
Baranova, V. A.
[1] An estimate of the coefficient c_4 of univalent functions depending on $|c_2|$. *Mat. Zametki*, **12** (1972), 127–130 (in Russian) = *Math. Notes*, **12** (1972), 510–512.
Bazilevich, I. E.
[1] Zum Koeffizientenproblem der schlichten Funktionen. *Mat. Sb.*, **1** (43) (1936), 211–228.
[2] Sur les théorèmes de Koebe–Bieberbach. *Mat. Sb.*, **1** (43) (1936), 283–292.
[3] Supplement to the papers "Zum Koeffizientenproblem der schlichten Funktionen" and "Sur les théorèmes de Koebe–Bieberbach". *Mat. Sb.*, **2** (44) (1937), 689–698 (in Russian).
[4] On distortion theorems and coefficients of univalent functions. *Mat. Sb.*, **28** (70) (1951), 147–164 (in Russian).
[5] On distortion theorems in the theory of univalent functions. *Mat. Sb.*, **28** (70) (1951), 283–292 (in Russian).
[6] On a case of integrability by quadratures of the equation of Loewner–Kufarev. *Mat. Sb.*, **37** (79) (1955), 471–476 (in Russian).
[7] Asymptotic property of the derivatives of a class of functions regular in the disk, in *Studies on Modern Problems in the Theory of Functions of a Complex Variable* (Gos. Izdat. Fiz.-Mat. Lit.: Moscow, 1961), 216–219 (in Russian).
[8] Generalization of an integral formula for a subclass of univalent functions. *Mat. Sb.*, **64** (106) (1964), 628–630 (in Russian).
[9] Coefficient dispersion of univalent functions. *Mat. Sb.*, **68** (110) (1965), 549–560 (in Russian) = *Amer. Math. Soc. Transl.* (2), **71** (1968), 168–180.
[10] On a univalence criterion for regular functions and the dispersion of their coefficients. *Mat. Sb.*, **74** (116) (1967), 133–146 (in Russian) = *Math. USSR-Sb.*, **3** (1967), 123–137.
[11] Supplement to the article "On a univalence criterion for regular functions", *Mat. Sb.*, **84** (126) (1971), 630–632 (in Russian) = *Math. USSR-Sb.*, **13** (1971), 626–630.
Becker, J.
[1] Löwnersche Differentialgleichung und quasikonform fortsetzbare schlichte Funktionen. *J. Reine Angew. Math.*, **255** (1972), 23–43.
[2] Löwnersche Differentialgleichung und Schlichtheitskriterien. *Math. Ann.*, **202** (1973), 321–335.
[3] Über die Lösungsstruktur einer Differentialgleichung in der konformen Abbildung. *J. Reine Angew. Math.*, **285** (1976), 66–74.
Bernardi, S. D.
[1] A survey of the development of the theory of schlicht functions. *Duke Math. J.*, **19** (1952), 263–287.
[2] *Bibliography of Schlicht Functions*. Courant Institute of Math. Sciences, New York Univ., 1966; Part II, *ibid.*, 1977. (Reprinted by Mariner Publishing Co.: Tampa, Florida, 1983; Part III added.)

Beurling, A.
[1] Ensembles exceptionnels. *Acta Math.*, **72** (1940), 1–13.
Bieberbach, L.
[1] Über einen Satz des Herrn Carathéodory. *Nachr. Königl. Ges. Wiss. Göttingen, Math.-Phys. Kl.*, 1913, 552–560.
[2] Über einige Extremalprobleme im Gebiete der konformen Abbildung. *Math. Ann.*, **77** (1915–1916), 153–172.
[3] Über die Koeffizienten derjenigen Potenzreihen, welche eine schlichte Abbildung des Einheitskreises vermitteln. *S.-B. Preuss. Akad. Wiss.*, 1916, 940–955.
[4] Aufstellung und Beweis des Drehungssatzes für schlichte konforme Abbildungen. *Math. Z.*, **4** (1919), 295–305.
[5] *Lehrbuch der Funktionentheorie*. Band I: Elemente der Funktionentheorie (B. G. Teubner: Leipzig, 1921); Band II: Moderne Funktionentheorie (Zweite Auflage, B. G. Teubner: Leipzig, 1931). (Reprinted by Johnson Reprint Corp.: New York, 1968.)
Bielecki, A., Krzyż, J., and Lewandowski, Z.
[1] On typically-real functions with a preassigned second coefficient. *Bull. Acad. Polon. Sci.*, **10** (1962), 205–208.
Bielecki, A. and Lewandowski, Z.
[1] Sur un théorème concernant les fonctions univalentes linéairement accessibles de M. Biernacki. *Ann. Polon. Math.*, **12** (1962), 61–63.
Biernacki, M.
[1] Sur quelques majorantes de la théorie des fonctions univalentes. *C. R. Acad. Sci. Paris*, **201** (1935), 256–258.
[2] Sur les fonctions univalentes. *Mathematica (Cluj)*, **12** (1936), 49–64.
[3] Sur la représentation conforme des domaines linéairement accessibles. *Prace Mat.-Fiz.*, **44** (1936), 293–314.
[4] *Les fonctions multivalentes*. Hermann: Paris, 1938.
[5] Sur les fonctions en moyenne multivalentes. *Bull. Soc. Math. France*, **70** (1946), 51–76.
[6] Sur les coefficients tayloriens des fonctions univalentes. *Bull. Acad. Polon. Sci.*, **4** (1956), 5–8.
[7] Sur l'intégrale des fonctions univalentes. *Bull. Acad. Polon. Sci.*, **8** (1960), 29–34.
Bombieri, E.
[1] Sul problema di Bieberbach per le funzioni univalenti. *Atti Accad. Naz. Lincei Rend. Cl. Sci. Fis. Mat. Natur.*, **35** (1963), 469–471.
[2] On functions which are regular and univalent in a half-plane. *Proc. London Math. Soc.*, **14A** (1964), 47–50.
[3] Sulla seconda variazione della funzione di Koebe. *Boll. Un. Mat. Ital.*, **22** (1967), 25–32.
[4] On the local maximum property of the Koebe function. *Invent. Math.*, **4** (1967), 26–67.
[5] A geometric approach to some coefficient inequalities for univalent functions. *Ann. Scuola Norm. Sup. Pisa*, **22** (1968), 377–397.
Brannan, D. A.
[1] Coefficient regions for univalent polynomials of small degree. *Mathematika*, **14** (1967), 165–169.
[2] On functions of bounded boundary rotation, I. *Proc. Edinburgh Math. Soc.*, **16** (1969), 339–347.
[3] On functions of bounded boundary rotation, II. *Bull. London Math. Soc.*, **1** (1969), 321–322.
[4] On univalent polynomials. *Glasgow Math. J.*, **11** (1970), 102–107.
[5] On coefficient problems for certain power series, in *Proceedings of the Symposium on Complex Analysis, Canterbury, 1973* (edited by J. Clunie and W. K. Hayman), London Math. Soc. Lecture Note Series, no. 12 (Cambridge University Press, 1974), 17–27.
Brannan, D. A. and Brickman, L.
[1] Coefficient regions for starlike polynomials. *Ann. Univ. Mariae Curie-Skłodowska*, **29** (1975), 15–21.
Brannan, D. A. and Clunie, J. G. (editors)
[1] *Aspects of Contemporary Complex Analysis*. Academic Press: London, 1980.
Brannan, D. A., Clunie, J. G., and Kirwan, W. E.
[1] On the coefficient problem for functions of bounded boundary rotation. *Ann. Acad. Sci. Fenn., Ser. AI*, no. 523 (1973), 18 pp.
Brannan, D. A. and Kirwan, W. E.
[1] A covering theorem for typically real functions. *Glasgow Math. J.*, **10** (1969), 153–155.

Brickman, L.

[1] Extreme points of the set of univalent functions. *Bull. Amer. Math. Soc.*, **76** (1970), 372–374.

[2] Φ-like analytic functions, I. *Bull. Amer. Math. Soc.*, **79** (1973), 555–558.

Brickman, L., Hallenbeck, D. J., MacGregor, T. H., and Wilken, D. R.

[1] Convex hulls and extreme points of families of starlike and convex mappings. *Trans. Amer. Math. Soc.*, **185** (1973), 413–428.

Brickman, L. and Leung, Y. J.

[1] Exposed points of the set of univalent functions. *J. London Math. Soc.*, to appear.

Brickman, L., Leung, Y. J., and Wilken, D. R.

[1] On extreme points and support points of the class *S*. *Ann. Univ. Mariae Curie-Skłodowska*, to appear.

Brickman, L., MacGregor, T. H., and Wilken, D. R.

[1] Convex hulls of some classical families of univalent functions. *Trans. Amer. Math. Soc.*, **156** (1971), 91–107.

Brickman, L. and Wilken, D.

[1] Support points of the set of univalent functions. *Proc. Amer. Math. Soc.*, **42** (1974), 523–528.

Brown, J. E.

[1] Geometric properties of a class of support points of univalent functions. *Trans. Amer. Math. Soc.*, **256** (1979), 371–382.

[2] Univalent functions maximizing $\mathrm{Re}\{a_3 + \lambda a_2\}$. *Illinois J. Math.*, **25** (1981), 446–454.

[3] Derivatives of close-to-convex functions, integral means and bounded mean oscillation. *Math. Z.*, **178** (1981), 353–358.

de Bruijn, N. G.

[1] Ein Satz über schlichte Funktionen. *Nederl. Akad. Wetensch. Proc.*, **44** (1941), 47–49 = *Indag. Math.*, **3** (1941), 8–10.

Bshouty, D. (Bishouty, D. H.)

[1] The Bieberbach conjecture for univalent functions with small second coefficients. *Math. Z.*, **149** (1976), 183–187.

[2] A note on Hadamard products of univalent functions. *Proc. Amer. Math. Soc.*, **80** (1980), 271–272.

[3] The Bieberbach conjecture for restricted initial coefficients. *Math. Z.*, **182** (1983), 149–158.

Caccioppoli, R.

[1] Sui funzionali lineari nel campo delle funzioni analitiche. *Atti Accad. Naz. Lincei Rend. Cl. Sci. Fis. Mat. Natur.*, **13** (1931), 263–266.

Campbell, D. M.

[1] Majorization–subordination theorems for locally univalent functions. *Bull. Amer. Math. Soc.*, **78** (1972), 535–538.

[2] Majorization–subordination theorems for locally univalent functions, II. *Canad. J. Math.*, **25** (1973), 420–425.

[3] Majorization–subordination theorems for locally univalent functions, III. *Trans. Amer. Math. Soc.*, **198** (1974), 297–306.

[4] A survey of properties of the convex combination of univalent functions. *Rocky Mountain J. Math.*, **5** (1975), 475–492.

Campbell, D. M., Clunie, J. G., and Hayman, W. K. (editors)

[1] Research problems in complex analysis, in *Aspects of Contemporary Complex Analysis* (edited by D. A. Brannan and J. G. Clunie; Academic Press: London, 1980), 527–572.

Campbell, D. M. and Lamoreaux, J.

[1] Continua in the plane with limit directions ± 1 are line segments, in *Complex Analysis: Proceedings of the S.U.N.Y. Brockport Conference* (Marcel Dekker, Inc.: New York, 1978), 49–52.

Campbell, D. M. and Singh, V.

[1] Valence properties of the solution of a differential equation. *Pacific J. Math.*, **84** (1979), 29–33.

Carathéodory, C.

[1] Über den Variabilitätsbereich der Fourier'schen Konstanten von positiven harmonischen Funktionen. *Rend. Circ. Mat. Palermo*, **32** (1911), 193–217.

[2] Untersuchungen über die konformen Abbildungen von festen und veränderlichen Gebieten. *Math. Ann.*, **72** (1912), 107–144.

[3] Zur Ränderzuordnung bei konformer Abbildung. *Nachr. Königl. Ges. Wiss. Göttingen, Math.-Phys. Kl.*, 1913, 509–518.

Cartan, H.

[1] Sur la possibilité d'étendre aux fonctions de plusieurs variables complexes la théorie des fonctions univalentes, in *Leçons sur les fonctions univalentes ou multivalentes* (P. Montel; Gauthier-Villars: Paris, 1933), appendix.

Chang, A., Schiffer, M. M., and Schober, G.

[1] On the second variation for univalent functions. *J. Analyse Math.*, **40** (1981), 203–238.

Charzyński, Z. and Schiffer, M.

[1] A new proof of the Bieberbach conjecture for the fourth coefficient. *Arch. Rational Mech. Anal.*, **5** (1960), 187–193.

[2] A geometric proof of the Bieberbach conjecture for the fourth coefficient. *Scripta Math.*, **25** (1960), 173–181.

Clunie, J.

[1] On schlicht functions. *Ann. of Math.*, **69** (1959), 511–519.

[2] On meromorphic schlicht functions. *J. London Math. Soc.*, **34** (1959), 215–216.

Clunie, J. and Duren, P. L.

[1] Addendum: An arclength problem for close-to-convex functions. *J. London Math. Soc.*, **41** (1966), 181–182.

Clunie, J. and Keogh, F. R.

[1] On starlike and convex schlicht functions. *J. London Math. Soc.*, **35** (1960), 229–233.

Clunie, J. and Pommerenke, Ch.

[1] On the coefficients of close-to-convex univalent functions. *J. London Math. Soc.*, **41** (1966), 161–165.

[2] On the coefficients of univalent functions. *Michigan Math. J.*, **14** (1967), 71–78.

Collingwood, E. F. and Lohwater, A. J.

[1] *The Theory of Cluster Sets*. Cambridge University Press: London and New York, 1966.

Coonce, H. B.

[1] A variational method for functions of bounded boundary rotation. *Trans. Amer. Math. Soc.*, **157** (1971), 39–51.

Courant, R.

[1] *Dirichlet's Principle, Conformal Mapping, and Minimal Surfaces* (with an appendix by M. Schiffer). Interscience: New York, 1950.

Cowling, V. F. and Royster, W. C.

[1] Domains of variability for univalent polynomials. *Proc. Amer. Math. Soc.*, **19** (1968), 767–772.

Crum, M. M.

[1] A property of schlicht functions. *J. London Math. Soc.*, **31** (1956), 493–494.

Curtiss, J. H.

[1] Faber polynomials and the Faber series. *Amer. Math. Monthly*, **78** (1971), 577–596.

Dieudonné, J.

[1] Sur le rayon d'univalence des polynômes. *C. R. Acad. Sci. Paris*, **192** (1931), 79–81.

[2] Sur les fonctions univalentes. *C. R. Acad. Sci. Paris*, **192** (1931), 1148–1150.

[3] Recherches sur quelques problèmes relatifs aux polynômes et aux fonctions bornées d'une variable complexe. *Ann. Sci. École Norm. Sup.*, **48** (1931), 247–358.

Dunford, N. and Schwartz, J. T.

[1] *Linear Operators*, part I. Interscience: New York, 1958.

Duren, P. L.

[1] Coefficient estimates for univalent functions. *Proc. Amer. Math. Soc.*, **13** (1962), 168–169.

[2] Distortion in certain conformal mappings of an annulus. *Michigan Math. J.*, **10** (1963), 431–441.

[3] Two inequalities involving elliptic functions. *Amer. Math. Monthly*, **70** (1963), 650–651.

[4] An arclength problem for close-to-convex functions. *J. London Math. Soc.*, **39** (1964), 757–761.

[5] On the Marx conjecture for starlike functions. *Trans. Amer. Math. Soc.*, **118** (1965), 331–337.

[6] *Theory of H^p Spaces*. Academic Press: New York, 1970.

[7] Coefficients of meromorphic schlicht functions. *Proc. Amer. Math. Soc.*, **28** (1971), 169–172.

[8] Estimation of coefficients of univalent functions by a Tauberian remainder theorem. *J. London Math. Soc.*, **8** (1974), 279–282.
[9] Asymptotic behavior of coefficients of univalent functions, in *Advances in Complex Function Theory, Maryland, 1973/74*, Lecture Notes in Math. No. 505 (Springer-Verlag, 1976), 17–23.
[10] Applications of the Garabedian–Schiffer inequality. *J. Analyse Math.*, **30** (1976), 141–149.
[11] Coefficients of univalent functions. *Bull. Amer. Math. Soc.*, **83** (1977), 891–911.
[12] Subordination, in *Complex Analysis, Kentucky, 1976*, Lecture Notes in Math. No. 599 (Springer-Verlag, 1977), 22–29.
[13] Extreme points of spaces of univalent functions, in *Linear Spaces and Approximation* (Birkhäuser Verlag: Basel, 1978), 471–477.
[14] Successive coefficients of univalent functions. *J. London Math. Soc.*, **19** (1979), 448–450.
[15] Arcs omitted by support points of univalent functions. *Comment. Math. Helv.*, **56** (1981), 352–365.

Duren, P. L. and Lehto, O.
[1] Schwarzian derivatives and homeomorphic extensions. *Ann. Acad. Sci. Fenn., Ser. AI*, no. 477 (1970), 11 pp.

Duren, P. L. and Leung, Y. J.
[1] Logarithmic coefficients of univalent functions. *J. Analyse Math.*, **36** (1979), 36–43.

Duren, P. L., Leung, Y. J., and Schiffer, M. M.
[1] Support points with maximum radial angle. *Complex Variables: Theory and Application*, **1** (1983), 263–277.

Duren, P. L. and McLaughlin, R.
[1] Two-slit mappings and the Marx conjecture. *Michigan Math. J.*, **19** (1972), 267–273.

Duren, P. L. and Schiffer, M.
[1] A variational method for functions schlicht in an annulus. *Arch. Rational Mech. Anal.*, **9** (1962), 260–272.
[2] The theory of the second variation in extremum problems for univalent functions. *J. Analyse Math.*, **10** (1962–63), 193–252.

Duren, P. L. and Schober, G.
[1] On a class of schlicht functions. *Michigan Math. J.*, **18** (1971), 353–356.
[2] Nonvanishing univalent functions. *Math. Z.*, **170** (1980), 195–216.
[3] Nonvanishing univalent functions, II. *Ann. Univ. Mariae Curie-Skłodowska*, to appear.

Duren, P. L., Shapiro, H. S., and Shields, A. L.
[1] Singular measures and domains not of Smirnov type. *Duke Math. J.*, **33** (1966), 247–254.

Dvořák, O.
[1] O funkcích prostých. *Časopis Pěst. Mat.*, **63** (1934), 9–16.

Ehrig, G.
[1] The Bieberbach conjecture for univalent functions with restricted second coefficients. *J. London Math. Soc.*, **8** (1974), 355–360.
[2] Coefficient estimates concerning the Bieberbach conjecture. *Math. Z.*, **140** (1974), 111–126.

Eilenberg, S.
[1] Sur quelques propriétés topologiques de la surface de sphère. *Fund. Math.*, **25** (1935), 267–272.

Eke, B. G.
[1] Remarks on Ahlfors' distortion theorem. *J. Analyse Math.*, **19** (1967), 97–134.
[2] The asymptotic behaviour of areally mean valent functions. *J. Analyse Math.*, **20** (1967), 147–212.

Epstein, B. and Schoenberg, I. J.
[1] On a conjecture concerning schlicht functions. *Bull. Amer. Math. Soc.*, **65** (1959), 273–275.

Essén, M. and Keogh, F. R.
[1] The Schwarzian derivative and estimates of functions analytic in the unit disc. *Math. Proc. Cambridge Philos. Soc.*, **78** (1975), 501–511.

Faber, G.
[1] Neuer Beweis eines Koebe–Bieberbachschen Satzes über konforme Abbildung. *S.-B. Bayer. Akad. Wiss. München*, 1916, 39–42.
[2] Über Potentialtheorie und konforme Abbildung. *S.-B. Bayer. Akad. Wiss. München*, 1920, 49–64.

Fekete, M. and Szegö, G.
[1] Eine Bemerkung über ungerade schlichte Funktionen. *J. London Math. Soc.*, **8** (1933), 85–89.

Feng, J. and MacGregor, T. H.
[1] Estimates on integral means of the derivatives of univalent functions. *J. Analyse Math.*, **29** (1976), 203–231.
FitzGerald, C. H.
[1] Quadratic inequalities and coefficient estimates for schlicht functions. *Arch. Rational Mech. Anal.*, **46** (1972), 356–368.
Friedland, S.
[1] On a conjecture of Robertson. *Arch. Rational Mech. Anal.*, **37** (1970), 255–261.
Friedland, S. and Schiffer, M.
[1] On coefficient regions of univalent functions. *J. Analyse Math.*, **31** (1977), 125–168.
Garabedian, P. R.
[1] Inequalities for the fifth coefficient. *Comm. Pure Appl. Math.*, **19** (1966), 199–214.
[2] An extension of Grunsky's inequalities bearing on the Bieberbach conjecture. *J. Analyse Math.*, **18** (1967), 81–97.
Garabedian, P. R., Ross, G. G., and Schiffer, M.
[1] On the Bieberbach conjecture for even *n*. *J. Math. Mech.*, **14** (1965), 975–989.
Garabedian, P. R. and Schiffer, M.
[1] A proof of the Bieberbach conjecture for the fourth coefficient. *J. Rational Mech. Anal.*, **4** (1955), 427–465.
[2] A coefficient inequality for schlicht functions. *Ann. of Math.*, **61** (1955), 116–136.
[3] The local maximum theorem for the coefficients of univalent functions. *Arch. Rational Mech. Anal.*, **26** (1967), 1–32.
Gehring, F. W. and Pommerenke, Ch.
[1] On the Nehari univalence criterion and quasidisks, to appear.
Gelfer, S. A.
[1] On the class of regular functions which do not take on any pair of values *w* and $-w$. *Mat. Sb.*, **19** (61) (1946), 33–46 (in Russian).
Goluzin, G. M.
[1] On distortion theorems in the theory of conformal mappings. *Mat. Sb.*, **1** (43) (1936), 127–135 (in Russian).
[2] Sur les théorèmes de rotation dans la théorie des fonctions univalentes. *Mat. Sb.*, **1** (43) (1936), 293–296.
[3] Some bounds on the coefficients of univalent functions. *Mat. Sb.*, **3** (45) (1938), 321–330 (in Russian).
[4] On the theory of univalent functions. *Mat. Sb.*, **6** (48) (1939), 383–388 (in Russian).
[5] Interior problems of the theory of schlicht functions. *Uspehi Mat. Nauk*, **6** (1939), 26–89; English transl. by T. C. Doyle, A. C. Schaeffer, and D. C. Spencer (O. N. R.: Washington, 1947); reprinted by University Microfilms, Ann Arbor, Mich., 1963.
[6] On *p*-valent functions. *Mat. Sb.*, **8** (50) (1940), 277–284 (in Russian).
[7] On the coefficients of univalent functions. *Mat. Sb.*, **12** (52) (1943), 40–47 (in Russian).
[8] On the theory of univalent functions. *Mat. Sb.*, **12** (52) (1943), 48–55 (in Russian).
[9] On the theory of univalent functions. *Mat. Sb.*, **18** (60) (1946), 167–179 (in Russian).
[10] On distortion theorems in the theory of conformal mapping. *Mat. Sb.*, **18** (60) (1946), 379–389 (in Russian).
[11] On distortion theorems in the theory of univalent functions. *Mat. Sb.*, **18** (60) (1946), 879–890 (in Russian).
[12] On distortion theorems and coefficients of univalent functions. *Mat. Sb.*, **19** (61) (1946), 183–202 (in Russian).
[13] A method of variation in conformal mapping, I. *Mat. Sb.*, **19** (61) (1946), 203–236 (in Russian).
[14] A method of variation in conformal mapping, II. *Mat. Sb.*, **21** (63) (1947), 83–117 (in Russian).
[15] A method of variation in conformal mapping, III. *Mat. Sb.*, **21** (63) (1947), 119–132 (in Russian).
[16] On the coefficients of univalent functions. *Mat. Sb.*, **22** (64) (1948), 373–380 (in Russian).
[17] On distortion theorems and coefficients of univalent functions. *Mat. Sb.*, **23** (65) (1948), 353–360 (in Russian).
[18] Some questions in the theory of univalent functions. *Trudy Mat. Inst. Steklov.*, **27** (1949), 1–112 (in Russian).
[19] On typically real functions. *Mat. Sb.*, **27** (69) (1950), 201–218 (in Russian).

[20] On subordinate univalent functions. *Trudy Mat. Inst. Steklov.*, **38** (1951), 68–71 (in Russian).
[21] On the theory of univalent functions. *Mat. Sb.*, **28** (70) (1951), 351–358 (in Russian).
[22] On majorants of subordinate analytic functions, I. *Mat. Sb.*, **29** (71) (1951), 209–224 (in Russian).
[23] On majorants of subordinate analytic functions, II. *Mat. Sb.*, **29** (71) (1951), 593–602 (in Russian).
[24] On the problem of coefficients of univalent functions. *Dokl. Akad. Nauk SSSR*, **81** (1951), 721–723 (in Russian).
[25] On the theory of univalent functions. *Mat. Sb.*, **28** (70) (1951), 351–358 (in Russian).
[26] On the theory of univalent functions. *Mat. Sb.*, **29** (71) (1951), 197–208 (in Russian).
[27] A method of variation in conformal mapping, IV. *Mat. Sb.*, **29** (71) (1951), 455–468 (in Russian).
[28] *Geometric Theory of Functions of a Complex Variable.* Moscow, 1952; German transl., Deutscher Verlag: Berlin, 1957; 2nd ed., Izdat. "Nauka": Moscow, 1966; English transl., Amer. Math. Soc., 1969.

Goodman, A. W.
[1] The rotation theorem for starlike univalent functions. *Proc. Amer. Math. Soc.*, **4** (1953), 278–286.
[2] Almost bounded functions. *Trans. Amer. Math. Soc.*, **78** (1955), 82–97.
[3] Univalent functions and nonanalytic curves. *Proc. Amer. Math. Soc.*, **8** (1957), 598–601.
[4] Analytic functions that take values in a convex region. *Proc. Amer. Math. Soc.*, **14** (1963), 60–64.
[5] The valence of sums and products. *Canad. J. Math.*, **20** (1968), 1173–1177.
[6] Open problems on univalent and multivalent functions. *Bull. Amer. Math. Soc.*, **74** (1968), 1035–1050.
[7] The domain covered by a typically-real function. *Proc. Amer. Math. Soc.*, **64** (1977), 233–237.
[8] An invitation to the study of univalent and multivalent functions. *Internat. J. Math. Math. Sci.*, **2** (1979), 163–186.
[9] *Univalent Functions*, Vols. I and II. Mariner Publishing Co.: Tampa, Florida, 1983.

Goodman, G. S.
[1] On the determination of univalent functions with prescribed initial coefficients. *Arch. Rational Mech. Anal.*, **24** (1967), 78–81.

Gorjaĭnov, V. V.
[1] Some properties of the solutions of the Loewner–Kufarev equation. *Dopovidi Akad. Nauk Ukrain. SSR, Ser. A*, 1978, 206–209 (in Ukrainian).
[2] On the parametric method in the theory of univalent functions. *Mat. Zametki*, **27** (1980), 559–568 (in Russian) = *Math. Notes*, **27** (1980), 275–279.

Grad, A.
[1] The region of values of the derivative of a schlicht function, in *Coefficient Regions for Schlicht Functions* (A. C. Schaeffer and D. C. Spencer; Amer. Math. Soc. Colloq. Publ., vol. 35, 1950), chapter xv.

Grinspan, A. Z.
[1] Logarithmic coefficients of functions in the class *S*. *Sibirsk. Mat. Ž.*, **13** (1972), 1145–1157 (in Russian) = *Siberian Math. J.*, **13** (1972), 793–801.
[2] Improved bounds for the difference of the moduli of adjacent coefficients of univalent functions, in *Some Questions in the Modern Theory of Functions* (Sib. Inst. Mat.: Novosibirsk, 1976), 41–45 (in Russian).
[3] On the Taylor coefficients of certain classes of univalent functions, in *Metric Questions in the Theory of Functions* (edited by G. D. Suvorov, "Naukova Dumka": Kiev, 1980), 28–32 (in Russian).
[4] On coefficients of powers of univalent functions. *Sibirsk. Mat. Ž.*, **22** (1981), no. 4 (128), 88–93 (in Russian) = *Siberian Math. J.*, **22** (1981), 551–555.

Gronwall, T. H.
[1] Some remarks on conformal representation. *Ann. of Math.*, **16** (1914–1915), 72–76.
[2] Sur la déformation dans la représentation conforme. *C. R. Acad. Sci. Paris*, **162** (1916), 249–252.
[3] Sur la déformation dans la représentation conforme sous des conditions restrictives. *C. R. Acad. Sci. Paris*, **162** (1916), 316–318.

[4] On the distortion in conformal mapping when the second coefficient in the mapping function has an assigned value. *Proc. Nat. Acad. Sci. U.S.A.*, **6** (1920), 300–302.

Grötzsch, H.

[1] Über einige Extremalprobleme der konformen Abbildung, I, II. *S.-B. Sächs. Akad. Wiss. Leipzig Math.-Natur. Kl.*, **80** (1928), 367–376; 497–502.

[2] Über die Verzerrung bei schlichter konformer Abbildung mehrfach zusammenhängender Bereiche, I, II, III. *S.-B. Sächs. Akad. Wiss. Leipzig Math.-Natur. Kl.*, **81** (1929), 38–47; 217–221; **83** (1931), 283–297.

[3] Verallgemeinerung eines Bieberbachschen Satzes. *Jber. Deutsch. Math.-Verein.*, **43** (1934), 143–145.

Grunsky, H.

[1] Neue Abschätzungen zur konformen Abbildung ein- und mehrfach zusammenhängender Bereiche. *Schr. Math. Inst. u. Inst. Angew. Math. Univ. Berlin*, **1** (1932), 95–140.

[2] Zwei Bemerkungen zur konformen Abbildung. *Jber. Deutsch. Math.-Verein.*, **43** (1934), 140–143.

[3] Koeffizientenbedingungen für schlicht abbildende meromorphe Funktionen. *Math. Z.*, **45** (1939), 29–61.

Haario, H.

[1] On coefficient bodies of univalent functions. *Ann. Acad. Sci. Fenn., Ser. AI. Math. Diss.*, no. 22 (1978), 49 pp.

Hallenbeck, D. J. and MacGregor, T. H.

[1] Subordination and extreme-point theory. *Pacific J. Math.*, **50** (1974), 455–468.

Hamilton, D. H.

[1] The successive coefficients of univalent functions. *J. London Math. Soc.*, **25** (1982), 122–138.

[2] The extreme points of Σ. *Proc. Amer. Math. Soc.*, **85** (1982), 393–396.

[3] On Littlewood's conjecture for univalent functions. *Proc. Amer. Math. Soc.*, **86** (1982), 32–36.

Hardy, G. H.

[1] *Divergent Series.* Oxford University Press: London, 1949.

Haslam-Jones, U. S.

[1] Tangential properties of a plane set of points. *Quart. J. Math.*, **7** (1936), 116–123.

Hayman, W. K.

[1] The asymptotic behaviour of *p*-valent functions. *Proc. London Math. Soc.*, **5** (1955), 257–284.

[2] *Multivalent Functions.* Cambridge University Press, 1958.

[3] Bounds for the large coefficients of univalent functions. *Ann. Acad. Sci. Fenn., Ser. AI*, no. 250 (1958), 13 pp.

[4] On the coefficients of univalent functions. *Proc. Cambridge Philos. Soc.*, **55** (1959), 373–374.

[5] On successive coefficients of univalent functions. *J. London Math. Soc.*, **38** (1963), 228–243.

[6] Coefficient problems for univalent functions and related function classes. *J. London Math. Soc.*, **40** (1965), 385–406.

[7] *Research Problems in Function Theory.* University of London, Athlone Press, 1967.

[8] Research problems in function theory, in *Proceedings of the Symposium on Complex Analysis, Canterbury, 1973* (edited by J. Clunie and W. K. Hayman), London Math. Soc. Lecture Note Series, no. 12 (Cambridge University Press, 1974), 143–180.

[9] The logarithmic derivative of multivalent functions. *Michigan Math. J.*, **27** (1980), 149–179.

Hengartner, W. and Schober, G.

[1] Extreme points for some classes of univalent functions. *Trans. Amer. Math. Soc.*, **185** (1973), 265–270.

[2] Compact families of univalent functions and their support points. *Michigan Math. J.*, **21** (1974), 205–217.

[3] Propriétés des points d'appui des familles compactes de fonctions univalentes. *C. R. Acad. Sci. Paris*, **279** (1974), 551–553.

[4] Some new properties of support points for compact families of univalent functions in the unit disk. *Michigan Math. J.*, **23** (1976), 207–216.

Herglotz, G.

[1] Über Potenzreihen mit positivem, reelen Teil im Einheitskreis. *S.-B. Sächs. Akad. Wiss. Leipzig Math.-Natur. Kl.*, **63** (1911), 501–511.

Hille, E.

[1] Remarks on a paper by Zeev Nehari. *Bull. Amer. Math. Soc.*, **55** (1949), 552–553.

Hille, E. and Phillips, R. S.
[1] *Functional Analysis and Semi-groups*. Amer. Math. Soc. Colloq. Publ., vol. 31, revised ed., 1957.
Horowitz, D.
[1] A refinement for coefficient estimates of univalent functions. *Proc. Amer. Math. Soc.*, **54** (1976), 176–178.
[2] Coefficient estimates for univalent polynomials. *J. Analyse Math.*, **31** (1977), 112–124.
[3] A further refinement for coefficient estimates of univalent functions. *Proc. Amer. Math. Soc.*, **71** (1978), 217–221.
Hummel, J. A.
[1] The coefficient regions of starlike functions. *Pacific J. Math.*, **7** (1957), 1381–1389.
[2] A variational method for starlike functions. *Proc. Amer. Math. Soc.*, **9** (1958), 82–87.
[3] Extremal problems in the class of starlike functions. *Proc. Amer. Math. Soc.*, **11** (1960), 741–749.
[4] The Grunsky coefficients of a schlicht function. *Proc. Amer. Math. Soc.*, **15** (1964), 142–150.
[5] Bounds for the coefficient body of univalent functions. *Arch. Rational Mech. Anal.*, **36** (1970), 128–134.
[6] *Lectures on Variational Methods in the Theory of Univalent Functions*. Lecture Notes, Univ. of Maryland, 1972.
[7] Inequalities of Grunsky type for Aharonov pairs. *J. Analyse Math.*, **25** (1972), 217–257.
[8] The Marx conjecture for starlike functions. *Michigan Math. J.*, **19** (1972), 257–266.
[9] A variational method for Gelfer functions. *J. Analyse Math.*, **30** (1976), 271–280.
[10] The Marx conjecture for starlike functions, II. *Ann. Univ. Mariae Curie-Skłodowska*, to appear.
Hummel, J. and Schiffer, M.
[1] Coefficient inequalities for Bieberbach–Eilenberg functions. *Arch. Rational Mech. Anal.*, **32** (1969), 87–99.
[2] Variational methods for Bieberbach–Eilenberg functions and for pairs. *Ann. Acad. Sci. Fenn., Ser. AI*, **3** (1977), 3–42.
Ilina, L. P.
[1] On the relative growth of adjacent coefficients of univalent functions. *Mat. Zametki*, **4** (1968), 715–722 (in Russian) = *Math. Notes*, **4** (1968), 918–922.
[2] Estimates for the coefficients of univalent functions in terms of the second coefficient. *Mat. Zametki*, **13** (1973), 351–357 (in Russian) = *Math. Notes*, **13** (1973), 215–218.
Ilina, L. P. and Kolomoiceva, Z. D.
[1] On a bound for $|c_4|$ depending on $|c_2|$ for the class *S*. *Vestnik Leningrad. Univ.*, **29** (1974), no. 1, 27–30 (in Russian).
Jenkins, J. A.
[1] On an inequality of Golusin. *Amer. J. Math.*, **73** (1951), 181–185.
[2] Symmetrization results for some conformal invariants. *Amer. J. Math.*, **75** (1953), 510–522.
[3] On a problem of Gronwall. *Ann. of Math.*, **59** (1954), 490–504.
[4] A general coefficient theorem. *Trans. Amer. Math. Soc.*, **77** (1954), 262–280.
[5] On Bieberbach–Eilenberg functions. *Trans. Amer. Math. Soc.*, **76** (1954), 389–396.
[6] On Bieberbach–Eilenberg functions, II. *Trans. Amer. Math. Soc.*, **78** (1955), 510–515.
[7] *Univalent Functions and Conformal Mapping*. Springer-Verlag: Berlin, 1958.
[8] On certain coefficients of univalent functions, in *Analytic Functions* (Princeton University Press, 1960), 159–194.
[9] On certain coefficients of univalent functions, II. *Trans. Amer. Math. Soc.*, **96** (1960), 534–545.
[10] On univalent functions with real coefficients. *Ann. of Math.*, **71** (1960), 1–15.
[11] An extension of the general coefficient theorem. *Trans. Amer. Math. Soc.*, **95** (1960), 387–407.
[12] Some problems for typically real functions. *Canad. J. Math.*, **13** (1961), 299–304.
[13] Some area theorems and a special coefficient theorem. *Illinois J. Math.*, **8** (1964), 80–99.
[14] On Bieberbach–Eilenberg functions, III. *Trans. Amer. Math. Soc.*, **119** (1965), 195–215.
[15] On certain extremal problems for the coefficients of univalent functions. *J. Analyse Math.*, **18** (1967), 173–184.
[16] A remark on "pairs" of regular functions. *Proc. Amer. Math. Soc.*, **31** (1972), 119–121.
Jensen, G.
[1] Quadratic differentials, in *Univalent Functions* (Ch. Pommerenke; Vandenhoeck and Ruprecht: Göttingen, 1975), 205–260.

Joh, K.
[1] Theorems on "schlicht" functions, III. *Proc. Phys. Math. Soc. Japan*, **21** (1939), 191–208.
Julia, G.
[1] Une équation aux dérivées fonctionelles liée à la représentation conforme. *Ann. Sci. École Norm. Sup.*, **39** (1922), 1–28.
Kamockiĭ, V. I.
[1] Application of the area theorem for investigation of the class $S(\alpha)$. *Mat. Zametki*, **28** (1980), 695–706 (in Russian) = *Math. Notes*, **28** (1980), 803–808.
[2] Estimate of integral means in the class $S(\alpha)$. *Sibirsk. Mat. Ž.*, **21** (1980), no. 1 (119), 211–215 (in Russian).
Kaplan, W.
[1] Close-to-convex schlicht functions. *Michigan Math. J.*, **1** (1952), 169–185.
Keogh, F. R.
[1] A subordination property of univalent functions. *Bull. London Math. Soc.*, **3** (1971), 181–184.
Keogh, F. R. and Miller, S. S.
[1] On the coefficients of Bazilevič functions. *Proc. Amer. Math. Soc.*, **30** (1971), 492–496.
Kirwan, W. E.
[1] Extremal problems for the typically real functions. *Amer. J. Math.*, **88** (1966), 942–954.
[2] On the coefficients of functions with bounded boundary rotation. *Michigan Math. J.*, **15** (1968), 277–282.
[3] Extremal problems for functions with bounded boundary rotation. *Ann. Acad. Sci. Fenn., Ser. AI*, no. 595 (1975), 19 pp.
Kirwan, W. E. and Pell, R.
[1] Extremal properties of a class of slit conformal mappings. *Michigan Math. J.*, **25** (1978), 223–232.
[2] A note on a class of slit conformal mappings. *Canad. J. Math.*, **30** (1978), 1166–1173.
Kirwan, W. E. and Schober, G.
[1] On extreme points and support points for some families of univalent functions. *Duke Math. J.*, **42** (1975), 285–296.
[2] Extremal problems for meromorphic univalent functions. *J. Analyse Math.*, **30** (1976), 330–348.
[3] New inequalities from old ones. *Math. Z.*, **180** (1982), 19–40.
Klee, V. L.
[1] Extremal structure of convex sets, II. *Math. Z.*, **69** (1958), 90–104.
Knopp, K.
[1] *Theory and Application of Infinite Series*, 2nd English ed. Blackie and Son: Glasgow, 1951.
Koebe, P.
[1] Über die Uniformisierung beliebiger analytischer Kurven. *Nachr. Akad. Wiss. Göttingen, Math.-Phys. Kl.*, 1907, 191–210.
[2] Über die Uniformisierung der algebraischen Kurven durch automorphe Funktionen mit imaginärer Substitutionsgruppe. *Nachr. Akad. Wiss. Göttingen, Math.-Phys. Kl.*, 1909, 68–76.
[3] Ränderzuordnung bei konformer Abbildung. *Nachr. Königl. Ges. Wiss. Göttingen, Math.-Phys. Kl.*, 1913, 286–288.
Komatu, Y.
[1] On convolution of power series. *Kōdai Math. Sem. Rep.*, **10** (1958), 141–144.
Korickiĭ, G. V.
[1] On the problem of curvature of level curves under univalent conformal mappings. *Uspehi Mat. Nauk*, **15** (95) (1960), 179–182 (in Russian).
Kössler, M.
[1] Simple polynomials. *Czechoslovak Math. J.*, **1** (76) (1951), 5–15.
Kövari, T. and Pommerenke, Ch.
[1] On Faber polynomials and Faber expansions. *Math. Z.*, **99** (1967), 193–206.
Kraus, W.
[1] Über den Zusammenhang einiger Charakteristiken eines einfach zusammenhängenden Bereiches mit der Kreisabbildung. *Mitt. Math. Sem. Giessen*, **21** (1932), 1–28.
Krzyż, J.
[1] On the maximum modulus of univalent functions. *Bull. Acad. Polon. Sci.*, **3** (1955), 203–206.

[2] The radius of close-to-convexity within the family of univalent functions. *Bull. Acad. Polon. Sci.*, **10** (1962), 201–204.
[3] Some remarks on close-to-convex functions. *Bull. Acad. Polon. Sci.*, **12** (1964), 25–28.
Krzyż, J. and Lewandowski, Z.
[1] On the integral of univalent functions. *Bull. Acad. Polon. Sci.*, **11** (1963), 447–448.
Kubota, Y.
[1] On extremal problems which correspond to algebraic univalent functions. *Kōdai Math. Sem. Rep.*, **25** (1973), 412–428.
[2] A coefficient inequality for certain meromorphic univalent functions. *Kōdai Math. Sem. Rep.*, **26** (1974–75), 85–94.
[3] On the fourth coefficient of meromorphic univalent functions. *Kōdai Math. Sem. Rep.*, **26** (1974–75), 267–288.
Kufarev, P. P.
[1] On one-parameter families of analytic functions. *Mat. Sb.*, **13** (55) (1943), 87–118 (in Russian).
[2] A remark on integrals of the Loewner equation. *Dokl. Akad. Nauk SSSR*, **57** (1947), 655–656 (in Russian).
[3] A theorem on solutions of a differential equation. *Uchen. Zap. Tomsk. Gos. Univ.*, **5** (1947), 20–21 (in Russian).
[4] A remark on extremal problems in the theory of univalent functions. *Uchen. Zap. Tomsk. Gos. Univ.*, **14** (1951), 3–7 (in Russian).
[5] On a property of extremal domains in the problem of coefficients. *Dokl. Akad. Nauk SSSR*, **97** (1954), 391–393 (in Russian).
[6] On a method for investigation of extremal problems in the theory of univalent functions. *Dokl. Akad. Nauk SSSR*, **107** (1956), 633–635 (in Russian).
Kühnau, R.
[1] Über vier Klassen schlichter Funktionen. *Math. Nachr.*, **50** (1971), 17–26.
Kung Sun
[1] The function $k(t)$ in Golusin and Löwner's differential equation. *Acta Math. Sinica*, **3** (1953), 225–230 (in Chinese).
[2] Contributions to the theory of schlicht functions. II. The coefficient problem. *Sci. Sinica*, **4** (1955), 359–373.
Landau, E.
[1] *Darstellung und Begründung einiger neuerer Ergebnisse der Funktionentheorie*. Julius Springer: Berlin, 1916; Zweite Auflage, 1926. (Reprinted by Chelsea: New York, 1946.)
[2] Einige Bemerkungen über schlichte Abbildung. *Jber. Deutsch. Math.-Verein.*, **34** (1925–26), 239–243.
[3] Über die Blochsche Konstante und zwei verwandte Weltkonstanten. *Math. Z.*, **30** (1929), 608–634.
[4] Über schlichte Funktionen. *Math. Z.*, **30** (1929), 635–638.
[5] Über ungerade schlichte Funktionen. *Math. Z.*, **37** (1933), 33–35.
Lebedev, N. A.
[1] Some estimates and extremal problems in the theory of conformal mapping. Dissertation, Leningrad State Univ., 1951 (in Russian).
[2] A method of variation in conformal mapping. *Dokl. Akad. Nauk SSSR*, **76** (1951), 25–27 (in Russian).
[3] Majorizing region for the expression $I = \ln(z^\lambda f'(z)^{1-\lambda}/f(z)^\lambda)$. *Vestnik Leningrad. Univ.*, **10** (1955), no. 8, 29–41 (in Russian) = *Amer. Math. Soc. Transl.* (2), **22** (1962), 43–57.
[4] Some estimates for functions regular and univalent in the disk. *Vestnik Leningrad. Univ.*, **10** (1955), no. 11, 3–21 (in Russian) = *Amer. Math. Soc. Transl.* (2), **22** (1962), 59–80.
[5] Some consequences of Grunsky's inequality. *Vestnik Leningrad. Univ.*, **27** (1972), no. 7, 45–55 (in Russian).
[6] On Hayman's regularity theorem. *Zap. Naučn. Sem. Leningrad. Otdel. Mat. Inst. Steklov. (LOMI)*, **44** (1974), 93–99 (in Russian).
[7] *The Area Principle in the Theory of Univalent Functions*. Izdat. "Nauka": Moscow, 1975 (in Russian).
Lebedev, N. A. and Milin, I. M.
[1] On the coefficients of certain classes of analytic functions. *Mat. Sb.*, **28** (70) (1951), 359–400 (in Russian).
[2] An inequality. *Vestnik Leningrad. Univ.*, **20** (1965), no. 19, 157–158 (in Russian).

Leeman, G. B.
[1] Constrained extremal problems for families of Stieltjes integrals. *Arch. Rational Mech. Anal.*, **52** (1973), 350–357.
[2] The constrained coefficient problem for typically real functions. *Trans. Amer. Math. Soc.*, **186** (1973), 177–189.
[3] Some regularity theorems for typically real functions. *Proc. Amer. Math. Soc.*, **40** (1973), 191–198.
[4] A new proof for an inequality of Jenkins. *Proc. Amer. Math. Soc.*, **54** (1976), 114–116.
[5] The seventh coefficient of odd symmetric univalent functions. *Duke Math. J.*, **43** (1976), 301–307.
Lehto, O.
[1] On the distortion of conformal mappings with bounded boundary rotation. *Ann. Acad. Sci. Fenn., Ser. AI*, no. 124 (1952), 14 pp.
Leung, Y. J.
[1] Successive coefficients of starlike functions. *Bull. London Math. Soc.*, **10** (1978), 193–196.
[2] Robertson's conjecture on the coefficients of close-to-convex functions. *Proc. Amer. Math. Soc.*, **76** (1979), 89–94.
[3] Integral means of the derivatives of some univalent functions. *Bull. London Math. Soc.*, **11** (1979), 289–294.
Levin, V. I.
[1] Bemerkung zu den schlichten Abbildungen des Einheitskreises. *Jber. Deutsch. Math.-Verein.*, **42** (1933), 68–70.
[2] Ein Beitrag zum Koeffizientenproblem der schlichten Funktionen. *Math. Z.*, **38** (1933), 306–311.
[3] Some remarks on the coefficients of schlicht functions. *Proc. London Math. Soc.*, **39** (1935), 467–480.
Lewandowski, Z.
[1] Quelques remarques sur les théorèmes de Schild relatifs à une classe de fonctions univalentes. *Ann. Univ. Mariae Curie-Skłodowska*, **9** (1955), 149–155.
[2] Sur l'identité de certaines classes de fonctions univalentes, I, II. *Ann. Univ. Mariae Curie-Skłodowska*, **12** (1958), 131–146; **14** (1960), 19–46.
[3] Sur les majorantes des fonctions holomorphes dans le cercle $|z| < 1$. *Ann. Univ. Mariae Curie-Skłodowska*, **15** (1961), 5–11.
[4] On a problem of M. Biernacki. *Ann. Univ. Mariae Curie-Skłodowska*, **17** (1963), 39–41.
[5] Some results concerning univalent majorants. *Ann. Univ. Mariae Curie-Skłodowska*, **18** (1964), 13–18.
Lindelöf, E.
[1] Mémoire sur certaines inégalités dans la théorie des fonctions monogènes et sur quelques propriétés nouvelles de ces fonctions dans le voisinage d'un point singulier essentiel. *Acta Soc. Sci. Fenn.*, **35** (1909), no. 7, 1–35.
Littlewood, J. E.
[1] On inequalities in the theory of functions. *Proc. London Math. Soc.*, **23** (1925), 481–519.
[2] On the coefficients of schlicht functions. *Quart. J. Math.*, **9** (1938), 14–20.
[3] *Lectures on the Theory of Functions.* Oxford University Press: London, 1944.
Littlewood, J. E. and Paley, R. E. A. C.
[1] A proof that an odd schlicht function has bounded coefficients. *J. London Math. Soc.*, **7** (1932), 167–169.
Loewner, C. (Löwner, K.)
[1] Untersuchungen über die Verzerrung bei konformen Abbildungen des Einheitskreises $|z| < 1$, die durch Funktionen mit nicht verschwindender Ableitung geliefert werden. *Ber. Verh. Sächs. Ges. Wiss. Leipzig*, **69** (1917), 89–106.
[2] Über Extremumsätze bei der konformen Abbildung des Äusseren des Einheitskreises. *Math. Z.*, **3** (1919), 65–77.
[3] Untersuchungen über schlichte konforme Abbildungen des Einheitskreises, I. *Math. Ann.*, **89** (1923), 103–121.
Loewner, C. and Netanyahu, E.
[1] On some compositions of Hadamard type in classes of analytic functions. *Bull. Amer. Math. Soc.*, **65** (1959), 284–286.
Lohwater, A. J., Piranian, G., and Rudin, W.
[1] On the derivative of a schlicht function. *Math. Scand.*, **3** (1955), 103–106.

Lonka, H. and Tammi, O.
[1] On the use of step-functions in extremum problems of the class with bounded boundary rotation. *Ann. Acad. Sci. Fenn., Ser. AI*, no. 418 (1968), 18 pp.
Lucas, K. W.
[1] On successive coefficients of areally mean p-valent functions. *J. London Math. Soc.*, **44** (1969), 631–642.
MacGregor, T. H.
[1] A covering theorem for convex functions. *Proc. Amer. Math. Soc.*, **15** (1964), 310.
[2] Majorization by univalent functions. *Duke Math. J.*, **34** (1967), 95–102.
[3] An inequality concerning analytic functions with a positive real part. *Canad. J. Math.*, **21** (1969), 1172–1177.
[4] The univalence of a linear combination of convex mappings. *J. London Math. Soc.*, **44** (1969), 210–212.
[5] Applications of extreme-point theory to univalent functions. *Michigan Math. J.*, **19** (1972), 361–376.
[6] Geometric problems in complex analysis. *Amer. Math. Monthly*, **79** (1972), 447–468.
Marden, M.
[1] *Geometry of Polynomials*, 2nd ed. American Math. Soc.: Providence, R.I., 1966.
Marty, F.
[1] Sur les dérivées seconde et troisième d'une fonction holomorphe et univalente dans le cercle unité. *C. R. Acad. Sci. Paris*, **194** (1932), 1308–1310.
[2] Sur le module des coefficients de MacLaurin d'une fonction univalente. *C. R. Acad. Sci. Paris*, **198** (1934), 1569–1571.
Marx, A.
[1] Untersuchungen über schlichte Abbildungen. *Math. Ann.*, **107** (1932), 40–67.
McLaughlin, R.
[1] On the Marx conjecture for starlike functions of order α. *Trans. Amer. Math. Soc.*, **142** (1969), 249–256.
[2] Extremalprobleme für eine Familie schlichter Funktionen. *Math. Z.*, **118** (1970), 320–330.
[3] Some extremal problems for functions univalent in an annulus. *Math. Scand.*, **28** (1971), 129–138.
[4] Extremal problems for a class of symmetric functions. *Arch. Rational Mech. Anal.*, **44** (1972), 310–319.
Michel, C.
[1] Eine Bemerkung zu schlichten Polynomen. *Bull. Acad. Polon. Sci.*, **18** (1970), 513–519.
Milin, I. M.
[1] The area method in the theory of univalent functions. *Dokl. Akad. Nauk SSSR*, **154** (1964), 264–267 (in Russian) = *Soviet Math. Dokl.*, **5** (1964), 78–81.
[2] Estimation of coefficients of univalent functions. *Dokl. Akad. Nauk SSSR*, **160** (1965), 769–771 (in Russian) = *Soviet Math. Dokl.*, **6** (1965), 196–198.
[3] On the coefficients of univalent functions. *Dokl. Akad. Nauk SSSR*, **176** (1967), 1015–1018 (in Russian) = *Soviet Math. Dokl.*, **8** (1967), 1255–1258.
[4] Adjacent coefficients of univalent functions. *Dokl. Akad. Nauk SSSR*, **180** (1968), 1294–1297 (in Russian) = *Soviet Math. Dokl.*, **9** (1968), 762–765.
[5] Hayman's regularity theorem for the coefficients of univalent functions. *Dokl. Akad. Nauk SSSR*, **192** (1970), 738–741 (in Russian) = *Soviet Math. Dokl.*, **11** (1970), 724–728.
[6] *Univalent Functions and Orthonormal Systems*. Izdat. "Nauka": Moscow, 1971 (in Russian); English transl., Amer. Math. Soc., Providence, R.I., 1977.
[7] On a property of the logarithmic coefficients of univalent functions, in *Metric Questions in the Theory of Functions* (edited by G. D. Suvorov; "Naukova Dumka": Kiev, 1980), 86–90 (in Russian).
Milin, V. I.
[1] Estimation of coefficients of odd univalent functions, in *Metric Questions in the Theory of Functions* (edited by G. D. Suvorov; "Naukova Dumka": Kiev, 1980), 78–86 (in Russian).
[2] Adjacent coefficients of odd univalent functions. *Sibirsk. Mat. Ž.*, **22** (1981), no. 2 (126), 149–157 (in Russian) = *Siberian Math. J.*, **22** (1981), 283–290.
Mocanu, P. T., Reade, M. O., and Złotkiewicz, E. J.
[1] On Bazilevič functions. *Proc. Amer. Math. Soc.*, **39** (1973), 173–174.
Mogk, E.
[1] Über ein Variationslemma von M. Schiffer. *Mitt. Math. Sem. Giessen*, **82** (1969), 42 pp.

Montel, P.
[1] *Leçons sur les fonctions univalentes ou multivalentes* (recueillies et rédigées par F. Marty, avec une note de H. Cartan). Gauthiers-Villars: Paris, 1933.
Nashed, M. Z.
[1] Differentiability and related properties of nonlinear operators: some aspects of the role of differentials in nonlinear functional analysis, in *Nonlinear Functional Analysis and Applications* (edited by L. B. Rall; Academic Press: New York, 1971), 103–309.
Nehari, Z.
[1] The Schwarzian derivative and schlicht functions. *Bull. Amer. Math. Soc.*, **55** (1949), 545–551.
[2] *Conformal Mapping*. McGraw-Hill: New York, 1952.
[3] Some inequalities in the theory of functions. *Trans. Amer. Math. Soc.*, **75** (1953), 256–286.
[4] Some criteria of univalence. *Proc. Amer. Math. Soc.*, **5** (1954), 700–704.
[5] On the coefficients of univalent functions. *Proc. Amer. Math. Soc.*, **8** (1957), 291–293.
[6] Inequalities for the coefficients of univalent functions. *Arch. Rational Mech. Anal.*, **34** (1969), 301–330.
[7] On the coefficients of Bieberbach–Eilenberg functions. *J. Analyse Math.*, **23** (1970), 297–303.
[8] A proof of $|a_4| \leq 4$ by Loewner's method, in *Proceedings of the Symposium on Complex Analysis, Canterbury, 1973* (edited by J. Clunie and W. K. Hayman), London Math. Soc. Lecture Note Series, no. 12 (Cambridge University Press, 1974), 107–110.
[9] A property of convex conformal maps. *J. Analyse Math.*, **30** (1976), 390–393.
[10] Univalence criteria depending on the Schwarzian derivative. *Illinois J. Math.*, **23** (1979), 345–351.
Nehari, Z. and Netanyahu, E.
[1] Coefficients of meromorphic schlicht functions. *Proc. Amer. Math. Soc.*, **8** (1957), 15–23.
Netanyahu, E.
[1] On univalent functions in the unit disk whose image contains a given disk. *J. Analyse Math.*, **23** (1970), 305–322.
Nevanlinna, R.
[1] Über die konforme Abbildung von Sterngebieten. *Översikt av Finska Vetenskaps-Soc. Förh.*, **63(A)**, no. 6 (1920–21), 1–21.
Nguyen Xuan Uy
[1] Removable sets of analytic functions satisfying a Lipschitz condition. *Ark. Mat.*, **17** (1979), 19–27.
Noonan, J. W.
[1] Asymptotic behavior of functions with bounded boundary rotation. *Trans. Amer. Math. Soc.*, **164** (1972), 397–410.
Noshiro, K.
[1] On the theory of schlicht functions. *J. Fac. Sci. Hokkaido Univ.*, **2** (1934–35), 129–155.
Obrock, A. E.
[1] The extremal functions for certain problems concerning schlicht functions. *Bull. Amer. Math. Soc.*, **71** (1965), 626–628.
[2] An inequality for certain schlicht functions. *Proc. Amer. Math. Soc.*, **17** (1966), 1250–1253.
Ogawa, S.
[1] A note on close-to-convex functions, I. *J. Nara Gakugei Univ.*, **8** (1959), 9–10.
Ozawa, M.
[1] On the Bieberbach conjecture for the sixth coefficient. *Kōdai Math. Sem. Rep.*, **21** (1969), 97–128.
[2] An elementary proof of the Bieberbach conjecture for the sixth coefficient. *Kōdai Math. Sem. Rep.*, **21** (1969), 129–132.
Ozawa, M. and Kubota, Y.
[1] On the eighth coefficient of univalent functions. *J. Analyse Math.*, **23** (1970), 323–352.
[2] Bieberbach conjecture for the eighth coefficient. *Kōdai Math. Sem. Rep.*, **24** (1972), 331–382.
[3] Bieberbach conjecture for the eighth coefficient, II. *Kōdai Math. Sem. Rep.*, **25** (1973), 257–288.
Paatero, V.
[1] Über die konforme Abbildung von Gebieten, deren Ränder von beschränkter Drehung sind. *Ann. Acad. Sci. Fenn., Ser. A*, **33** (1931), no. 9, 77 pp.

[2] Über Gebiete von beschränkter Randdrehung. *Ann. Acad. Sci. Fenn., Ser. A*, **37** (1933), no. 9, 20 pp.

Pearce, K.

[1] New support points of S and extreme points of HS. *Proc. Amer. Math. Soc.*, **81** (1981), 425–428.

Pederson, R. N.

[1] On unitary properties of Grunsky's matrix. *Arch. Rational Mech. Anal.*, **29** (1968), 370–377.

[2] A proof of the Bieberbach conjecture for the sixth coefficient. *Arch. Rational Mech. Anal.*, **31** (1968), 331–351.

[3] A note on the local coefficient problem. *Proc. Amer. Math. Soc.*, **20** (1969), 345–347.

Pederson, R. and Schiffer, M.

[1] A proof of the Bieberbach conjecture for the fifth coefficient. *Arch. Rational Mech. Anal.*, **45** (1972), 161–193.

Pell, R.

[1] Support point functions and the Loewner variation. *Pacific J. Math.*, **86** (1980), 561–564.

[2] A new proof of the Bieberbach conjecture for the third coefficient, *Complex Variables: Theory and Application*, **1** (1983), 151–154.

Peschl, E.

[1] Zur Theorie der schlichten Funktionen. *J. Reine Angew. Math.*, **176** (1937), 61–94.

Pfaltzgraff, J. A.

[1] On the Marx conjecture for a class of close-to-convex functions. *Michigan Math. J.*, **18** (1971), 275–278.

[2] Univalence of the integral of $f'(z)^\lambda$. *Bull. London Math. Soc.*, **7** (1975), 254–256.

Pfluger, A.

[1] *Lectures on Conformal Mapping*. Lecture Notes, Dept. of Mathematics, Indiana Univ., 1969.

[2] On the convexity of some sections of the nth coefficient body for schlicht functions, in *Some Problems of Mathematics and Mechanics* (Izdat. "Nauka": Leningrad, 1970), 233–241 (in Russian) = *Amer. Math. Soc. Transl.* (2), **104** (1976), 215–222.

[3] Lineare Extremalprobleme bei schlichten Funktionen. *Ann. Acad. Sci. Fenn., Ser AI*, no. 489 (1971), 32 pp.

[4] On a coefficient problem for schlicht functions, in *Advances in Complex Function Theory, Maryland, 1973/74*, Lecture Notes in Math. No. 505 (Springer-Verlag, 1976), 79–91.

[5] Some coefficient problems for starlike functions. *Ann. Acad. Sci. Fenn., Ser. AI*, **2** (1976), 383–396.

[6] On a coefficient inequality for schlicht functions, in *Romanian-Finnish Seminar in Complex Analysis, Bucharest, 1976*, Lecture Notes in Math. No. 743 (Springer-Verlag, 1979), 336–343.

[7] Über die Koeffizienten schlichter Funktionen. *Bonner Math. Schriften*, Sonderband Nr. 121 (Bonn, 1980), 41–61.

Pick, G.

[1] Über den Koebeschen Verzerrungssatz. *Ber. Verh. Königl. Ges. Wiss. Leipzig, Math.-Phys. Kl.*, **68** (1916), 58–64.

[2] Über die konforme Abbildung eines Kreises auf ein schlichtes und zugleich beschränktes Gebiet. *S.-B. Kaiserl. Akad. Wiss. Wien*, **126** (1917), 247–263.

Pinchuk, B.

[1] A variational method for functions of bounded boundary rotation. *Trans. Amer. Math. Soc.*, **138** (1969), 107–113.

Pokornyi, V. V.

[1] On some sufficient conditions for univalence. *Dokl. Akad. Nauk SSSR*, **79** (1951), 743–746 (in Russian).

Pólya, G. and Schoenberg, I. J.

[1] Remarks on de la Vallée Poussin means and convex conformal maps of the circle. *Pacific J. Math.*, **8** (1958), 295–334.

Pommerenke, Ch.

[1] Über einige Klassen meromorpher schlichter Funktionen. *Math. Z.*, **78** (1962), 263–284.

[2] On starlike and close-to-convex functions. *Proc. London Math. Soc.*, **13** (1963), 290–304.

[3] On meromorphic starlike functions. *Pacific J. Math.*, **13** (1963), 221–235.

[4] Über die Faberschen Polynome schlichter Funktionen. *Math. Z.*, **85** (1964), 197–208.

[5] Über die Subordination analytischer Funktionen. *J. Reine Angew. Math.*, **218** (1965), 159–173.

[6] On the Loewner differential equation. *Michigan Math. J.*, **13** (1966), 435–443.
[7] On the coefficients of univalent functions. *J. London Math. Soc.*, **42** (1967), 471–474.
[8] Relations between the coefficients of a univalent function. *Invent. Math.*, **3** (1967), 1–15.
[9] The Grunsky inequalities for univalent functions. Lecture Notes, Imperial College London, March 1969.
[10] On the Grunsky inequalities for univalent functions. *Arch. Rational Mech. Anal.*, **35** (1969), 234–244.
[11] On a variational method for univalent functions. *Michigan Math. J.*, **17** (1970), 1–3.
[12] Probleme aus der Funktionentheorie. *Jber. Deutsch. Math.-Verein.*, **73** (1971), 1–5.
[13] Problems in complex function theory. *Bull. London Math. Soc.*, **4** (1972), 354–366.
[14] *Univalent Functions* (with a chapter on quadratic differentials by G. Jensen). Vandenhoeck and Ruprecht: Göttingen, 1975.
[15] On the Becker univalence criterion. *Ann. Univ. Mariae Curie-Skłodowska*, to appear.

Poole, J. T.
[1] A note on variational methods. *Proc. Amer. Math. Soc.*, **15** (1964), 929–932.

Prawitz, H.
[1] Über Mittelwerte analytischer Funktionen. *Arkiv Mat. Astr. Fys.*, **20** (1927–28), no. 6, 1–12.

Privalov, I. I.
[1] On functions giving a univalent conformal mapping. *Mat. Sb.*, **31** (1924), 350–365 (in Russian).

Prohorov, D. V.
[1] The geometric characterization of certain classes of univalent functions. *Vestnik Leningrad. Univ.*, **29** (1974), no. 13, 51–55 (in Russian).

Quine, J. R.
[1] On the self-intersections of the image of the unit circle under a polynomial mapping. *Proc. Amer. Math. Soc.*, **39** (1973), 135–140.
[2] On univalent polynomials. *Proc. Amer. Math. Soc.*, **57** (1976), 75–78.

Reade, M. O.
[1] On close-to-convex univalent functions. *Michigan Math. J.*, **3** (1955–56), 59–62.

Reich, E.
[1] Schlicht functions with real coefficients. *Duke Math. J.*, **23** (1956), 421–427.

Remak, R.
[1] Über eine spezielle Klasse schlichter konformer Abbildungen des Einheitskreises. *Mathematica, Zutphen. B*, **11** (1943), 175–192; **12** (1943), 43–49.

Rényi, A.
[1] On the coefficients of schlicht functions. *Publ. Math. Debrecen*, **1** (1949), 18–23.

Riesz, F.
[1] Sur une inégalité de M. Littlewood dans la théorie des fonctions. *Proc. London Math. Soc.*, **23** (1925), 36–39.

Robertson, M. S.
[1] A note on schlicht polynomials. *Trans. Royal Soc. Canada*, **26** (1932), 43–48.
[2] On the coefficients of a typically-real function. *Bull. Amer. math. Soc.*, **41** (1935), 565–572.
[3] On the theory of univalent functions. *Ann. of Math.*, **37** (1936), 374–408.
[4] A remark on the odd schlicht functions. *Bull. Amer. Math. Soc.*, **42** (1936), 366–370.
[5] The partial sums of multivalently star-like functions. *Ann. of Math.*, **42** (1941), 829–838.
[6] Applications of a lemma of Fejér to typically real functions. *Proc. Amer. Math. Soc.*, **1** (1950), 555–561.
[7] Convolutions of schlicht functions. *Proc. Amer. Math. Soc.*, **13** (1962), 585–589.
[8] The generalized Bieberbach conjecture for subordinate functions. *Michigan Math. J.*, **12** (1965), 421–429.
[9] A generalization of the Bieberbach coefficient problem for univalent functions. *Michigan Math. J.*, **13** (1966), 185–192.
[10] Quasi-subordination and coefficient conjectures. *Bull. Amer. Math. Soc.*, **76** (1970), 1–9.
[11] Quasi-subordinate functions, in *Mathematical Essays Dedicated to A. J. Macintyre* (Ohio University Press: Athens, Ohio, 1970), 311–330.

Robinson, R. M.
[1] Bounded univalent functions. *Trans. Amer. Math. Soc.*, **52** (1942), 426–449.
[2] Univalent majorants. *Trans. Amer. Math. Soc.*, **61** (1947), 1–35.
[3] Extremal problems for star mappings. *Proc. Amer. Math. Soc.*, **6** (1955), 364–377.

Rogosinski, W.
[1] Über positive harmonische Sinusentwicklungen. *Jber. Deutsch. Math.-Verein.*, **40** (1931), 33–35.
[2] Über positive harmonische Entwicklungen und typisch-reelle Potenzreihen. *Math. Z.*, **35** (1932), 93–121.
[3] Zum Majorantenprinzip der Funktionentheorie. *Math. Z.*, **37** (1933), 210–236.
[4] Zum Schwarzschen Lemma. *Jber. Deutsch. Math.-Verein.*, **44** (1934), 258–261.
[5] On a theorem of Bieberbach–Eilenberg. *J. London Math. Soc.*, **14** (1939), 4–11.
[6] On subordinate functions. *Proc. Cambridge Philos. Soc.*, **35** (1939), 1–26.
[7] On the coefficients of subordinate functions. *Proc. London Math. Soc.*, **48** (1943), 48–82.
Royster, W. C.
[1] On the univalence of a certain integral. *Michigan Math. J.*, **12** (1965), 385–387.
Royster, W. C. and Suffridge, T. J.
[1] Typically real polynomials. *Publ. Math. Debrecen*, **17** (1970), 307–312.
Ruscheweyh, St.
[1] On the radius of univalence of the partial sums of convex functions. *Bull. London Math. Soc.*, **4** (1972), 367–369.
[2] Über die Faltung schlichter Funktionen. *Math. Z.*, **128** (1972), 85–92.
Ruscheweyh, St. and Sheil-Small, T.
[1] Hadamard products of schlicht functions and the Pólya–Schoenberg conjecture. *Comment. Math. Helv.*, **48** (1973), 119–135.
Ruscheweyh, St. and Wirths, K.-J.
[1] Über die Faltung schlichter Funktionen, II. *Math. Z.*, **131** (1973), 11–23.
Schaeffer, A. C., Schiffer, M., and Spencer, D. C.
[1] The coefficient regions of schlicht functions. *Duke Math. J.*, **16** (1949), 493–527.
Schaeffer, A. C. and Spencer, D. C.
[1] The coefficients of schlicht functions. *Duke Math. J.*, **10** (1943), 611–635.
[2] The coefficients of schlicht functions, II. *Duke Math. J.*, **12** (1945), 107–125.
[3] The coefficients of schlicht functions, III. *Proc. Nat. Acad. Sci. U.S.A.*, **32** (1946), 111–116.
[4] A variational method in conformal mapping. *Duke Math. J.*, **14** (1947), 949–966.
[5] The coefficients of schlicht functions, IV. *Proc. Nat. Acad. Sci. U.S.A.*, **35** (1949), 143–150.
[6] *Coefficient Regions for Schlicht Functions.* Amer. Math. Soc. Colloq. Publ., vol. 35, 1950.
Schiffer, M.
[1] Sur un principe nouveau pour l'évaluation des fonctions holomorphes. *Bull. Soc. Math. France*, **64** (1936), 231–240.
[2] Un calcul de variation pour une famille de fonctions univalentes. *C. R. Acad. Sci. Paris*, **205** (1937), 709–711.
[3] Sur un problème d'extrémum de la représentation conforme. *Bull. Soc. Math. France*, **66** (1938), 48–55.
[4] A method of variation within the family of simple functions. *Proc. London Math. Soc.*, **44** (1938), 432–449.
[5] On the coefficients of simple functions. *Proc. London Math. Soc.*, **44** (1938), 450–452.
[6] Sur la variation de la fonction de Green de domaines plans quelconques. *C. R. Acad. Sci. Paris*, **209** (1939), 980–982.
[7] Variation of the Green function and theory of the p-valued functions. *Amer. J. Math.*, **65** (1943), 341–360.
[8] Sur l'équation différentielle de M. Löwner. *C. R. Acad. Sci. Paris*, **221** (1945), 369–371.
[9] Hadamard's formula and variation of domain-functions. *Amer. J. Math.*, **68** (1946), 417–448.
[10] Faber polynomials in the theory of univalent functions. *Bull. Amer. Math. Soc.*, **54** (1948), 503–517.
[11] Some recent developments in the theory of conformal mapping, in *Dirichlet's Principle, Conformal Mapping, and Minimal Surfaces* (R. Courant; Interscience: New York, 1950), appendix.
[12] Applications of variational methods in the theory of conformal mapping, in *Calculus of Variations and its Applications*, Proc. Symposia Appl. Math. 8 (1958), 93–113 (for Amer. Math. Soc.: McGraw-Hill, New York).
[13] Extremum problems and variational methods in conformal mapping, in *Proc. International Congress of Mathematicians, Edinburgh, 1958*, 211–231.
[14] Univalent functions whose n first coefficients are real. *J. Analyse Math.*, **18** (1967), 329–349.

[15] On the coefficient problem for univalent functions. *Trans. Amer. Math. Soc.*, **134** (1968), 95–101.

[16] Some distortion theorems in the theory of conformal mapping. *Atti Accad. Naz. Lincei Mem.*, **10** (1970), 3–19.

[17] Inequalities in the theory of univalent functions, in *Inequalities*, vol. III (edited by O. Shisha; Academic Press: New York, 1972), 311–319.

Schiffer, M. and Schmidt, H. G.

[1] A new set of coefficient inequalities for univalent functions. *Arch. Rational Mech. Anal.*, **42** (1971), 346–368.

Schiffer, M. and Tammi, O.

[1] On the fourth coefficient of univalent functions with bounded boundary rotation. *Ann. Acad. Sci. Fenn., Ser. AI*, no. 396 (1967), 26 pp.

Schmidt, H. G.

[1] Some examples of the method of quadratic differentials in the theory of univalent functions. Matematisk Institut, Aarhus Universitet, Preprint No. 35 (1970), 64 pp.

Schober, G.

[1] *Univalent Functions—Selected Topics.* Lecture Notes in Math. No. 478, Springer-Verlag, 1975.

[2] Coefficients of inverses of meromorphic univalent functions. *Proc. Amer. Math. Soc.*, **67** (1977), 111–116.

Schur, I.

[1] Bemerkungen zur Theorie der beschränkten Bilinearformen mit unendlich vielen Veränderlichen. *J. Reine Angew. Math.*, **140** (1911), 1–28.

[2] Über Potenzreihen, die im Innern des Einheitskreises beschränkt sind. *J. Reine Angew. Math.*, **147** (1917), 205–232; **148** (1918), 122–145.

[3] On Faber polynomials. *Amer. J. Math.*, **67** (1945), 33–41.

Shah Tao-shing

[1] The principle of area in the theory of univalent functions. *Acta Math. Sinica*, **3** (1953), 208–212 (in Chinese).

[2] Goluzin's number $(3 - \sqrt{5})/2$ is the radius of superiority in subordination. *Science Record*, **1** (1957), no. 4, 25–28.

[3] On the radius of superiority in subordination. *Science Record*, **1** (1957), no. 5, 53–57.

Sheil-Small, T.

[1] On convex univalent functions. *J. London Math. Soc.*, **1** (1969), 483–492.

[2] On Bazilevič functions. *Quart. J. Math. Oxford*, **23** (1972), 135–142.

[3] On linearly accessible univalent functions. *J. London Math. Soc.*, **6** (1973), 385–398.

[4] On the convolution of analytic functions. *J. Reine Angew. Math.*, **258** (1973), 137–152.

[5] The Hadamard product and linear transformations of classes of analytic functions. *J. Analyse Math.*, **34** (1978), 204–239.

Shirokov, N. A.

[1] Hayman's regularity theorem. *Zap. Naučn. Sem. Leningrad. Otdel. Mat. Inst. Steklov. (LOMI)*, **24** (1972), 182–200 (in Russian) = *J. Soviet Math.*, **2** (1974), 693–708.

Singh, R.

[1] On Bazilevič functions. *Proc. Amer. Math. Soc.*, **38** (1973), 261–271.

Smirnov, V. I. and Lebedev, N. A.

[1] *Constructive Theory of Functions of a Complex Variable.* Izdat. "Nauka": Moscow and Leningrad, 1964; English transl., M.I.T. Press: Cambridge, Mass., 1968.

Špaček, L.

[1] Příspěvek k teorii funkcí prostých. *Časopis Pěst. Mat.*, **62** (1933), 12–19.

Spencer, D. C.

[1] Some remarks concerning the coefficients of schlicht functions. *J. Math. Phys.*, **21** (1942), 63–68.

[2] Some problems in conformal mapping. *Bull. Amer. Math. Soc.*, **53** (1947), 417–439.

Springer, G.

[1] The coefficient problem for schlicht mappings of the exterior of the unit circle. *Trans. Amer. Math. Soc.*, **70** (1951), 421–450.

[2] Extreme Punkte der konvexen Hülle schlichter Funktionen. *Math. Ann.*, **129** (1955), 230–232.

Srebro, U.

[1] Is the slit of a rational slit mapping in S straight? *Proc. Amer. Math. Soc.*, to appear.

Stroganoff, W.
[1] Über den arc $f'(z)$ unter der Bedingung, dass $f(z)$ die konforme Abbildung eines sternartigen Gebietes auf das Innere des Einheitskreises der z-Ebene liefert. *Trudy Mat. Inst. Steklov.*, **5** (1934), 247–258.
Strohhäcker, E.
[1] Beiträge zur Theorie der schlichten Funktionen. *Math. Z.*, **37** (1933), 356–380.
Struik, D. J.
[1] *Lectures on Classical Differential Geometry.* Addison-Wesley: Cambridge, Mass., 1950.
Styer, D. and Wright, D.
[1] On the valence of the sum of two convex functions. *Proc. Amer. Math. Soc.*, **37** (1973), 511–516.
Suffridge, T. J.
[1] Convolutions of convex functions. *J. Math. Mech.*, **15** (1966), 795–804.
[2] On univalent polynomials. *J. London Math. Soc.*, **44** (1969), 496–504.
[3] Some remarks on convex maps of the unit disk. *Duke Math. J.*, **37** (1970), 775–777.
[4] Extreme points in a class of polynomials having univalent sequential limits. *Trans. Amer. Math. Soc.*, **163** (1972), 225–237.
[5] Starlike functions as limits of polynomials, in *Advances in Complex Function Theory, Maryland, 1973/74*, Lecture Notes in Math. No. 505 (Springer-Verlag, 1976), 164–203.
Szász, O.
[1] Über Funktionen, die den Einheitskreis schlicht abbilden. *Jber. Deutsch. Math.-Verein.*, **42** (1933), 73–75.
Szegö, G.
[1] Aufgabe 2, Konforme Abbildung; Lösung der Aufgabe 2. *Jber. Deutsch. Math.-Verein.*, **31** (1922), Abt. 2, 42; *Ibid.*, **32** (1923), Abt. 2, 45.
[2] Über eine Extremalaufgabe aus der Theorie der schlichten Abbildungen. *S.-B. Berlin. Math. Ges.*, **22** (1923), 38–47.
[3] Zur Theorie der schlichten Abbildungen. *Math. Ann.*, **100** (1928), 188–211.
Tammi, O.
[1] *Extremum Problems for Bounded Univalent Functions, I, II.* Lecture Notes in Math. Nos. 646 and 913, Springer-Verlag, 1978 and 1982.
Teichmüller, O.
[1] Ungleichungen zwischen den Koeffizienten schlichter Funktionen. *S.-B. Preuss. Akad. Wiss. Phys.-Math. Kl.*, 1938, 363–375.
Thomas, D. K.
[1] On Bazilevič functions. *Trans. Amer. Math. Soc.*, **132** (1968), 353–361.
Titchmarsh, E. C.
[1] *The Theory of Functions*, 2nd ed. Oxford University Press, 1939.
Toeplitz, O.
[1] Die linearen vollkommenen Räume der Funktionentheorie. *Comment. Math. Helv.*, **23** (1949), 222–242.
Trimble, S. Y.
[1] A coefficient inequality for convex univalent functions. *Proc. Amer. Math. Soc.*, **48** (1975), 266–267.
Tsao, A.
[1] Disproof of a coefficient conjecture for meromorphic univalent functions. *Trans. Amer. Math. Soc.*, **274** (1982), 783–796.
Tsuji, M.
[1] *Potential Theory in Modern Function Theory.* Maruzen: Tokyo, 1959.
Warschawski, S. E.
[1] On the higher derivatives at the boundary in conformal mapping. *Trans. Amer. Math. Soc.*, **38** (1935), 310–340.
Wilken, D. R.
[1] The integral means of close-to-convex functions. *Michigan Math. J.*, **19** (1972), 377–379.
Wolibner, W.
[1] Sur certaines conditions nécessaires et suffisantes pour qu'une fonction analytique soit univalente. *Colloq. Math.*, **2** (1949–51), 249–253.
Zamorski, J.
[1] On Bazilevič schlicht functions. *Ann. Polon. Math.*, **12** (1962), 83–90.

List of Symbols

$\mathbb{C}$	complex plane, 1
$\hat{\mathbb{C}}$	extended complex plane, 6
$\mathbb{D}$	unit disk, 1
$\mathbb{\Delta}$	exterior of unit disk, 14
$\mathbb{T}$	unit circle, 1
$\mathbb{R}$	real line, 218
$\bar{E}$	closure of E, 1
$\mathring{E}$	interior of E, 1
∂E	boundary of E, 1
$d(E)$	diameter of E, 294
$\rho(E)$	conformal radius of E, 294
S	normalized univalent functions, 9
S_R	univalent functions with real coefficients, 55
$S^{(2)}$	odd univalent functions, 28
$S^{(m)}$	functions with m-fold symmetry, 28
S_α	functions with Hayman index α, 188
$S(A_n)$	functions with initial coefficients A_n, 334
C	convex functions, 40
S^*	starlike functions, 40
K	close-to-convex functions, 46
K_0	normalized close-to-convex functions, 46
T	typically real functions, 55
V_k	functions of bounded boundary rotation, 269
P	functions with positive real part, 40
P_R	subclass of P with real coefficients, 56
Σ	functions univalent in $\mathbb{\Delta}$, 28
Σ'	nonvanishing functions in Σ, 28
Σ_0	functions in Σ with $b_0 = 0$, 29
$\tilde{\Sigma}$	full mappings, 29
Σ^*	starlike functions in Σ, 137
$\mathscr{H}(D)$	analytic functions in D, 37
$\mathscr{A}$	$\mathscr{H}(\mathbb{D})$, 275
H^p	Hardy space, 61
ℓ^p	sequence space, 129

L^p	Lebesgue space, 215
$k(z)$	Koebe function, 26
$\ell(z)$	linear function, 45
$\Delta\psi$	Laplacian of ψ, 16
$\mathbb{V}_n$	coefficient region, 334
$\mathbb{V}_n(A_k)$	section of coefficient region, 346
$f \prec g$	subordination, 190
$f \ll g$	coefficient domination, 238
$f \circ g$	composition, 11
$f * g$	convolution, 246
$f \circledast g$	integral convolution, 247
$M_\infty(r, f)$	maximum modulus, 60
$M_p(r, f)$	integral mean, 37
$\{f, z\}$	Schwarzian derivative, 258
$\mathcal{S}$	Schwarzian derivative operator, 258
u^*	Baernstein star-function, 215
$\delta(\cdot\,; f)$	Gâteaux differential, 276
$\ell(\cdot\,; f)$	Fréchet differential, 276

Index

Grundlehren der mathematischen Wissenschaften

Continued from page ii

206. André: Homologie des Algébres Commutatives
207. Donoghue: Monotone Matrix Functions and Analytic Continuation
208. Lacey: The Isometric Theory of Classical Banach Spaces
209. Ringel: Map Color Theorem
210. Gihman/Skorohod: The Theory of Stochastic Processes I
211. Comfort/Negrepontis: The Theory of Ultrafilters
212. Switzer: Algebraic Topology—Homotopy and Homology
213. Shafarevich: Basic Algebraic Geometry
214. van der Waerden: Group Theory and Quantum Mechanics
215. Schaefer: Banach Lattices and Positive Operators
216. Pólya/Szegö: Problems and Theorems in Analysis II
217. Stenström: Rings of Quotients
218. Gihman/Skorohod: The Theory of Stochastic Process II
219. Duvant/Lions: Inequalities in Mechanics and Physics
220. Kirillov: Elements of the Theory of Representations
221. Mumford: Algebraic Geometry I: Complex Projective Varieties
222. Lang: Introduction to Modular Forms
223. Bergh/Löfström: Interpolation Spaces. An Introduction
224. Gilbarg/Trudinger: Elliptic Partial Differential Equations of Second Order
225. Schütte: Proof Theory
226. Karoubi: K-Theory, An Introduction
227. Grauert/Remmert: Theorie der Steinschen Räume
228. Segal/Kunze: Integrals and Operators
229. Hasse: Number Theory
230. Klingenberg: Lectures on Closed Geodesics
231. Lang: Elliptic Curves: Diophantine Analysis
232. Gihman/Skorohod: The Theory of Stochastic Processes III
233. Stroock/Varadhan: Multi-dimensional Diffusion Processes
234. Aigner: Combinatorial Theory
235. Dynkin/Yushkevich: Markov Control Processes and Their Applications
236. Grauert/Remmert: Theory of Stein Spaces
237. Köthe: Topological Vector-Spaces II
238. Graham/McGehee: Essays in Commutative Harmonic Analysis
239. Elliott: Probabilistic Number Theory I
240. Elliott: Probabilistic Number Theory II
241. Rudin: Function Theory in the Unit Ball of C^n
242. Blackburn/Huppert: Finite Groups I
243. Blackburn/Huppert: Finite Groups II
244. Kubert/Lang: Modular Units
245. Cornfeld/Fomin/Sinai: Ergodic Theory
246. Naimark: Theory of Group Representations
247. Suzuki: Group Theory I
248. Suzuki: Group Theory II
249. Chung: Lectures from Markov Processes to Brownian Motion
250. Arnold: Geometrical Methods in the Theory of Ordinary Differential Equations
251. Chow/Hale: Methods of Bifurcation Theory
252. Aubin: Nonlinear Analysis on Manifolds, Monge—Ampère Equations
253. Dwork: Lectures on p-adic Differential Equations
254. Freitag: Siegelsche Modulfunktionen
255. Lang: Complex Multiplication

Made in the USA
Monee, IL
13 September 2026

62976972R00221